Creativity Models for Innovation in Management and Engineering

Carolina Machado
University of Minho, Portugal

J. Paulo Davim
University of Aveiro, Portugal

IGI Global
PUBLISHER of TIMELY KNOWLEDGE

A volume in the Advances in Logistics, Operations, and Management Science (ALOMS) Book Series

Published in the United States of America by
IGI Global
Business Science Reference (an imprint of IGI Global)
701 E. Chocolate Avenue
Hershey PA, USA 17033
Tel: 717-533-8845
Fax: 717-533-8661
E-mail: cust@igi-global.com
Web site: http://www.igi-global.com

Library of Congress Cataloging-in-Publication Data

Names: Machado, Carolina, 1965- editor. | Davim, J. Paulo, editor.
Title: Creativity models for innovation in management and engineering / Carolina Machado, and J. Paulo Davim, editor.
Description: Hershey, PA : Business Science Reference, [2022] | Includes bibliographical references and index. | Summary: "This book introduces new lines of research about creativity and innovation in management and engineering areas, including models, theories and tools presented and discussed by the different contributions that allow management to take a more creative, innovative and strategic role in organizations"-- Provided by publisher.
Identifiers: LCCN 2021054757 (print) | LCCN 2021054758 (ebook) | ISBN 9781668423394 (hardcover) | ISBN 9781668423400 (paperback) | ISBN 9781668423417 (ebook)
Subjects: LCSH: Creative ability in business. | Creative ability in technology. | Diffusion of innovations.
Classification: LCC HD53 .C75195 2022 (print) | LCC HD53 (ebook) | DDC 658.4/063--dc23/eng/20211117
LC record available at https://lccn.loc.gov/2021054757
LC ebook record available at https://lccn.loc.gov/2021054758

This book is published in the IGI Global book series Advances in Logistics, Operations, and Management Science (ALOMS) (ISSN: 2327-350X; eISSN: 2327-3518)

British Cataloguing in Publication Data
A Cataloguing in Publication record for this book is available from the British Library.

Advances in Logistics, Operations, and Management Science (ALOMS) Book Series

John Wang
Montclair State University, USA

ISSN:2327-350X
EISSN:2327-3518

Mission

Operations research and management science continue to influence business processes, administration, and management information systems, particularly in covering the application methods for decision-making processes. New case studies and applications on management science, operations management, social sciences, and other behavioral sciences have been incorporated into business and organizations real-world objectives.

The **Advances in Logistics, Operations, and Management Science** (ALOMS) Book Series provides a collection of reference publications on the current trends, applications, theories, and practices in the management science field. Providing relevant and current research, this series and its individual publications would be useful for academics, researchers, scholars, and practitioners interested in improving decision making models and business functions.

Coverage

- Political Science
- Computing and information technologies
- Operations Management
- Decision analysis and decision support
- Organizational Behavior
- Finance
- Production Management
- Services management
- Risk Management
- Information Management

IGI Global is currently accepting manuscripts for publication within this series. To submit a proposal for a volume in this series, please contact our Acquisition Editors at Acquisitions@igi-global.com or visit: http://www.igi-global.com/publish/.

The Advances in Logistics, Operations, and Management Science (ALOMS) Book Series (ISSN 2327-350X) is published by IGI Global, 701 E. Chocolate Avenue, Hershey, PA 17033-1240, USA, www.igi-global.com. This series is composed of titles available for purchase individually; each title is edited to be contextually exclusive from any other title within the series. For pricing and ordering information please visit http://www.igi-global.com/book-series/advances-logistics-operations-management-science/37170. Postmaster: Send all address changes to above address.

Titles in this Series

For a list of additional titles in this series, please visit: http://www.igi-global.com/book-series/advances-logistics-operations-management-science/37170

Moving From R&D Development to Production and Sales in Start-Up Companies Emerging Research and Opportunities
Amiram Porath (APC, Israel)
Business Science Reference • © 2022 • 300pp • H/C (ISBN: 9781799856856) • US $215.00

Advancing Interpersonal Emotion Regulation and Social Regulation
Donta S. Harper (Argosy University, USA)
Information Science Reference • © 2022 • 384pp • H/C (ISBN: 9781668424780) • US $215.00

Handbook of Research on Activating Middle Executives' Agency to Lead and Manage During Times of Crisis
Ann-Marie Wilmot (The University of the West Indies, Mona, Jamaica) and Canute Sylvester Thompson (The University of the West Indies, Mona, Jamaica)
Business Science Reference • © 2022 • 415pp • H/C (ISBN: 9781668443316) • US $325.00

Business Models to Promote Technology, Culture, and Leadership in Post-COVID-19 Organizations
Joseph Robert Heinzman (Western Governors University, USA) Debbie S. Judge (Western Governors University, USA) Vincent Franklin (Western Governors University, USA) and Jim A. McCleskey (Western Governors University, USA & Houston Community College System, USA)
Business Science Reference • © 2022 • 300pp • H/C (ISBN: 9781668443583) • US $250.00

Global Risk and Contingency Management Research in Times of Crisis
Narasimha Rao Vajjhala (University of New York Tirana, Tirana, Albania) and Kenneth David Strang (W3-Research, USA & RMIT University, Australia)
Business Science Reference • © 2022 • 356pp • H/C (ISBN: 9781668452790) • US $270.00

Applied Guide for Event Study Research in Supply Chain Management
Lincoln C. Wood (University of Otago, New Zealand & Curtin University, Australia)
Information Science Reference • © 2022 • 261pp • H/C (ISBN: 9781799889694) • US $215.00

Black Women's Formal and Informal Ways of Leadership
Mesha Garner (Illinois State University, USA)
Information Science Reference • © 2022 • 300pp • H/C (ISBN: 9781668438275) • US $215.00

IGI Global
PUBLISHER of TIMELY KNOWLEDGE

701 East Chocolate Avenue, Hershey, PA 17033, USA
Tel: 717-533-8845 x100 • Fax: 717-533-8661
E-Mail: cust@igi-global.com • www.igi-global.com

List of Reviewers

Paulina Aburto-De La Llave, *Syngenta, Mexico*
Connie Atristain-Suárez, *Universidad Panamericana, Mexico*
Eglantina Hysa, *Epoka University, Albania*
Abiodun S. Isiaka, *University of Regina, Canada*
José Joaquim Machado, *Polytechnic Institute of Cávado and Ave, Portugal*
Nima Norouzi, *Bournemouth University, UK*
Mirela Panait, *Petroleum-Gas University of Ploiesti, Romania*
Lukman Raimi, *Universiti Brunei Darussalam, Brunei*
Eduardo Guilherme Satolo, *São Paulo State University, Brazil*
Kaveh Sheibani, *ORLab Analytics, Canada*
Andrea Uribe-Palau, *Universidad Panamericana, Mexico*

Table of Contents

Detailed Table of Contents

In recent years there has been a growth of interest in the development of systematic search methods for solving problems in operational research and artificial intelligence. This chapter introduces a new idea for the integration of approaches for hard combinatorial optimisation problems. The proposed methodology evaluates objects in a way that combines fuzzy reasoning with a greedy mechanism. In other words, a fuzzy solution space is exploited using greedy methods. This seems to be superior to the standard greedy version. The chapter consists of two main parts. The first part focuses on description of the theory and mathematics of the so-called fuzzy greedy evaluation concept. The second part demonstrates through computational experiments the effectiveness and efficiency of the proposed method for a number of complex problems in engineering management and beyond.

Whereas the COVID-19 pandemic tested the solidity, agility, and resilience of organizations, it as well enhanced a refined debate on the conceptual frameworks that have traditionally been guiding the managerial decisions and organizational structures, policies, and practices. This chapter aims at exploring the impact of the COVID-19 pandemic on knowledge, technology, creativity, and innovation management research, highlighting creativity as the core vector to understand the reconfiguration of the renewed corporate structures and processes. By literature review, it identifies key concepts, assumptions, and theoretical constructs, aiming at highlighting creativity as the core asset to understand how the worldwide organizations have been able to overcome the twofold challenges and opportunities of the recent

Mexican consumers generally recognize the wood furniture industry in Mexico, assign it an authenticity value, and associate it with a positive image. When customers perceive that the company is committed to creating superior value for the customer through the MO and EO, they show loyalty and purchase preference behaviours.

CSR practices derive directly from strategic actions of companies but, more indirectly, from other factors that influence organizational behavior, such as culture. But talking about culture is something too comprehensive since we have national and organizational culture. At the level of CSR practices, will the framework of values of the leadership and the employees of an organization or the norms and values of the country where it operates be more influential? What is the role played by an innovative culture in the implementation of CSR? How can companies promote corporate social innovation? To answer these questions, this chapter will present some studies and discussions to contribute to the reflection of this issue not only to sensitize organizations to the importance of assuming CSR behaviors in an innovative way but also to analyze the role that culture has at this level.

The aim of the study is to demonstrate the application of exergoeconomic analysis as a cost control mechanism in manufacturing operations with a focus on industrial beverage carbonation system. Exergoeconomic analysis is a thermodynamic tool that can identify cost rate of exergy destruction (hidden cost) associated with a machine or a system that cannot be identified using typical cost management techniques applied in industries. Exergy analysis was performed to examine the performance parameters of each unit in the system using mass and energy balance. Cost rate of exergy destruction was performed using the thermoeconomic analysis of energy systems software. The study reviews that the total hidden cost in the carbonation system was $777.31/hr, while the total cost rate related to the investment and maintenance of the system was $45.13/hr. The study concluded that reduction of exergy destruction in the mixer and other subsystems within the carbonation system is very crucial to the improvement of the cost rate in beverage production.

Preface

Looking to provide a channel of communication to disseminate the knowledge of how to manage in a highly competitive, creative, and innovative environment, between academics/researchers and managers the subjects developed in this book, entitled *Creativity Models for Innovation in Management and Engineering,* assume a particular emphasis in nowadays organizations.

Looking not only to answer, but above all, to anticipate to the markets and environment challenges, this book, focusing on the areas of creativity and innovation in management and engineering, seeks cultural and geographic diversity in studies of organization management and engineering, as well as uses that have a special impact on organizational sustainability, organizational communications, change processes, work practices, and creative and innovative strategies, reflecting the diversity of societal and infrastructural conditions.

In todays' competitive environments only the most creative and innovative organizations are able to survive. Indeed, instead of react to the market challenges, creative and innovative organizations act in a proactive way, establishing and developing strategies able to leverage, continuously, their creativity and their innovative abilities to attain long term success and competitiveness.

Whether in the profit or non-profit sectors, or in the public or private sector, this book is designed to improve the knowledge, abilities, skills, and effectiveness, of all those that in their daily life are involved in creativity and innovative management and engineering.

Able to be used by academics, researchers, human resources managers, managers, engineers, and other professionals in related matters with creative and innovative management and engineering, this book highlights the theoretical and practical implications of a more creative and innovative management and engineering, in order to contribute to a better understanding of the subject.

To discuss the issues related with creativity and innovation in management and engineering is essential because:

- It contributes to the development of new lines of research about creativity and innovation in management and engineering areas;
- The strategies, policies, models, theories, and tools presented and discussed by the different contributions present in this book allow management to take a more creative, innovative and strategic role in organizations.
- In a complex and increasingly competitive world, the development of sophisticated analyzes necessary for efficient and effective decision-making, using creative and innovative information, tools, and mechanisms, is fundamental for strategic action compatible with a greater capacity to

adapt, but above all, anticipation of an increasingly aggressive and competitive environment, with a view to improving business, performance and organizational growth.

Looking to contribute to the increase of research related to these new trends and developments in the field of creativity and innovation management and engineering, the present book, divided into 10 chapters, covers in Chapter 1 "An Integrated Intelligent Computation for Complex Problems in Engineering Management" while Chapter 2 discusses "Creativity as the Ultimate Asset in the COVID-19 Reconfigured Organizational Environment." Chapter 3 focuses on "Thermodynamics and Economics Analogies," and Chapter 4 deals with "Quality Function Deployment in Agribusiness Supply Chain in the Food Sector: Has Its Potential Been Thoroughly Exploited?" Chapter 5 contains information about "CSR Programs of Financial Institutions: Development-Oriented Issues or Just Greenwashing?" while Chapter 6 highlights "Leadership and Organizational Culture in Football: A Coaches' Perspective." Chapter 7 focuses on "Market and Entrepreneurial Orientation Strategies in the Wood Furniture Industry in Mexico" while Chapter 8 discusses "The Influence of Culture on Innovation and CSR Practices." Finally, Chapter 9 speaks about "Exergoeconomic Analysis as a Cost Control Mechanism in Manufacturing Operations-Focus on Industrial Beverage Carbonation System", and Chapter 10 develops a reflection about "To Build a Future Abroad: The Strength of Women as an Example of Tenacity and Achievement of Goals."

This subject, with a specific emphasis on creativity and innovation in the Management and Engineering areas, will be critical and of great usefulness for all those - managers, engineers, researchers, human resource managers, academics, as well as professionals from different areas - that in their daily, academic and professional routines need to lead with creative and innovative management and engineering issues.

Carolina Machado
University of Minho, Portugal

J. Paulo Davim
University of Aveiro, Portugal

Acknowledgment

The editors acknowledge their gratitude to IGI Global for this opportunity and for their professional support. Finally, we would like to thank to all chapter authors for their interest and availability to work on this project.

Carolina Machado
University of Minho, Portugal

J. Paulo Davim
University of Aveiro, Portugal

Introduction

Focusing on the latest and more recent research findings that are occurring in the fields of Management and Engineering in different countries, *Creativity Models for Innovation in Management and Engineering*, organized in its ten chapters, looks to help researchers and practitioners to select among the different options, models, strategies and policies, the more relevant priorities to manage competitive, creative, and innovative organizations.

Indeed, both managers, engineers, researchers, human resource managers, academics, as well as professionals from different areas that in their daily, academic, and professional routines, need to lead with creative and innovative management and engineering issues need to obtain tools that allow them to know and better understand the organizations they have to manage, making them more competitive and capable of being successful in the aggressive markets in which they operate. Highlighting the main ideas, tools and strategies studied by different researchers from different countries, the various chapters that make up this book are seen as critical and very useful contributions for all these professionals.

In other words, this book, organized in ten chapters looks to:

- Highlight the latest and more recent research findings about creativity and innovation issues, that are occurring in the field of management and engineering, in different countries;
- Show in what ways companies around the world are facing today's challenges and developments in creative and innovative management and engineering;
- Share knowledge and insights about creativity and innovation in modern and competitive organizations, on an international scale;
- Help researchers and practitioners to select among the different options, models and strategies the more relevant policies and practices to manage competitive, creative and innovative organizations.

From the chapters presented in *Creativity Models for Innovation in Management and Engineering*, it turns out that there are many challenges that organizations are facing in the environment in which they are involved, characterized by multiple and continuous transformations and challenges. The conclusions they reached allow us to draw huge and extremely critical contributions, undoubtedly of great use to all those who in their daily and professional lives seek to deal with and manage competitive, innovative, socially responsible, and truly sustainable organizations.

Finally, it is important to highlight that this subject of creativity and innovation, with a specific emphasis in the Management and Engineering areas, will be critical and of great usefulness for final undergraduate engineering and management courses or as a subject on management and engineering at the postgraduate level. Also, this book can serve as a useful reference for academics, researchers, managers, engineers, and other professionals in related matters with creative and innovative management and engineering.

We therefore hope that all potential users and readers of this book will consider it as a useful tool capable of leading them to a successful management of increasingly creative and innovative organizations.

Carolina Machado
University of Minho, Portugal

J. Paulo Davim
University of Aveiro, Portugal

Chapter 1
An Integrated Intelligent Computation for Complex Problems in Engineering Management

Kaveh Sheibani
ORLab Analytics, Canada

ABSTRACT

In recent years there has been a growth of interest in the development of systematic search methods for solving problems in operational research and artificial intelligence. This chapter introduces a new idea for the integration of approaches for hard combinatorial optimisation problems. The proposed methodology evaluates objects in a way that combines fuzzy reasoning with a greedy mechanism. In other words, a fuzzy solution space is exploited using greedy methods. This seems to be superior to the standard greedy version. The chapter consists of two main parts. The first part focuses on description of the theory and mathematics of the so-called fuzzy greedy evaluation concept. The second part demonstrates through computational experiments the effectiveness and efficiency of the proposed method for a number of complex problems in engineering management and beyond.

INTRODUCTION

In recent years, there has been a growth of interest in the development of systematic search methods for solving problems in operational research and artificial intelligence. Metaheuristics that are employed as strategies in optimisation are a fairly young research field. They are approaches that organise an interaction between solution improvement procedures and higher-level tactics in order to create a process capable of escaping from premature local optima and performing a robust search of a solution space. A metaheuristic can be viewed as a generic approach, for a type of hard optimisation problem. It is applicable to a wide set of different optimisation problems, with relatively few modifications needed to apply it to a specific problem. A much newer area of research is the hybridisation of metaheuristics. It

DOI: 10.4018/978-1-6684-2339-4.ch001

has become evident that a skilled combination of general ideas from different methods can provide an efficient performance and high flexibility.

The use of search techniques on a solution space are central to the design of metaheuristics. Indeed, adopting a robust search technique significantly improves the overall performance. In this chapter, we introduce a new idea for the integration of approaches for hard combinatorial optimisation problems. The proposed methodology evaluates objects in a way that combines fuzzy reasoning with a greedy mechanism. In other words, we exploit a fuzzy solution space (fuzzy set) using greedy methods. Our methodology also attempts to adapt its knowledge from previous experiments, thereby improving the exploration of the promising areas of the search space. The effectiveness and efficiency of this so-called fuzzy greedy evaluation concept are investigated within search, optimisation and learning systems for hard combinatorial optimisation problems. For this purpose, the travelling salesman problem (TSP) and the flow- shop scheduling problem (FSP) have been selected for investigation as two of the most widely studied problems in the field of combinatorial optimisation.

The chapter consists of two main parts. The first part focuses on description of the theory and mathematics of the so-called fuzzy greedy evaluation concept. The second part demonstrates through computational experiments, the effectiveness and efficiency of the proposed concept for hard combinatorial optimisation problems. The text contains an extensive bibliography, which covers many relevant books and significant papers.

FUZZY GREEDY EVALUATION

Introduction

In recent decades there has been a growth of interest in methods for finding optimal solutions to a class of problems called combinatorial optimization. The subject is very wide and many books and articles have been published on its various aspects. Typical examples of combinatorial optimization problems are the travelling salesman (Lawler, 1985; Gutin and Punnen, 2002; Gutin, 2013), variants of the assignment problem (Cela *et al*, 2014), scheduling problems (Pinedo *et al*, 2015), the set covering (Beasley, 1990; Boschetti and Maniezzo, 2015), and vehicle routing problems (Coelho and Laporte, 2015). Combinatorial problems are normally easy to describe but difficult to solve (Korte and Vygen, 2012). The foundations for the theory of computational complexity are to be found in Cook's (1971) seminal paper. In his paper, Cook attempted to classify problems in practice as easy or hard. A problem is called easy if an algorithm can be developed which solves the problem to optimality in a polynomial-time (Du and Pardalos, 2005). A problem is called hard or intractable if such efficient algorithms do not exist or the solution cannot be found within a reasonable computational time bound.

Due to the practical importance of combinatorial optimization problems, many methods have been developed for solving them (Pulleyblank (2014); Martello and Ries (2015); Nishi *et al*, (2017)). These methods can be classified as either exact or approximate. Exact methods guarantee to find an optimal solution in a bounded amount of time. Of course, for those combinatorial optimization problems which belong to the class NP-hard (Garey and Johnson, 1979), exact methods need an exponential amount of time. Thus, approximation methods which usually called heuristics, are often considered to be the only practical tools available to solve hard combinatorial optimization problems.

This section overviews a new idea, so-called fuzzy greedy evaluation concept for the integration of approaches for hard combinatorial optimization problems. The section also provides a comprehensive discussion on mathematics of the proposed methodology. In this presentation, the proposed method can be seen as a generic heuristic for the integration of approaches to hard combinatorial optimization problems. Furthermore, the current work shows another application of the proposed method with different objectives in addition to the previously published literature on the same subject. For example, it shows that how different adaptations of the proposed method can be yielded to different results.

The section consists of two main parts. The first part focuses on description of the theory and mathematics of the so-called fuzzy greedy evaluation concept. The second part demonstrates computational examples of the proposed concept for hard combinatorial optimization problems. The concluding remarks contain some suggestions for further research.

Fuzzy Sets

In many real-world problems fuzzy sets allow us to represent vague concepts expressed in natural language. The membership function of a fuzzy set A can be denoted by $\mu_A: X \rightarrow [0,1]$. Each fuzzy set should be uniquely defined by one particular membership function. Consider a fuzzy set where membership function is defined in Equation (1). This is one of the general formulae of a parameterized family of membership functions described in Klir and Yuan (1995).

$$\mu_A\left(x\right) = \frac{1}{1 + p\left(x - r\right)^2} \tag{1}$$

This fuzzy set expresses, in a particular form, the general concept of a class of real numbers that are close to r. When the non-negative parameter p increases, the graph of $\mu_A(x)$ becomes narrower. The function has the following properties: $\mu_A(r) = 1$ and $\mu_A(x) < 1$ for all $x \neq r$. More detailed discussions of fuzzy sets and their properties can be found in (Klir and Yuan, 1995; Zimmermann, 2001; Wang and Klir, 2013).

Alternative Fuzzy Membership Functions

As we mentioned above, the fuzzy membership functions can have very different shapes of their graphs, in the context of a particular application. We can also mention the following membership functions, which are described in Klir and Yuan (1995).

$$\mu A_1(x) = \begin{cases} \rho(x - r) + 1, & x \in \left[r - 1/\rho, r\right] \\ \rho(r - x) + 1, & x \in \left[r, r + 1/\rho\right] \\ 0 & \text{otherwise} \end{cases} \tag{2}$$

$$\mu A_2(x) = \begin{cases} \left(1+\cos(\rho\pi(x-r))\right)/2, & x \in [r - 1/\rho, r + 1/\rho] \\ 0 & \text{otherwise} \end{cases} \tag{3}$$

$$\mu A_3(x) = e^{-|\rho(x-r)|} \tag{4}$$

Greedy Algorithms

The greedy method is an approach to algorithm design which is particularly suited for solving certain optimization problems. Such algorithms make choices according to what appears to be the best choice at the moment, without any consideration of any previous choices or future consequences. Greedy algorithms do not necessarily yield optimal solutions, but for some problems they do.

The greedy paradigm is often used in solving combinatorial optimization problems (Cook *et al*, 1998). This phenomenon can be explained by the fact that, although it is widely assumed that greedy algorithms rarely produce optimal solutions, they often provide solutions that are significantly better than the worst case (Vince, 2002; Curtis, 2003; Bang-Jensen *et al*, 2004). Hence, a natural question to pose is: can we characterize the type of optimization problems for which greedy algorithms produce optimal solutions? In a sense, the theory of matroids (Whitney, 1935; Edmonds, 1971), and later greedoids (Korte and Lovasz, 1981; Korte and Lovasz, 1983; Korte *et al*, 1991) are useful in determining when the greedy method yields optimal solutions. Vince (2002) put greedy algorithms into a simple framework, which explains when they do not work for all linear objective functions. Curtis (2003) addressed principles for the classification of those greedy algorithms that do provide optimal solutions to optimization problems. In extreme cases, for certain cases of optimization problems, a greedy algorithm may produce the unique worst possible solution (Gutin and Yeo, 2002; Gutin *et al*, 2002). Bang-Jensen *et al* (2004) characterized certain cases in which the greedy algorithm fails.

Because of the simplicity and versatility of greedy algorithms, they are frequently applied in many various areas such as combinatorial optimization problems. A variety of applications is addressed in (Cormen *et al*, 1990; Curtis 2003). For example, the greedy algorithm gives optimal solutions for Dijkstra's algorithm for shortest path (Dijkstra, 1959) and minimal spanning tree (Kruskal, 1956; Prim, 1957). More detailed discussions on the subject can be found in (Horowitz and Sahni, 1978; Papadimitriou and Steiglitz, 1982; Wilson, 1985; Grimaldi, 1989; Cook *et al*, 1998).

Fuzzy Greedy Evaluation

Consider combinatorial optimization problems, each of which is associated with a discrete solution space X, a feasible space S with the property $S \subseteq X$ that is defined by the problem constraints, and an objective function f: $X \rightarrow \Re$. In the case of minimization, the aim is to find a feasible solution $\mathbf{x}^* \in S$ such that $f(\mathbf{x}^*) \leq f(\mathbf{x})$, $\forall \mathbf{x} \in S$, where, $\mathbf{x} = (x_1,\ldots,x_n)$ is a vector of decision variables (solution). A cost function $c(x)$, $\forall x \in \{x_1,\ldots,x_n\}$ is defined for each specific problem. This function represents the incremental increase in the value of the objective function due to the incorporation of an element 'x' into the solution under construction. We refer to $c(x)$ as a greedy evaluation function. Let the set of candidate elements be the set of all such elements that can be incorporated into the partial solution under construction without causing infeasibility. Therefore, the selection of the next element 'x' for incorporation into the partial solution under construction is determined by the evaluation of all candidate elements according to our greedy evaluation function.

In the standard greedy approach, the greedy evaluation function denotes a degree of priority for incorporating the corresponding element 'x' into the solution under construction. We now describe an alternative approach. We will treat the set X as a fuzzy set with a well-defined membership function $\mu(x)$, the form of which is given by Equation (5).

$$\mu\left(x\right) = \frac{1}{1 + \lambda^2 \left(\left(\frac{1-\lambda}{\lambda}\right)x - \theta\right)^2} \tag{5}$$

The variable $x \in X$ corresponds to one of the variables in the definition of our combinatorial optimization problem. The parameter θ is a basic measure for evaluating the priority to be assigned to variable x. The parameter λ is a tuning parameter that is chosen by experimentation such that $0 \leq \lambda < 1$. This parameter will be seen to play an important part in the proposed approach.

The evaluation function in Equation (5) has the following properties: $\mu(\lambda\theta / (1 - \lambda)) = 1$ and $0 < \mu(x) < 1$ for all $x \neq \lambda\theta / (1 - \lambda)$. This implies that x closest numerically to $\lambda\theta / (1 - \lambda)$ should be given higher priority. Let x_{min} and x_{max} be the smallest and the largest values of x, respectively. We define a small enough value $\lambda_{min} = x_{min} / (x_{min} + \theta)$ for $\theta > 0$ and any $\lambda \leq \lambda_{min}$, for which inequality $x \geq \lambda_{min}\theta / (1 - \lambda_{min})$ holds. We also define a big enough value $\lambda_{max} = x_{max} / (x_{max} + \theta)$ for $\theta > 0$ and any $\lambda \geq \lambda_{max}$, for which inequality $x \leq \lambda_{max}\theta / (1 - \lambda_{max})$ holds.

Equation (5) is a modification of Equation (1), which is one of the general formulas of the families of fuzzy membership functions. We will refer to this function as a fuzzy greedy evaluation function, and it will replace the role of the greedy evaluation function in determining the degree of priority assigned to an element 'x'. Figure 1 shows examples of the proposed fuzzy greedy evaluation function on different values of λ.

Figure 1. Distribution of $\mu(x)$ with θ = 10, and λ = 0.4, 0.5 and 0.6.

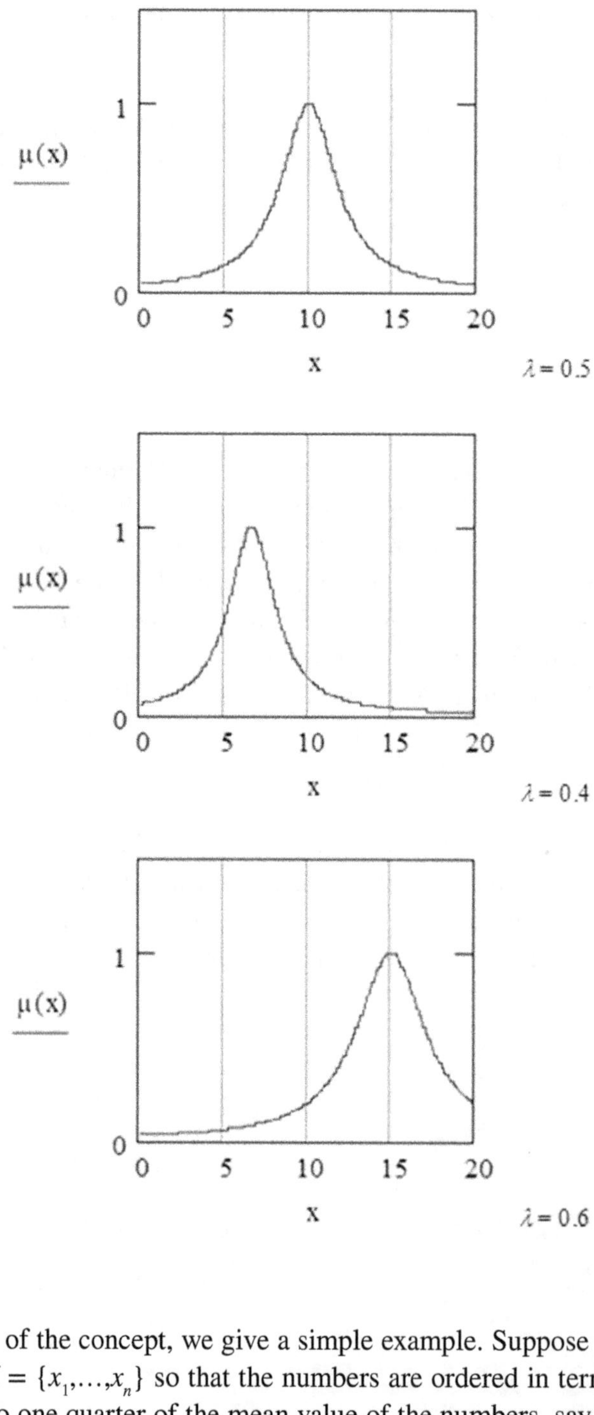

To convey the nature of the concept, we give a simple example. Suppose we want to build a list of a given set of integers $X = \{x_1,\ldots,x_n\}$ so that the numbers are ordered in terms of their closeness to a number which is equal to one quarter of the mean value of the numbers, say $\theta/4$, where $\theta = (x_1+\ldots+x_n)/n$. We will set the tuning parameter λ in Equation (5) to 0.2; for this λ value, the equation shows that $\mu(x)$ is maximized when x takes the value $\theta/4$. Clearly, if element 1 has a value x_1 and element 2 has a value x_2 and if $|x_1 - \theta/4| < |x_2 - \theta/4|$, then element 1 will be of higher priority than element 2. Thus, it is only necessary to sort the list of candidate elements by descending order of their membership grades, corresponding to $\mu(x)$. Much more will be said below about the role of this parameter and how its value should be set in the context of particular combinatorial optimization problems.

Fuzzy Greedy Search

The fuzzy set theory has been applied in an attempt to answer this significant question in context of the greedy algorithms: what is the best choice at present? We propose that the fuzzy greedy evaluation function $\mu(x)$ in Equation (5) can be employed to identify a suitable choice. We introduce a new algorithm for evaluating the candidate elements by using $\mu(x)$ at each stage of the classical greedy algorithm. We have discussed the greedy parameter λ, which needs to be tuned. As a simple procedure, the algorithm can be invoked for a fixed number of times for any given problem instance. This number, say t, depends on whether we decide to let λ take all possible values to $k = 2, 3, 4,\ldots$ decimal places over the range from λ_{min} to λ_{max}. We will call these procedures as the fuzzy greedy search heuristic. The steps are as follows:

Algorithm 1. Fuzzy Greedy Search Heuristic.

```
WHILE (λmin ≤ λ up to k decimal places ≤ λmax) DO
BEGIN
                WHILE (X is not empty) DO
                BEGIN
1.          Choose an {x} which maximizes μ(x);
2.          X = X - {x};
3.          Incorporate {x} to build a solution;
                END
                Keep current best_solution;
END
RETURN best_solution;
```

Suppose $X = \{1,2,\ldots,10\}$. List the elements in order of their closeness to the mean of the elements of X. Examine the performance of the fuzzy greedy algorithm using different values of λ equal to λ_{min}, 0.4, 0.5, 0.6 and λ_{max}.

Let $\theta = 11/2$ as the mean of the elements of X. From Equation (5), it is obvious that a value of $\lambda = 0.5$ will give $\mu(\theta) = 1$ so that using Algorithm (1) will give an optimal list. λ_{min} is equal to 2/13. λ_{max} is equal to 20/31. Using the fuzzy greedy search algorithm we get the ordered lists shown in Table 1.

Table 1. Performance of the fuzzy greedy algorithm using different values of λ.

$\lambda = \lambda_{min}$	$\lambda = 0.4$	$\lambda = 0.5$	$\lambda = 0.6$	$\lambda = \lambda_{max}$
1	4	5	8	10
2	3	6	9	9
3	5	4	7	8
4	2	7	10	7
5	6	3	6	6
6	1	8	5	5
7	7	2	4	4
8	8	9	3	3
9	9	1	2	2
10	10	10	1	1

In this trivial example, it was intuitively clear what the optimal choice of λ should be. However, for the combinatorial optimization problems, this will not be the case; suitable values of λ obtain by experimentation. The computational complexity of the fuzzy greedy algorithm is the same as that of the straightforward greedy algorithm. This can be viewed as a transformation of the exchange sort algorithm known as *bubble sort*. The computational complexity of the algorithm is $O(n^2)$ when bubble sort is used. This can be improved to $O(n \log n)$ if *merge sort* is used. More detailed and comprehensive discussions of sorting methods can be found in (Horowitz and Sahni, 1978; Neapolitan and Naimipour, 2003).

Sheibani (2019) provides annotated bibliographies of the fuzzy greedy search heuristic, including tutorials, surveys, conference proceedings, theses, and applications.

A HYBRID METAHEURISTIC FOR THE TRAVELLING SALESMAN PROBLEM

Introduction

Combinatorial optimization problems are normally easy to describe but difficult to solve. To illustrate this point, consider the travelling salesman problem (TSP). There are many variations of the TSP Gutin and Punnen (2002). In this chapter, the Euclidean TSP is considered. This problem can be represented by n cities numbered $1,2,\ldots,n$ and a Euclidean distance ($c_{i,j}$) between any two cities i and j. The goal in the TSP is to find a tour, which visits each city exactly once and is of minimum length. As the starting point is arbitrary and the distance between every pair of cities is the same, there are clearly $(n-1)!/2$ feasible solutions. The TSP is a classical NP-hard optimization problem. Hence, approximation methods are generally considered to be the only practical way to solve most real-life problems. In recent years, there has been a growth of interest in the development of systematic search methods for solving such problems in operational research. A much newer area of research is the hybridization of metaheuristics Sheibani (2004). The use of search techniques on a solution space is central to the design of a solution method. Indeed, adopting a robust search technique significantly improves the overall performance.

The systematic study of the TSP as a combinatorial optimization problem began with the work of Dantzig *et al.* (1954). Various approaches have been proposed for the problem as demonstrated in Merz and Freisleben (2001); Schmitt and Amini (1998); Laporte and Palekar (2002) and Laporte (2010). We will now discuss some of those methods that are most relevant to our discussion. The first efforts to find approximate solutions to the TSP by using genetic algorithms (GA) were those of Goldberg and Lingle (1985), using partially mapped crossover (PMX); Grefenstette *et al.* (1985), using greedy crossover; Davis (1985), using order crossover (OX); and Oliver *et al.* (1987), using cycle crossover (CX). Other related works include Cheng and Gen (1994), using greedy selection crossover (GSX); Chatterjee *et al.* (1996) using a non-random initial population obtained by the nearest neighbour heuristic, and an asexual scheme in the new population generations; Qu and Sun (1999), using a modification of the GSX, and a measure to prevent premature convergence. Ha *et al.* (2020) considered a special application of the travelling salesman problem. A more recent study by Valdez (2020) demonstrates a comparison of a number of metaheuristics for solving the travelling salesman problem.

In this chapter, we introduce a new adaptive fuzzy greedy search operator for a hybrid meta-heuristic, which is a combination of GA and greedy randomized adaptive search procedures (GRASP) (Feo and Resende, 1995) to find near-optimum solutions for the TSP. The concluding remarks contain some suggestions for further research.

A Hybrid Metaheuristic for TSP

The aim of the design of a hybrid method is to combine the strengths of some different techniques in order to improve the efficiency of a single approach (Sheibani, 2010). In this chapter, we propose a hybrid metaheuristic for the TSP. The proposed method is based on the use of GA and some of the ideas of the GRASP. The methodology uses the solutions from the construction phase of GRASP as the initial population of GA. A new adaptive fuzzy greedy search operator is developed for this hybrid method. Figure 2 illustrates the proposed hybrid method in pseudo-code (the variable *P*(*t*) represents a set of population members at generation *t*).

Figure 2. The proposed hybrid metaheuristic in pseudo-code

```
BEGIN
t←0;
P(t) ←Ø;
Solution ←Ø;
WHILE (P(t) is not complete) DO
P(t) ← GRASP_construction(seed);
evaluate(P(t));
WHILE (not termination condition) DO
BEGIN
t←t+1;
P(t) ← select(P(t - 1));
recombine(P(t));
evaluate(P(t));
END
END
```

In order to apply our hybrid method to the TSP, a solution is represented by a string of integers. Each is a sequence of n arranged cities, numbered 1 to n, which represents the order of the cities on a circle.

In the proposed hybrid metaheuristic we build a feasible solution, which is represented as a string of integers (chromosome). The proposed method uses a set of candidate cities, each of which can be incorporated into the partial solution under construction without causing infeasibility. The priority of the cities in the list is determined according to an evaluation function as shown in Equation (5). Here, x is the distance between any two cities ($c_{i,j}$). The parameter θ is a basic measure for evaluating the priority to be assigned to x, which is the average distance between successive cities in the current best solution obtained by the algorithm. This parameter will be seen to play an adaptive role, in that good choices made at previous stages (giving rise to the best solution so far) will also influence future choices. The parameter λ is a greedy tuning parameter that is chosen by experimentation such that $0 \leq \lambda < 1$ to adjust θ.

The evaluation function $\mu(x)$ has the following properties: $\mu(\lambda\theta / (1 - \lambda)) = 1$ and $0 < \mu(x) < 1$ for all $x \neq \lambda\theta / (1 - \lambda)$. This implies that the choice of next city to travel has distance closest numerically to $\lambda\theta / (1 - \lambda)$. For more details on the fuzzy greedy evaluation methodology, we refer the interested reader to (Sheibani, 2004-2021).

Initial Population Generation

The initial population was generated using the construction phase of GRASP. This process will be referred to as the GRASP-link. In this phase, a candidate list is formed from the possible cities (i.e. those not yet visited). Then we select two of the candidate cities from the list, which have maximum μ values in order to build the restricted candidate list (RCL). One of these elements in the RCL is selected at random to incorporate into the partial solution under construction. This process is repeated until a full TSP tour is formed.

Selection Scheme

A mixed strategy based on the roulette wheel selection and the elitist replacement was adopted. The roulette wheel method uses a probability distribution for selection of a chromosome, which is proportional to its fitness ($1/c$), where c is the tour length. Elitist replacement puts the fittest chromosome in the current population directly into the next generation.

An Adaptive Fuzzy Greedy Search Operator

A new search operator (called FGSX) was proposed by modifying the method of Qu and Sun (1999). The operating principle of FGSX is shown schematically in Figure 3. Let P_1 and P_2 be two randomly selected chromosomes from the previous generation. Each is a sequence of 9 arranged cities, numbered 1 to 9, which represents the order of the cities on a circle. First, we arbitrarily select a city; say 4, as the starting point in the offspring O_1. Then, we duplicate all cities in the selected parent chromosomes, which have not been incorporated in the offspring O_1 (between two cut points marked by '½') as shown under the heading "Duplication" in the figure. This guarantees that the next two possible candidate cities have not already been incorporated in the offspring under construction. The next city in the offspring is determined as shown under the heading "Selection". Assume that $\mu(c_{4,3})$ is greater then $\mu(c_{4,5})$, indicating that the choice of travel from city 4 to 3 is more suitable than 5, so we should select city 3 as the second city in the offspring O_1. The process is continued until a completely new offspring is formed. It is important to note that the FGSX can be adaptive in the sense that it attempts to learn from the best solution obtained in the previous generation by updating the parameter θ at each generation.

Figure 3. An illustration of the proposed FGSX

Parents

P_1: (1 2 3 4 5 6 7 8 9)
P_2: (4 3 2 1 5 6 7 8 9)

Duplication

1 2 3 4 | 5 6 7 8 9 1 2 3 | 4 5 6 7 8 9
4 | 3 2 1 5 6 7 8 9 | 4 3 2 1 5 6 7 8 9

Selection

Assuming that $\mu(c_{4,5}) < \mu(c_{4,3})$,
thus, city 3 is selected for offspring O_1

Offspring

O_1: (4 3 - - - - - - -)

Mutation Operator

We used a simple reordering operator as a mutation by selecting two points along the length of the single chromosome at random and then reversing the order of the sub-sequence between these two points. We should also mention that this reordering operator is commonly known as a 2-opt move in the TSP literature, which involves the reversal of a tour segment. The operating principle is shown schematically in Figure 4.

Figure 4. A simple reordering operator

Mutation

$$S: (1\ 2 \mid 3\ 4\ 5\ 6\ 7 \mid 8\ 9) \rightarrow S': (1\ 2 \mid 7\ 6\ 5\ 4\ 3 \mid 8\ 9)$$

Parameter Setting

The proposed method was implemented in C++. The experimental results were obtained for 10 replications with different seeds common in all problem instances for the random number generation. The random number generator used the standard C++ library function. The test problems that were chosen are part of the extensive set of standard benchmark problems available in TSPLIB (Reinelt, 1991). The performance is measured for both the solution quality and algorithm computational time (CPU time) on a 2.80 GHz processor. The solution quality is determined with the percentage deviation of the obtained solution (tour length) from the best-known solution taken from TSPLIB.

The performance of a GA greatly depends on the structure of the problem considered, architecture of the algorithm and the settings for the algorithm parameters such as population size, crossover and mutation rates. Finding universal parameter values to prevent premature convergence still remains an unsolved problem (Back (1992) and Yun and Gen (2003)). Our experimentation showed that good performance of the proposed hybrid method was obtained when setting the genetic parameters as follows: population size *popsize* = 45, crossover probability $P_c = 0.45$, mutation probability $P_m = 0.55$ and the number of iterations to be 50,000 for the termination condition of the algorithm. At the end of this section we briefly discuss the dependency between the values of genetic parameters and the rate of convergence to good quality local optima.

We introduced the tuning parameter λ in addition to the above GA's parameters. The performance of the proposed method is sensitive to the chosen value of λ. Extensive experimentation showed that good performance is obtained when setting λ to 0.1, 0.2, or 0.3. We found that usually this range of λ values gave the best solution. Figure 5 illustrates the effect of different values of λ on the computational performance of the proposed method.

Figure 5. A comparison of the proposed method on the eil51 problem with different values of λ

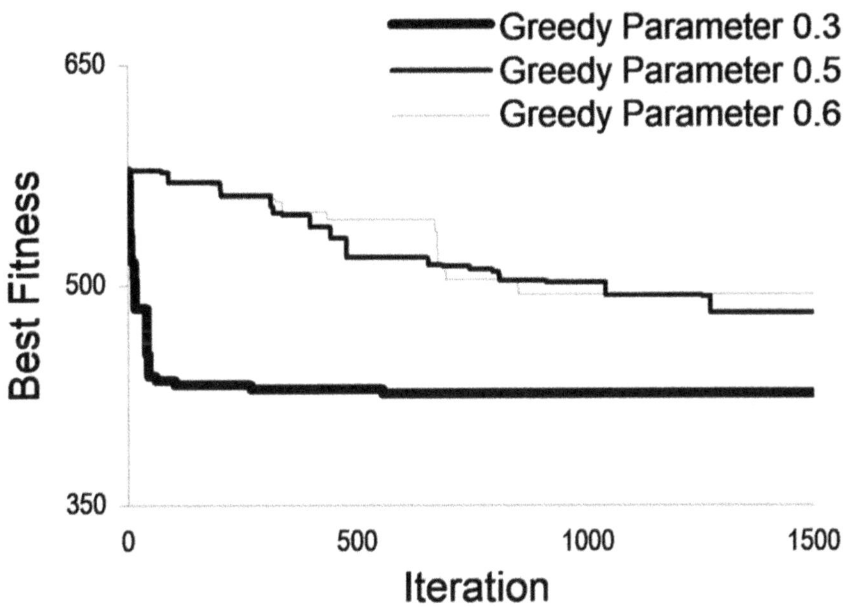

We evaluated the performance of the proposed method using random and the GRASP-link initializations; all other parameter settings were the same. A comparison of the results is tabulated in Table 2.

Table 2. Comparison of two different initialization methods

	GRASP-Link		*Rand-Initial*	
Problem	*Mean Error%*	*Variance*	*Mean Error%*	*Variance*
eil51	0.11	0.01	0.51	0.35
eil76	0.89	0.69	0.53	0.20
kroA100	0.22	0.05	0.31	0.08
lin105	0.08	0.02	0.74	1.15
bier127	0.83	0.14	1.36	0.31
ch130	1.47	0.41	2.36	0.24
kroA150	1.37	0.26	1.88	1.09
tsp225	1.86	1.29	2.89	2.20
Average	**0.86**	**0.36**	**1.33**	**0.70**

Given that the initialization process is computationally inexpensive, compared to the rest of the GA, the results show the value of using an effective initialization process. The evolution of solutions using these two methods is illustrated in Figure 6. The results indicate that for a small number of generations, the GRASP-Link initialization leads to better performance than a random initialization.

Figure 6. Two different initializations for the eil76 problem

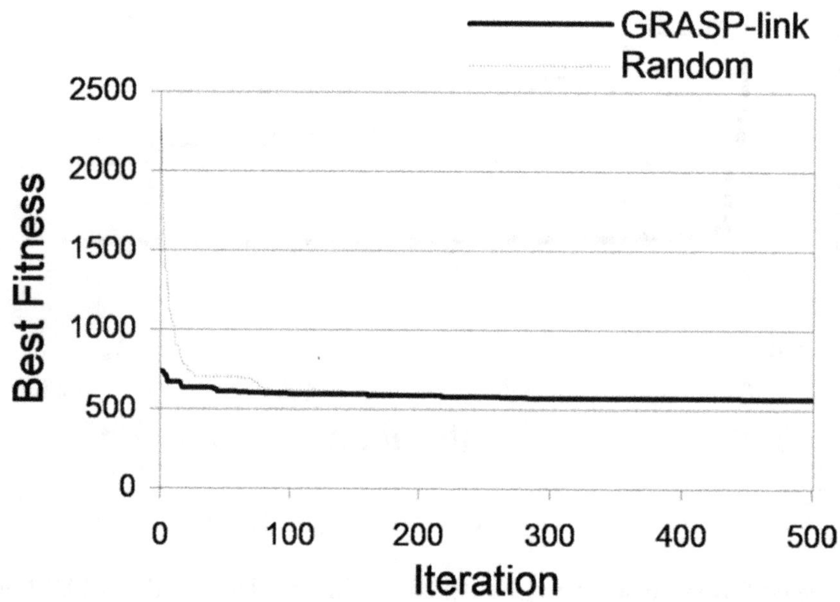

A comparison of the proposed operator FGSX with the simplified GSX illustrated in Table 3. It should be noted that GSX has been the most effective operator in comparison with PMX, OX, and CX for the TSP problems Qu and Sun (1999). On average, the proposed FGSX has an error of 0.14% less than that of GSX.

Table 3. Comparison of the FGSX with the simplified GSX

Problem	FGSX		GSX	
	Mean Error%	Variance	Mean Error%	Variance
eil51	0.11	0.01	0.44	0.33
eil76	0.89	0.69	0.89	0.69
kroA100	0.22	0.05	0.42	0.69
kroB100	0.91	0.07	0.91	0.07
kroC100	0.59	0.22	0.66	0.56
kroD100	1.29	0.63	1.18	0.35
kroE100	0.42	0.05	0.49	0.09
rd100	0.70	0.68	1.00	0.98
eil101	1.46	0.58	1.46	0.58
lin105	0.08	0.02	0.27	0.17
bier127	0.83	0.14	1.19	0.67
ch130	1.47	0.41	1.70	0.63
ch150	0.58	0.09	0.87	0.40
kroA150	1.37	0.26	1.63	0.41
kroB150	1.62	1.05	1.62	1.05
tsp225	1.86	1.29	1.86	1.29
Average	**0.90**	**0.39**	**1.04**	**0.56**

Sensitivity Analysis

Given the results of Table 3, it might be reasonable to plan experimentation in such a way that larger problems are given a larger number of iterations. In order to keep the computational effort approximately unchanged, we keep the maximum number of tour evaluations constant; this would mean simultaneously reducing the size of the population and rising the number of iterations. Some bigger problems with different parameter values but with the identical tour requirement were considered. The results are tabulated in Table 4.

Table 4. Comparison with different values of the parameters

Problem	Popsize	Iteration	Mean Error%	Variance	CPU sec
lin318	25	90,000	2.37	0.72	2858
	45	50,000	2.33	0.31	2919
pcb442	25	90,000	2.26	0.20	5476
	45	50,000	3.55	1.14	5580
vm1084	15	150,000	3.75	0.31	32270
	45	50,000	6.99	1.79	33580

The results reported in Table 4 show that the quality of solutions for the two larger problems significantly improved by increasing the number of iterations and reducing the population size. This experimentation thus leads us to set the parameters to the values that depend on the size of the problem.

A HYBRID METAHEURISTIC FOR THE PERMUTATION FLOW-SHOP SCHEDULING

Introduction

The aim of the design of a hybrid approach is to combine the strengths of some different techniques in order to improve the efficiency of a single method (Sheibani, 2011). In this chapter, we propose an evolutionary approach which is an integration of the fuzzy greedy heuristic into genetic algorithms to combinatorial optimization problems with specific reference to the flow-shop scheduling problem (FSP).

The FSP problem can be stated as follows: there are n jobs to be processed by m machines in an identical sequence on each machine. The usual objective is to minimize the completion time of the last job to leave the system, commonly termed the makespan (C_{max}). The FSP with $m > 2$ belongs to the class of combinatorial optimization problems known to be NP-hard in the strong sense (Garey *et al*, 1976). Hence, approximation methods are generally considered to be the only practical way to solve most real-life problems. A discussion of the computational complexity of the FSP can be found in (Gonzalez and Sahni, 1978; Garey and Johnson, 1979; Brucker, 2001). There are many variations of the FSP (Pinedo *et al*, 2015). In this chapter, the permutation flow-shop scheduling problem (PFSP) with makespan criterion is considered. In the PFSP after completing a job on one machine, the job is processed on the next machine or joins a queue if the machine is busy. All queues are assumed to operate under the first-in-first-out (FIFO) discipline. It means that a job cannot pass another job while waiting in a queue. Therefore, clearly there are $n!$ possible schedules.

Metaheuristics have been popular approaches for this problem (Varadharajan and Rajendran, 2005; Grabowski and Wodecki, 2004; Rajendran and Zigler, 2004; Ruiz and Stützle, 2007; Costa *et al* 2015). We will now discuss some of those methods that are most relevant to our discussion. Reeves (1995) developed a genetic algorithm which the initial population was basically generated randomly, except for one member that was constructed by the NEH heuristic (Nawaz *et al*, 1983). The selection scheme used a rank-based method with a uniform distribution. The algorithm uses the one-point order crossover operator called C1 and a shift mutation with an adaptive rate. Chen *et al* (1995) proposed a very simple genetic algorithm. In this algorithm the initial population was generated in two phases. In the first phase a number of individuals were constructed by the CDS heuristic (Campbell *et al*, 1970), except for one member, which was constructed by Dannenbring's (1977) heuristic. In the second phase, the members produced in the first phase were modified by a pairwise exchange procedure to produce additional members. The deviation of each member's makespan from that of the largest makespan in the population was utilized to define the fitness of a member. They applied the partially mapped crossover operator (PMX) of Goldberg and Lingle (1985) in the rest of the algorithm but did not use mutation. In their implemented algorithm, the solutions become stable after 20 generations. Murata *et al* (1996) developed a GA with two hybrid versions, using SA and local search (LS), in order to improve the overall performance. In their proposed GA, the initial population was randomly generated; a probability distribution was adopted as a selection scheme; a two-point crossover operator, and the shift mutation operator were applied at each generation, both with probability one. They concluded that the proposed hybrid versions outperformed all of the implemented pure methods i.e., GA, SA, and LS each used separately. Reeves and Yamada (1998) also developed a hybrid GA. The main feature of their proposed method was a local-search-based crossover operator, which they called the multi-step crossover fusion operator (MSXF). For another hybrid GA, we can mention the proposed algorithm of Wang and Zheng (2003). In this algorithm, the initial population included the solution obtained by the NEH heuristic. They also replaced an individual from the current generation by a solution that had been produced by means of an SA algorithm. This seemed to act like a mutation operator in their proposed hybrid algorithm. Etiler *et al* (2004) applied the linear order crossover operator (LOX) proposed by Falkenauer and Bouffouix (1991), and the shift mutation operator. Iyer and Saxena (2004), who used a randomly generated initial population and modified the one-point order crossover operator (OX) (Davis, 1985) in a way which preserves the largest common sub-sequence in the parent's chromosomes. They also employed the pairwise exchange mutation operator. A more recent study by Sioud and Gagné (2018) proposed an enhanced migrating bird optimization algorithm for the permutation flow-shop problem with sequence dependent setup times minimizing the makespan. A state of the art review of intelligent scheduling methods can be found in Zarandi *et al.* (2020).

Representation of Solution

In order to apply our GA to the PFSP, a solution is represented by a string of integer numbers, each of which represents the job in that position of the sequence.

Fuzzy Greedy Initialization

In genetic algorithms, the initial population is usually chosen randomly among the possible individuals. There is evidence that better starting solutions usually yield better final solutions (Sheibani, 2005). The aim of the design of a hybrid method is to combine the strengths of some different techniques in order to improve the efficiency of a single approach. In practice, the use of heuristics for producing an initial population for a GA seems to be effective for exploring the most promising regions of the search space. We developed a GA by applying a constructive heuristic for initialization.

The performance of a GA is often sensitive to the quality of its initial population (Ahuja and Orlin, 1997). In this implementation, we employed the FGH heuristic for generating the initial population. We randomly chose sufficient values of λ in the FGH heuristic's effective range, including λ_{max} to construct a sequence (it should be noted that the FGH heuristic reduces to the NEH heuristic for $\lambda \geq \lambda_{max}$, see Sheibani (2010). This process is repeated until the initial population is completed. We will refer to this process as the FGH initialization. The steps are shown in Algorithm 2. The FGH heuristic for the PFSP consists of two phases: arranging the jobs in priority order and then constructing a sequence by a simple job insertion principle similar to the NEH heuristic (Nawaz *et al*, 1983). The priority of the jobs is determined according to Equation (5). Here, x is a generic variable associated with the data defining a particular instance of the PFSP. The parameter θ is a basic measure for evaluating the priority to be assigned to x. The parameter λ is a tuning parameter that is chosen by experimentation in the heuristic's effective range between λ_{min} and λ_{max} to adjustθ. Let $P_{i,j}$ be the processing time of job j on machine i. We represent x with x_j as the sum of the processing times of job j on each machine, and θ as the average of the sums of the processing times of the jobs on each machine, through the following equations:

$$x_j = \sum_{i=1}^{m} P_{ij} \tag{6}$$

$$\theta = \frac{1}{n}\sum_{j=1}^{n} x_j \tag{7}$$

Algorithm 2. The FGH initialization for the PFSP.

```
WHILE (initial population is not completed) DO
BEGIN
1.          λ = Rand (λmin, λmax).
2.          Calculate μ(xj) for each job j.
3.          Arrange the jobs by descending order of μ(xj).
4.          Select the next job and insert it in all possible positions in the
partial sequence and keep the best one (i.e. minimum makespan) as the current
partial sequence.
5.          Repeat step 4 until all jobs are scheduled.
6.          Add current solution to initial population.
END
RETURN initial_population;
```

Figure 7 exemplifies the effect of different values of λ on the computational performance of the proposed heuristic. For more details on the FGH heuristic for the PFSP, we refer the interested reader to Sheibani (2010 and 2017).

Figure 7. The FGH heuristic on a PFSP instance.

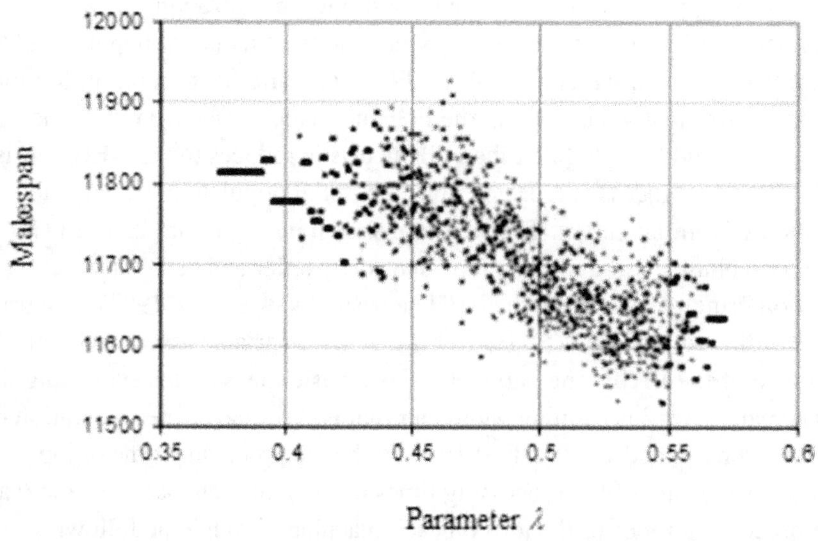

Selection Scheme

A mixed strategy based on the roulette wheel selection and the elitist replacement was adopted. The roulette wheel method uses a probability distribution for selection of a chromosome, which is proportional to its fitness, here in this implementation we have considered as $1/C_{max}$. Elitist replacement puts the fittest chromosome in the current population directly into the next generation.

Crossover Operator

There are several crossover operators that can be used for sequencing problems (Belfiore *et al*, 1998). Some good examples are OX and LOX crossover operators. In this experimentation, we used PMX operator. The operating principle is as follows: We build the offspring by randomly choosing a sub-sequence from one parent and preserving the order and positions of as many jobs as possible from the other parent, in which each job of the sub-sequence from the alternate parent is mapped to the position held by this job in the first parent. The remaining jobs are inherited from the alternate parent. The operating principle of the PMX is shown schematically in Figure 8. Let P_1 and P_2 be two selected chromosomes from the previous generation. Each is a string of 10 arranged jobs, numbered 0 to 9, which represents the order of the jobs on a sequence. In the figure, a randomly selected sub-sequence is shown between two cut points marked by '½' in the chromosomes.

Figure 8. The partially mapped crossover (PMX).

Parents

P_1:(0 1 2 3 | 4 5 6 7 | 8 9)

P_2:(4 3 2 1 | 0 9 8 7 | 6 5)

Offspring

O_1:(4 1 2 3 | 0 9 8 7 | 6 5)

O_2:(0 3 2 1 | 4 5 6 7 | 8 9)

Mutation Operator

We used the simple pairwise exchange operator as a mutation by selecting two points along the length of the single chromosome at random and then jobs at these two points are exchanged. The operating principle is shown schematically in Figure 9.

Figure 9. The simple pairwise exchange operator.

Mutation

(0123456789) → (7123456089)

Parameter Setting

The proposed GA was implemented in C++. The test problems are part of an extensive set of standard benchmark instances (Taillard, 1993). We evaluate the performance of the developed GA against its simple version (i.e. starting off with a random initial population). The solution quality is measured by the percentage deviation of the obtained solution from the best-known solution.

The performance of a GA greatly depends on the structure of the problem considered, architecture of the algorithm and the settings for the algorithm parameters such as population size, crossover and mutation rates. Finding universal parameter values to prevent premature convergence still remains an unsolved problem. Our experimentation showed that good performance of the proposed GA is obtained when setting the genetic parameters as follows: population size 50, crossover probability $P_c = 0.6$ and mutation probability $P_m = 0.4$. The termination criterion is 50,000 iterations. In this experimentation, we performed 10 replications with different seeds common for all instances. The random number generator used the standard C++ library function.

Effectiveness of the Initial Population

We examined the performance of the proposed method against its simple version (i.e. starting off with a random initial population), all other settings were the same. Table 5 summarizes the computational results.

Table 5. Comparison of two different initialization methods

Problem (Job × Machine)		FGH-Initialization Mean Error%	FGH-Initialization Variance	Random-Initialization Mean Error%	Random-Initialization Variance
20 × 5	ta001	0	0	1.34	0.19
	ta002	0.33	0.03	0.28	0.05
	ta003	0.71	0.24	1.35	0.32
	ta004	0.64	0.02	1.47	0.35
	ta005	1.14	0	1.06	0.05
	ta006	0	0	1.19	0.03
	ta007	0.57	0.07	1.37	0
	ta008	0.20	0.07	0.61	0.14
	ta009	1.42	0.05	1.58	1.15
	ta010	0.27	0.27	1.13	1.14
50 × 20	ta051	3.33	0.06	5.32	0.80
	ta052	3.87	0.12	6.34	0.48
	ta053	4.93	0.21	7.22	1.73
	ta054	4.15	0.20	5.56	0.91
	ta055	4.73	0.24	6.03	1.10
	ta056	3.95	0.04	5.80	1.12
	ta057	4.47	0.27	6.12	0.64
	ta058	4.86	0.28	6.48	0.56
	ta059	4.01	0.05	6.57	0.64
	ta060	3.97	0.06	5.57	1.27
200 × 10	ta091	0.69	0	1.71	0.15
	ta092	0.92	0.01	2.70	0.13
	ta093	1.03	0	2.07	0.05
	ta094	0.68	0.21	1.21	0.10
	ta095	0.51	0.01	2.80	0.33
	ta096	0.75	0.03	2.37	0.32
	ta097	0.69	0.02	2.17	0.08
	ta098	0.74	0.01	2.11	0.04
	ta099	0.59	0.02	2.18	0.17
	ta100	0.89	0.02	2.19	0.17
Average		**1.83**	**0.09**	**3.13**	**0.47**

It is easy to observe that the proposed FGH initialization method has a very significant influence on the quality of the overall solution. Given that the initialization process was computationally inexpensive compared to the rest of the GA, this shows the value of using an effective initialization process.

Computational Analysis

We investigated the number of generations that the proposed GA requires to converge to the local optima obtained in our results in which 50,000 iterations were performed for each replication. We observed that in 26.33% of the cases (out of a total of 300 independent runs corresponding to the 10 instances from each group of problems considered), the fittest initial solution generated showed no improvement even through 50,000 iterations. In our opinion, this could be due in part to the high quality initial solutions that are obtained using the FGH heuristic. Another possible explanation is that the recombination method used was not sufficiently effective. Alternative recombination methods could be investigated. We also examined the performance of the proposed hybrid method when the maximum number of makespan evaluations (*popsize* × *iterations*) is reduced by a factor of 50. We changed the population size to 10 and the number of iterations to 5000, giving 50,000 evaluations of makespan in total. The results are reported in Table 6.

Table 6. The developed GA with the different setups

Problem Size (Job × Machine)	*Problem*	*Popsize*	P_c	P_m	*Iteration*	*Best Error%*	*Mean Error%*	*Worst Error%*	*Variance*
20×5	ta001-010	10	0.6	0.4	5000	0.41	1.00	1.82	0.30
50×20	ta051-060	10	0.6	0.4	5000	3.96	4.74	5.34	0.25
200×10	ta091-100	10	0.6	0.4	5000	0.62	0.94	1.10	0.03
Average						**1.66**	**2.23**	**2.75**	**0.19**
20×5	ta001-010	50	0.6	0.4	50,000	0.31	0.53	0.90	0.07
50×20	ta051-060	50	0.6	0.4	50,000	3.70	4.23	4.86	0.15
200×10	ta091-100	50	0.6	0.4	50,000	0.52	0.75	1.00	0.03
Average						**1.51**	**1.83**	**2.25**	**0.09**

The results reported in Table 3 show that the quality of solutions for the smaller problem significantly improved by increasing the number of iterations and the population size. This experimentation thus leads us to set the parameters to the values that depend on the size of the problem. Given these results, it might be reasonable to plan experimentations in such a way that larger problems are given a larger number of iterations.

Concluding Remarks

This chapter proposed a new idea for the integration of approaches for hard combinatorial optimisation problems. The effectiveness and efficiency of this so-called fuzzy greedy evaluation concept have been investigated within search, optimisation and learning systems for hard combinatorial optimisation problems. In this context, a hybrid genetic algorithm has been developed for the travelling salesman problem

(TSP), and a hybrid genetic algorithm has been developed for the permutation flow-shop scheduling problem (PFSP).

The proposed methodology has the potential for application to other combinatorial optimization problems, if a suitable evaluation measure can be properly defined based on global information for a given problem solution. For future research we believe that the following topics are potentially useful: (1) developing efficient adaptations of the proposed heuristic, (2) extending the proposed method to other variations or objectives, and (3) developing efficient methods using the fuzzy greedy evaluation concept in other areas of optimization.

REFERENCES

Ahuja, R. K., & Orlin, J. B. (1997). Developing fitter genetic algorithms. *INFORMS Journal on Computing*, *9*(3), 251–253. doi:10.1287/ijoc.9.3.251

Back, T. (1992), The interaction of mutation rate, selection, and self-adaptation within a genetic algorithm. *Proceedings of the 2nd International Conference on Parallel Problem Solving From Nature*, 85-94.

Baker, K. R. (1974). *Introduction to Sequencing and Scheduling*. John Wiley & Sons.

Bang-Jensen, J., Gutin, G., & Yeo, A. (2004). When the greedy algorithm fails. *Discrete Optimization*, *1*(2), 121–127. doi:10.1016/j.disopt.2004.03.007

Beasley, J. E. (1990). An algorithm for set covering problems. *European Journal of Operational Research*, *31*(1), 85–93. doi:10.1016/0377-2217(87)90141-X

Belfiore, N. P., Esposito, A., & Vincent, T. L. (1998). A theoretical and experimental study of crossover operators of genetic algorithms. *Journal of Optimization Theory and Applications*, *99*(2), 271–302. doi:10.1023/A:1021766025497

Boschetti, M., & Maniezzo, V. (2015). A set covering based matheuristic for a real-world city logistics problem. *International Transactions in Operational Research*, *22*(1), 169–195. doi:10.1111/itor.12110

Brucker, P. (2001). *Scheduling Algorithms* (3rd ed.). Springer-Verlag. doi:10.1007/978-3-662-04550-3

Campbell, H. G., Dudek, R. A., & Smith, M. L. (1970). A heuristic algorithm for the n-job m-machine sequencing problem. *Management Science*, *16*(10), B630–B637. doi:10.1287/mnsc.16.10.B630

Cela, E., Deineko, V. G., & Woeginger, G. J. (2014). Linearizable special cases of the QAP. *Journal of Combinatorial Optimization*, 1–11.

Chatterjee, S., Carrera, C., & Lynch, L. A. (1996). Genetic algorithms and traveling salesman problems. *European Journal of Operational Research*, *93*(3), 490–510. doi:10.1016/0377-2217(95)00077-1

Chen, C. L., Vempati, V. S., & Aljaber, N. (1995). An application of genetic algorithms for flow shop problems. *European Journal of Operational Research*, *80*(2), 389–396. doi:10.1016/0377-2217(93)E0228-P

Cheng, R., & Gen, M. (1994). Crossover on intensive search and traveling salesman problem. *Computers & Industrial Engineering*, *27*(1-4), 485–488. doi:10.1016/0360-8352(94)90340-9

Coelho, L. C., & Laporte, G. (2015). Classification, models and exact algorithms for multi-compartment delivery problems. *European Journal of Operational Research*, *242*(3), 854–864. doi:10.1016/j.ejor.2014.10.059

Cook, S. A. (1971). The complexity of theorem-proving procedures. *Proceedings of the 3rd Annual ACM Symposium on Theory of Computing*, 151-158. 10.1145/800157.805047

Cook, W. J., Cunningham, W. H., Pulleyblank, W. R., & Schrijver, A. (1998). *Combinatorial Optimization*. John Wiley & Sons.

Cormen, T., Leiserson, C. E., & Rivest, R. L. (1990). *Introduction to Algorithms*. The MIT Press.

Costa, A., Cappadonna, F. A., & Fichera, S. (2015). A hybrid genetic algorithm for minimizing makespan in a flow-shop sequence-dependent group scheduling problem. *Journal of Intelligent Manufacturing*, 1–15.

Curtis, S. A. (2003). The classification of greedy algorithms. *Science of Computer Programming*, *49*(1-3), 125–157. doi:10.1016/j.scico.2003.09.001

Dannenbring, D. G. (1977). An evaluation of flow shop sequencing heuristics. *Management Science*, *23*(11), 1174–1182. doi:10.1287/mnsc.23.11.1174

Dantzig, G. B., Fulkerson, D. R., & Johnson, S. M. (1954). Solution of a large scale traveling salesman problem. *Operations Research*, *2*, 393–410.

Davis, L. (1985). Applying adaptive algorithms to epistatic domains. *Proceedings of the 9th International Joint Conference on Artificial Intelligence*, 162–164.

Dijkstra, E. W. (1959). A note on two problems in connexion with graphs. *Numerische Mathematik*, *1*(1), 269–271. doi:10.1007/BF01386390

Du, D. Z., & Pardalos, P. M. (Eds.). (2005). *Handbook of Combinatorial Optimization*. Springer-Verlag. doi:10.1007/b102533

Edmonds, J. (1971). Matroids and the greedy algorithm. *Mathematical Programming*, *1*(1), 126–136. doi:10.1007/BF01584082

Etiler, O., Toklu, B., Atak, M., & Wilson, J. M. (2004). A genetic algorithm for flow shop scheduling problems. *The Journal of the Operational Research Society*, *55*(8), 830–835. doi:10.1057/palgrave.jors.2601766

Falkenauer, E., & Bouffouix, S. (1991). A genetic algorithm for job shop. *Proceedings of the IEEE International Conference on Robotics and Automation*, 824–829.

Feo, T. A., & Resende, M. G. C. (1995). Greedy randomized adaptive search procedures. *Journal of Global Optimization*, *6*(2), 109–133. doi:10.1007/BF01096763

French, S. (1982). *Sequencing and Scheduling: An Introduction to the Mathematics of the Job-Shop*. Ellis Horwood.

Garey, M. R., & Johnson, D. S. (1979). *Computers and Intractability a Guide to the Theory of NP-Completeness*. W.H. Freeman and Company.

Garey, M. R., Johnson, D. S., & Sethi, R. (1976). The complexity of flowshop and jobshop scheduling. *Mathematics of Operations Research, 1*(2), 117–129. doi:10.1287/moor.1.2.117

Goldberg, D. E., & Lingle, R. (1985). Alleles, loci and the travelling salesman problem. In *Proceedings of the 2nd International Conference on Genetic Algorithms*. Lawrence Erlbaum.

Gonzalez, T., & Sahni, S. (1978). Flowshop and Jobshop Schedules: Complexity and Approximation. *Operations Research, 26*(1), 36–52. doi:10.1287/opre.26.1.36

Grabowski, J., & Wodecki, M. (2004). A very fast tabu search algorithm for the permutation flow shop problem with makespan criterion. *Computers & Operations Research, 31*(11), 1891–1909. doi:10.1016/S0305-0548(03)00145-X

Grefenstette, J. J., Gopal, R., Rosmaita, B., & Van Gucht, D. (1985), Genetic algorithms for the traveling salesman problem. *Proceedings the 2nd International Conference on Genetic Algorithms*, 160-168.

Grimaldi, R. P. (1989). *Discrete and Combinatorial Mathematics*. Addison-Wesley.

Gutin, G. (2013). Section 4.6 Traveling Salesman Problems. Handbook of Graph Theory, 336.

Gutin, G., & Punnen, A. P. (2002). *The Traveling Salesman Problem and Its Variations*. Kluwer Academic Publishers.

Gutin, G., & Yeo, A. (2002). Anti-matroids. *Operations Research Letters, 30*(2), 97–99. doi:10.1016/S0167-6377(02)00117-7

Gutin, G., Yeo, A., & Zverovich, A. (2002). Traveling salesman should not be greedy: Domination analysis of greedy-type heuristics for the TSP. *Discrete Applied Mathematics, 117*(1-3), 81–86. doi:10.1016/S0166-218X(01)00195-0

Ha, Q. M., Deville, Y., Pham, Q. D., & Hà, M. H. (2020). A hybrid genetic algorithm for the traveling salesman problem with drone. *Journal of Heuristics, 26*(2), 219–247. doi:10.100710732-019-09431-y

Ho, J. C. (1995). Flowshop sequencing with mean flowtime objective. *European Journal of Operational Research, 81*(3), 571–578. doi:10.1016/0377-2217(93)E0353-Y

Holland, J. H. (1975). *Adaptation in natural and artificial systems: an introductory analysis with applications to biology, control, and artificial intelligence*. University of Michigan Press.

Horowitz, E., & Sahni, S. (1978). *Fundamentals of Computer Algorithms*. Computer Science Press.

Iyer, S. K., & Saxena, B. (2004). Improved genetic algorithm for the permutation flowshop scheduling problem. *Computers & Operations Research, 31*(4), 593–606. doi:10.1016/S0305-0548(03)00016-9

Johnson, S. M. (1954). Optimal Two-and-Three-Stage Production Schedules with Set-up Times Included. *Naval Research Logistics, 1*(1), 61–68. doi:10.1002/nav.3800010110

Klir, G. J., & Yuan, B. (1995). *Fuzzy Sets and Fuzzy Logic: Theory and Applications*. Prentice Hall.

Korte, B., & Lovasz, L. (1981). Mathematical structures underlying greedy algorithms. In Fundamentals of Computation Theory. Lecture Notes in Computer Science. Springer-Verlag. doi:10.1007/3-540-10854-8_22

Korte, B., & Lovasz, L. (1983). Structure properties of greedoids. *Combinatorica, 3*(3-4), 359–374. doi:10.1007/BF02579192

Korte, B., Lovasz, L., & Schrader, R. (1991). *Greedoids*. Springer-Verlag. doi:10.1007/978-3-642-58191-5

Korte, B., & Vygen, J. (2012). *Combinatorial optimization: theory and algorithms. Algorithms and Combinatorics* (Vol. 21). Springer-Verlag Berlin Heidelberg. doi:10.1007/978-3-642-24488-9

Kruskal, J. B. (1956). On the shortest spanning subtree and the traveling salesman problem. *Proceedings of the American Mathematical Society, 7*(1), 48–50. doi:10.1090/S0002-9939-1956-0078686-7

Laporte, G. (2010). A concise guide to the Traveling Salesman Problem. *The Journal of the Operational Research Society, 61*(1), 35–40. doi:10.1057/jors.2009.76

Laporte, G., & Palekar, U. (2002). Some applications of the clustered travelling salesman problem. *The Journal of the Operational Research Society, 53*(9), 972–976. doi:10.1057/palgrave.jors.2601420

Lawler, E. L. (1985). *The Traveling Salesman Problem: A Guided Tour of Combinatorial Optimization*. John Wiley & Sons.

Liu, J., & Reeves, C. R. (2001). Constructive and Composite Heuristic Solutions to the P//åCi Scheduling Problem. *European Journal of Operational Research, 132*(2), 439–452. doi:10.1016/S0377-2217(00)00137-5

Martello, S., & Ries, B. (2015). Advances in Combinatorial Optimization. *Discrete Applied Mathematics, 196*(C), 1–3. doi:10.1016/j.dam.2015.08.011

Merz, P., & Freisleben, B. (2001). Memetic algorithms for the traveling salesman problem. *Complex Systems, 13*, 297–345.

Michalewicz, Z. (1996). *Genetic algorithms + data structures = evolution programs* (3rd ed.). Springer-Verlag. doi:10.1007/978-3-662-03315-9

Mohanty, S., Patra, P. K., Mohapatra, S., & Ray, M. (2017). MPSO: A Novel Meta-Heuristics for Load Balancing in Cloud Computing. *International Journal of Applied Evolutionary Computation, 8*(1), 1–25. doi:10.4018/ijaec.2017010101

Murata, T., Ishibuchi, H., & Tanaka, H. (1996). Genetic algorithms for flowshop scheduling problems. *Computers & Industrial Engineering, 30*(4), 1061–1071. doi:10.1016/0360-8352(96)00053-8

Nawaz, M., Enscore, E. E. Jr, & Ham, I. (1983). A heuristic algorithm for the m-machine, n-job flow-shop sequencing problem. *Omega International Journal of Management Science, 11*(1), 91–95. doi:10.1016/0305-0483(83)90088-9

Neapolitan, R. E., & Naimipour, K. (2003). *Foundations of Algorithms Using C++ Pseudo-code*. Jones & Bartlett Publishers.

Nishi, T., Ohno, A., Inuiguchi, M., Takahashi, S., & Ueda, K. (2017). A Combined column generation and heuristics for railway short-term rolling stock planning with regular inspection constraints. *Computers & Operations Research*, *81*, 14–25. doi:10.1016/j.cor.2016.11.025

Oliver, I. M., Smith, D. J., & Holland, J. R. C. (1987), A study of permutation crossover operators on the traveling salesman problem. *Proceedings the 2nd International Conference on Genetic Algorithms*, 224-230.

Papadimitriou, C. D., & Steiglitz, K. (1982). *Combinatorial Optimization: Algorithms and Complexity*. Prentice-Hall.

Pinedo, M., Zacharias, C., & Zhu, N. (2015). Scheduling in the service industries: An overview. *Journal of Systems Science and Systems Engineering*, *24*(1), 1–48. doi:10.100711518-015-5266-0

Prim, R. C. (1957). Shortest connection networks and some generalizations. *The Bell System Technical Journal*, *36*(6), 1389–1401. doi:10.1002/j.1538-7305.1957.tb01515.x

Pulleyblank, W. R. (Ed.). (2014). *Progress in combinatorial optimization*. Academic Press.

Qu, L., & Sun, R. (1999). A synergetic approach to genetic algorithms for solving travelling salesman problem. *Information Sciences*, *117*(3-4), 267–283. doi:10.1016/S0020-0255(99)00026-2

Rajendran, C., & Ziegler, H. (2004). Ant-colony algorithms for permutation flowshop scheduling to minimize makespan/total flowtime of jobs. *European Journal of Operational Research*, *155*(2), 426–438. doi:10.1016/S0377-2217(02)00908-6

Reeves, C. R. (1995). A genetic algorithm for flowshop sequencing. *Computers & Operations Research*, *22*(1), 5–13. doi:10.1016/0305-0548(93)E0014-K

Reeves, C. R. (1997). Genetic algorithms for the operations researcher. *INFORMS Journal on Computing*, *9*(3), 231–250. doi:10.1287/ijoc.9.3.231

Reeves, C. R., & Yamada, T. (1998). Genetic algorithms, path relinking, and the flow-shop sequencing prob-lem. *Evolutionary Computation*, *6*(1), 45–60. doi:10.1162/evco.1998.6.1.45 PMID:10021740

Reinelt, G. (1991). TSPLIB - a traveling salesman problem library. *ORSA Journal on Computing*, *3*(4), 376–384. doi:10.1287/ijoc.3.4.376

Ruiz, R., & Maroto, C. (2005). A comprehensive review and evaluation of permutation flowshop heuristics. *European Journal of Operational Research*, *165*(2), 479–494. doi:10.1016/j.ejor.2004.04.017

Ruiz, R., Maroto, C., & Alcaraz, J. (2006). Two new robust genetic algorithms for the flowshop scheduling problem. *Omega the International Journal of Management Science*, *34*(5), 461–476. doi:10.1016/j.omega.2004.12.006

Ruiz, R., & Stützle, T. (2007). A simple and effective iterated greedy algorithm for the permutation flow shop problem. *European Journal of Operational Research*, *177*(3), 2033–2049. doi:10.1016/j.ejor.2005.12.009

Schmitt, L. J., & Amini, M. M. (1998). Performance characteristics of alternative genetic algorithmic approaches to the traveling salesman problem using path representation: An empirical study. *European Journal of Operational Research*, *108*(3), 551–570. doi:10.1016/S0377-2217(97)00206-3

Sheibani, K. (2004). An adaptive greedy genetic algorithm using fuzzy reasoning for the travelling salesman problem. *Proc. Hybrid Metaheuristics, 16th European Conf. on Artificial Intelligence*, 31-49.

Sheibani, K. (2005). *Fuzzy Greedy Evaluation in Search, Optimisation, and Learning* [PhD thesis]. London Metropolitan University, London, UK.

Sheibani, K. (2006). *Fuzzy greedy search and job-shop problem*. Proc. the UK Planning & Scheduling Special Interest Group.

Sheibani, K. (2008). Fuzzy Greedy Search in Combinatorial Optimisation. Tadbir Institute for Operational Research, Systems Design, and Financial Services.

Sheibani, K. (2010). A fuzzy greedy heuristic for permutation flow-shop scheduling. *The Journal of the Operational Research Society*, *61*(5), 813–818. doi:10.1057/jors.2008.194

Sheibani, K. (2010). An evolutionary approach using fuzzy greedy initialization to permutation flow-shop scheduling with the makespan criterion. *Proc. IEEE Intl. Conf. on Industrial Engineering and Engineering Management*, 638-641. 10.1109/ICIEEM.2010.5646535

Sheibani, K. (2010). The fuzzy greedy search in combinatorial optimization with specific reference to the travelling salesman problem. *Proc. IEEE Intl. Conf. on Industrial Engineering and Engineering Management*, 1367-1370. 10.1109/IEEM.2010.5674393

Sheibani, K. (2011). A hybrid metaheuristic using fuzzy greedy search operator for combinatorial optimization with specific reference to the travelling salesman problem. *Iranian Journal of Operations Research*, *2*(2), 63–71.

Sheibani, K. (2013). A constructive heuristic for permutation flow-shop scheduling with the makespan criterion: in a particular case where the sum of the processing times of each job is the same on each machine. *Proc. 5th Intl. Conf. on Applied Operational Research*, 18-24.

Sheibani, K. (2017). Fuzzy greedy search heuristic for combinatorial optimization with specific application to machine scheduling. *International Journal of Management Science and Engineering Management*, *12*(2), 148–153. doi:10.1080/17509653.2016.1172993

Sheibani, K. (2018). A very fast heuristic for combinatorial optimization with specific application to priority rule sequencing in operations management: Fuzzy greedy search. *International Journal of Operations Research and Information Systems*, *9*(3), 77–89. doi:10.4018/IJORIS.2018070104

Sheibani, K. (2019). An integrated heuristic for machine sequencing with specific reference to the permutation flow-shop scheduling problem. *International Journal of Strategic Engineering*, *2*(2), 1–8. doi:10.4018/IJoSE.2019070101

Sheibani, K. (2019). An annotated bibliography of fuzzy greedy search. *Operational Research and Management Science Letters*, *2*(1), 2–5.

Sheibani, K. (2021). Fuzzy greedy search: an algorithmic approach for combinatorial optimisation. In V. González-Prida & M. C. Carnero (Eds.), *Advanced Models and Tools for Effective Decision Making Under Uncertainty and Risk Contexts* (pp. 254–275). IGI Global. doi:10.4018/978-1-7998-3246-1.ch010

Sioud, A., & Gagné, C. (2018). Enhanced migrating birds optimization algorithm for the permutation flow shop problem with sequence dependent setup times. *European Journal of Operational Research, 264*(1), 66–73. doi:10.1016/j.ejor.2017.06.027

Sule, D. R. (1997). *Industrial Scheduling*. PWS Publishing Company.

Taillard, E. (1993). Benchmarks for basic scheduling problems. *European Journal of Operational Research, 64*(2), 278–285. doi:10.1016/0377-2217(93)90182-M

Tang, L., & Liu, J. (2002). A modified genetic algorithm for the flow shop sequencing problem to minimize mean flow time. *Journal of Intelligent Manufacturing, 13*(1), 61–67. doi:10.1023/A:1013681029869

Tucker, A. (1995). *Applied Combinatorics*. John Wiley & Sons.

Valdez, F., Moreno, F., & Melin, P. (2020). A Comparison of ACO, GA and SA for Solving the TSP Problem. In *Hybrid Intelligent Systems in Control, Pattern Recognition and Medicine* (pp. 181–189). Springer. doi:10.1007/978-3-030-34135-0_13

Varadharajan, T. K., & Rajendran, C. (2005). A multi-objective simulated-annealing algorithm for scheduling in flowshops to minimize the makespan and total flowtime of jobs. *European Journal of Operational Research, 167*(3), 772–795. doi:10.1016/j.ejor.2004.07.020

Vince, A. (2002). A framework for the greedy algorithm. *Discrete Applied Mathematics, 121*(1-3), 247–260. doi:10.1016/S0166-218X(01)00362-6

Wang, L., & Zheng, D. Z. (2003). An effective hybrid heuristic for flow shop scheduling. *International Journal of Advanced Manufacturing Technology, 21*(1), 38–44. doi:10.1007001700300005

Wang, Z., & Klir, G. (2013). *Fuzzy measure theory*. Springer Science & Business Media.

Whitney, H. (1935). On the abstract properties of linear dependence. *American Journal of Mathematics, 57*(3), 509–533. doi:10.2307/2371182

Wilson, R. J. (1985). *Introduction to Graph Theory* (3rd ed.). Longman.

Yun, Y., & Gen, M. (2003). Performance analysis of adaptive genetic algorithms with fuzzy logic and heuristics. *Fuzzy Optimization and Decision Making, 2*(2), 161–175. doi:10.1023/A:1023499201829

Zarandi, M. H. F., Asl, A. A. S., Sotudian, S., & Castillo, O. (2020). A state of the art review of intelligent scheduling. *Artificial Intelligence Review, 53*(1), 501–593. doi:10.100710462-018-9667-6

Zimmermann, H. J. (2001). *Fuzzy Set Theory and its Applications* (4th ed.). Springer-Verlag. doi:10.1007/978-94-010-0646-0

Chapter 2
Creativity as the Ultimate Asset in the COVID-19 Reconfigured Organizational Environment

Diana Santos Fernandes
School of Economics and Management, University of Minho, Portugal

Carolina Feliciana Machado
https://orcid.org/0000-0002-9685-1576
School of Economics and Management, University of Minho, Portugal & Interdisciplinary Centre of Social Sciences (CICS), University of Minho, Portugal

ABSTRACT

Whereas the COVID-19 pandemic tested the solidity, agility, and resilience of organizations, it as well enhanced a refined debate on the conceptual frameworks that have traditionally been guiding the managerial decisions and organizational structures, policies, and practices. This chapter aims at exploring the impact of the COVID-19 pandemic on knowledge, technology, creativity, and innovation management research, highlighting creativity as the core vector to understand the reconfiguration of the renewed corporate structures and processes. By literature review, it identifies key concepts, assumptions, and theoretical constructs, aiming at highlighting creativity as the core asset to understand how the worldwide organizations have been able to overcome the twofold challenges and opportunities of the recent environmental conditions, defending that creativity hence emerged as the core asset so that organizations could test and reinforce their resilience, boosting overall performance via transversal dynamics to all the organization's structure, stakeholders, policies, and practices.

INTRODUCTION

As of the year 2020, the world has been challenged with the wide-ranging impacts of the Covid-19 pandemic (George et al., 2020). We have been assisting to a reconfiguration of the worldwide dynamics

DOI: 10.4018/978-1-6684-2339-4.ch002

into paths never imagined because as of that year we have been assisting to a phenomenon of global scale which crystalized into the translation of a health crisis into an economic, financial, and social one.

The truth is that the economic world as we know has faced a clear and profound redefinition, marked by deep transformations and creative solutions to the refined dynamics it faced. It is possible to point out that the widespread geographical proliferation of the virus stimulated social distancing as both a public health need, but also as a subsequent daily habit, which has ended up by conducting to the shutdown of businesses, events, and corporate offices, henceforth, notably affording negative consequences on the so far fueled undercurrents on the global financial markets. Not disregarding this economic impact of the Coronavirus, we can derive a second vector to explain its imprint on the renewed organizational subtleties, if we acknowledge the exponential rate at which the virus had been spreading, and for sure as well the delicate and sharp uncertainty on its current and future situation. This way, we can argue that it climaxed into a generalized feeling of constant panic by the worldwide population, by which people had been impelled to design and adopt safe and preventive habits not only in their daily routines (and in those, we stress their work dynamics) but also in their overall patterns of consumption and investment. And this alteration is assumed as wide-ranging not only in its geographical scope, but also on the actors it impacted, given that both consumers, investors, international trade partners, employers, and employees have been profoundly affected.

Based on the above mentioned, it is possible to defend that, whereas this pandemic has undeniably and profoundly tested the solidity, agility, and resilience of organizations, it as well keeps stressing the urgency of a refined academic debate and reviewed managerial approaches in what concerns to the core assumptions fundamental to the conceptual frameworks that have traditionally been guiding the managerial decisions, corporate structural configurations, and organizational policies and practices, and this based on the premise that, with this new global context, both new-fangled problems and challenges arose, nevertheless, as well revigorated opportunities have been afforded to the organizational environment, in both professional and personal spheres affecting its overall stakeholders.

Henceforth, the present chapter assumes both academic and managerial relevance, based on the argument (as already exposed) that it is consensual that, in recent times, the world has been experiencing uncertain and turbulent environmental conditions, which impact the financial, economic, social, and political scopes. As so, the worldwide complex interdependencies have been stressfully challenged with increased competition, shortages on the demand on certain products/services and economic sectors, constraints on the supply of certain goods and services, volatile dynamics on pricing, and unpredictable technological changes, coupled and worsened with the Covid-19 pandemic. This to say, the present chapter aims at providing not only a review and exposition around the consequences this pandemic has addressed to the sustainability of businesses but, and subsequently, it aims as well at providing useful insights on which assets have assumed with amplified significance as a means to strengthen organizational resilience to overcome those reconfigured challenges, this way highlighting Creativity as the ultimate organizational asset to strive on this new-fangled global corporate environment. This also to say, stressing the relevance of Creativity as the ultimate organizational asset to boost organizational performance, premise that we sustain to have gain *momentum* and fundament in the current world, due to the twofold challenges and opportunities brought about by the worldwide impacts of the Covid-19 pandemic on the corporate level.

The chapter subsequently emerges as a result of the above-mentioned reflection, so, it develops as a thinking product acknowledging that the conditions which mark the refined worldwide dynamics require all stakeholders to realize the urgency of encouraging all organizational actors to be creative, hence imprinting Creativity on the overall organizational structure, strategies, policies, tools, and practices

(Amabile, 1988; Dnishev & Alzhanova, 2016; Shalley & Gilson, 2004). Empirical evidence shows that employee creativity sharply contributes to organizational innovation, effectiveness, efficiency, and sustainability (Shalley et al., 2004), hence, this chapter will add to the literature by providing further insights which corroborate this line of argumentation, highlighting the impact of this organizational asset as a core vector for increased levels of corporate performance, sustaining the competitive advantage.

Henceforth, the chapter aims to explore the impact of the Covid-19 pandemic on Knowledge, Technology, Creativity, and Innovation Management research, highlighting Creativity as the core vector to understand the underlying subtleties and impacts, and how it has then affected the reconfiguration of the renewed corporate structures and dynamics. As so, we aim at weighting Creativity as the core asset to understand how the worldwide organizations have been able to overcome the twofold challenges and opportunities of the recent environmental conditions, defending that Creativity indeed emerged in this context as the core asset so that organizations could stress their resilience and boost overall performance. The chapter, thus, identifies key concepts, assumptions, and theoretical constructs, also exposing them on their managerial impacts, to then discuss how these both differ and interconnect, providing useful insights on how such areas of investigation and core organizational assets emerged and developed, on this reconfigured global reality.

The chapter, hence, articulates on the premise which postulates that the better we know Creativity, then the more understandable the theoretical construct and managerial application of Innovation will become; hence, it will be possible to achieve greater chances to boost organizational performance, in the first instance indeed because Innovation needs Creativity.

The argumentation will, therefore, be structured around two main points, which despite being different from each other and, thus, possible to be dissected and analyzed separately, assume as intrinsically connected in their theoretical construction and managerial implications. Therefore, the argumentation will develop as a continuous and logical cadence of ideas, so that all sections communicate with each other to climax by high pointing the relevance of Creativity as the ultimate corporate asset in this reconfigured pandemic organizational environment.

This to say, the first section of the chapter will provide a theoretical background review on the concepts of Knowledge, Creativity, and Innovation, specifying their idiosyncrasies but also exposing the main vectors in which they correspond. Framed by this background, in a second section, the chapter will explore further insights on Creativity, aiming at ascertaining what is possible to learn about Creativity from the organizations' response to Covid-19. This section will proceed with a literature review to theoretically address the twofold opportunities and challenges for Creativity in the organizational environment, specifically framed in the Covid-19 era.

Hence, in methodological terms, the chapter consists of theoretical work, adopting a descriptive exposition comparing and interconnecting different inputs and concepts from various disciplines and research fields, via a deductive analysis. This will be so, as the chapter will add to the existing research on the topic via presenting a systematic literature review of Knowledge, Innovation and Creativity theory in organizational contexts, and their potential towards boosting organizational performance, as a means to stress the organizational resilience in the turbulent times, which are a consensual characteristic of the current Covid-19 world.

This chapter will provide additional contribution to the current debate on the drivers of Organizational Performance by emphasizing the importance of both organizational as well as individual-level factors, this to say, by highlighting people as the key asset of all organizations to obtain a sustainable competitive advantage. We believe this consists of an important aspect to shed light on. Resulting from this line of thinking, this chapter will provide additional inputs which will highpoint that Creativity acts as a result of people skills and knowledge, being operationalized and boosted by the correct organizational structures, policies, and practices.

In conclusion, this chapter's originality stems from the fact that it consubstantiates an insightful and enriching contribution to the field of knowledge because it interrelates multidisciplinary contributions, interconnecting concepts that have traditionally been presented in the literature as often juxtaposed, which has confused both theoretical debates as well as empirical research and subsequent managerial decisions, hence providing useful insights to expose the conceptual and practical differences and correspondences between them. Therefore, this chapter will be providing an interesting contribution to literature, imprinting a more ambitious and rigorous investigation.

BACKGROUND

Knowledge

The impact caused by the accentuated evolution of information technology in society, as well as the profound modifications resulting from an economic model that preaches intense competitiveness, have been reflected in significant changes in the structuring paradigm of organizations and how knowledge has been understood in the development of new products, processes, and organizational structures (Silva, 2002). To create and stabilize a sustainable competitive advantage, organizations are increasingly investing in the application of concepts such as learning, knowledge, and competence. This logic is based on the argument which advocates that the ongoing process composed by the creation, acquisition, and application of knowledge by the organization on its essence represents an intangible resource capable of conferring a competitive advantage to this organizational structure, sustaining it, according to Shinyashiki et al. (2003). This knowledge is derived from the interactions developed in learning processes that occur in the organizational environment, reasoning that can be better and deeply understood, on its plenitude, by acknowledging that the more specific such knowledge demonstrates in face of the organization, the more it will become the foundation of its core assets and competencies, and, consequently, an important strategic pilar.

Organizational Knowledge is an invisible asset slowly accumulated over time, hence, impossible to be traded or easily imitated by competitors, as it represents the basis of the organization's history and culture. The more specificities it demonstrates towards the organization, the more it will become a strategic asset, considering that this knowledge is the foundation of the organization's core competencies since it belongs to its human capital. Therefore, in the current (and future) society, knowledge increasingly becomes a central asset. The basic economic resources start to count (besides capital, natural resources, and labor) on the contribution of the knowledge required for production and business processes. This logic is based on Drucker (1993), who maintains that value is created by productivity and the ability to innovate, apply knowledge to work, creating knowledge workers. Thus, Knowledge Management goes beyond investment in technology or innovation management. It necessarily involves understanding the characteristics and demands of the competitive environment and understanding the individual and collective needs associated with the creation and learning processes.

According to Abramowitz & David (1996), the present century has been characterized by the progressive intensity of knowledge in the productive system, from which the increasingly important role of knowledge production and learning stands out. Indeed, in the contemporary era, political and economic discourses increasingly emphasize a knowledge-based economy as the structural platform of the current world's interdependent and complex order, hence suggesting that the most promising strategy for economic growth will indeed crystalize to be the one that emphasizes the knowledge base at the very core of the economy. This line of reasoning, subsequently, culminates to propose a clear and deep relationship between knowledge creation and diffusion, and economic power, given that the essence at the core of such relationship indicates that the Innovation highway depends on the evolution of knowledge. The root for such an argument grounds on the premise which advocates that the nature of global economic growth has been changed (and, indeed, keeps constantly changing) by the speed of innovation, which has only turned possible by the rapidly and profoundly changing technology, which climaxed to the possibility of generating shorter product life cycles, and, simultaneously, a high rate in what concerns to the development of new goods. Deriving from such a refined scenario, organizations now need to ensure that their business strategies, development paths, and growth prospects are creative, innovative, competitive, and sustainable, so that they can assume as agents able to build and sustain competitive advantage, facing competitors, increasing their market share.

Not disregarding the above-mentioned scenario, the issue now emerges because, in the current global context, innovation, however, has become both growingly needed but also assumedly complex, given the fast and multifaced progress in the amount and variety of the available knowledge, but also due to the constant evolution of technology and to the fluctuating customer needs, factors that, on their very end, also end up by fostering the firstly mentioned one in regards to the amount and variety of knowledge. Also on the analysis of such context, it shall not be overlooked the increasing broad international competitive pressure, which has been graining superior levels of density and complexity, hence increasingly organized in a multifaceted interdependence network (Martins, 2010). And from such a line of thinking it is possible to derive the argument that, since innovation assumes as extremely dependent on the availability of knowledge, the complexity created by the explosion of its wealth, variety, and scope needs to be identified and managed to ensure successful innovation.

According to Winter (1987), there is a serious lack of appropriate terminology and conceptual schemes in the analysis of the role of knowledge in the economy. Knowledge and information are referred to in economic models in two different contexts, but we will only focus on the perspective that considers knowledge as an asset, conceiving it as an input (competence) and output (innovation) in the productive process. In this sense, the knowledge economy will focus on the conditions by which knowledge becomes a "normal commodity", something similar to a tangible product that can be produced and reproduced, having the ability to be transacted in markets. In articulation with this, there is the concept of "learning", a core process in which competencies and capabilities are acquired, allowing the learner to achieve individual or organizational objectives with a greater degree of success.

The set of definitions of Knowledge Management is profuse. Gloet & Terziovski (2004) describe the concept simultaneously as a process because it simultaneously consists of the formalization of, and of the access to experience, knowledge, and expertise that creates new capabilities, enables superior performance, encourages innovation, and supports customer value creation. This conceptualization encompasses several terms connected to the dynamics of knowledge, such as its creation, valuation, measurement, mapping, transportation, storage, sharing, and distribution. Darroch & McNaughton (2002) indicate that knowledge management refers to a management function that creates/locates knowledge, manages its flow, and ensures its efficient use for the long-term benefit of the organization. An organization that demonstrates competence in knowledge management will then take it on as its philosophy.

Based on the above mentioned, it becomes clear that the conceptual construct of knowledge, and the respective inputs which then derive into the concept of Knowledge Management, relate strongly and closely to the concept of Intellectual Capital. According to Klein (1998), this theoretical construct refers to the intellectual capital of companies, namely, their knowledge, experience, expertise, and various intangible assets, rather than their tangible physical and financial capital, which increasingly determines their competitive positions.

In fact, and following the inputs provided by Edvinsson (1997), Intellectual Capital refers to the conglomerate of knowledge and information, so in other words, it condenses, articulates, and synthetizes the junction of Human Capital and Structural Capital. Moreover, indeed as the term itself indicates, in its essence it refers primarily to the skills, experience, knowledge, and abilities held, developed, and retained by human beings – people consist, thus, on this conceptual construct's very core. As mentioned above, the concept can be detailed into two different, nevertheless articulated, components: Human Capital and Structural Capital. Human Capital refers, then, to a set of information which is acquired, improved, transmitted, and transformed into Knowledge by the individuals which, at their very core, entail a given organization, dynamics which is operationalized via a progressive and ongoing process all along with time. This to say, this capital culminates to assume as a critical vehicle for the creation and support of added value to this same organization, acting, subsequently, as well as a source of sustainable competitive advantage for it. Important to mention, such Knowledge can be tacit – arising from life-long experiences – or explicit – structured, encrypted, and stored. So, and deriving from this, it is noteworthy to highpoint that its transmission to other people is possible, via a myriad of forms, and all along with a temporal spectrum. In this sense, it is also important to mention that, over time, this information can be updated, transformed, and modified through training, experiences, research, exchanges, recycling, and, of course, also because of the challenges posed by the environment.

All people have an intellectual potential, whether in the degree of knowledge, creativity, competence, and/or skills, which can be similar or very different from each other. On the contrary, it is developed over time and as a function of the environment in which the person is inserted, the stimuli received, and the experiences he or she goes through.

Following the line of Kaya et al. (2010), within organizations, Intellectual Capital consists of the sum of everyone's knowledge. It constitutes the intellectual matter, that is, it concerns the knowledge, information, intellectual property, and experiences that can be used to generate wealth, add value to the organization, and enhance its added value. The Intellectual Capital that is in people's minds is currently considered the most important resource that an organization can have. Admitting the importance of Intellectual Capital, it is called the new competitive advantage, that is, a differential in obtaining success and resources from an organization. It can be said that it is through the intellectual capital that a company can develop and stand out in the market where it operates, however, this does not leave aside the other resources.

In the industrial period, only physical resources (land, capital, and labor) were considered as an organization's resources, but with the transition from the industrial period to the knowledge period, the intellect has also started to be valued as a differential for organizations.

Following the line of Koenig (1996), the management of Intellectual Capital within organizations has become crucial, revealing the need to seek resources to encourage employees to share knowledge, encompassing all levels throughout the organizational structure. In this way, the organization retains know-how and gets ideas for possible improvements and cost reduction. As Intellectual Capital is too complex to be measured, the concept of Knowledge Management was created to help organizations to use means to better organize this intellect, to systematize, codify, apply, and capitalize on it. Intellectual capital thus comes to be considered an organizational competitive advantage, since it presents rarity and value, and is thus difficult to imitate. It is important to emphasize that Intellectual Capital cannot be a concept thought of solely as a function of business interests. Therefore, it must also be thought of as a strategic way of human emancipation, since it is a resource, first and foremost, of human beings, of everyone.

Therefore, the strategic relevance of intellectual capital for innovation and creativity processes in companies becomes clear. We can better understand it by alluding to Barry & Brown (1985), who argues that a successful company must understand how people work and how communication and information technology can help them perform their tasks more effectively. Fleury & Oliveira (2001), in the same line of thought, emphasize that organizational learning and innovation have become key elements for the survival and success of organizations, as well as the main solutions to the problem of competitiveness. Thus, in the 21st century, knowledge and innovation management and the management of intellectual capital in organizations will become areas of vital strategic importance, which, therefore, should translate into tools of fundamental relevance in the new administrative models, both for organizational survival and for the maintenance of a sustainable competitive advantage.

CREATIVITY

In this context, the question that Morgan (1995) poses is whether there really exists a possibility of designing organizations that are as flexible, resilient, and resourceful as the functioning of the brain, thus able to create new forms of organization that spread brain-like capabilities throughout the company, in detriment of what has been done by perceiving knowledge and intelligence as a resource confined to special and confined units of the organization. In this logic, and according to Wigand et al. (1997), in this new business format, the division of labor should be increasingly dominated by cognitive rather than technical aspects. The value will therefore be created by transforming information into knowledge, and its application in the company, which will require a more holistic approach.

The premise underlying such argument is the one which postulates that the same organizational environment that is propitious to learning is also favorable to creativity, as highlighted by the works of Alencar (1998) and of Sternberg et al. (1997): work motivated by pleasure and satisfaction, comprehensive knowledge and experience, an organizational culture that includes predominant values for innovation and, of special relevance, the willingness to take risks and learn from one's own mistakes, since creativity implies dealing with the unknown.

In this logic of ideas, learning from experience, applying knowledge gained from experience, dealing with complex situations, solving problems even in the face of a scarcity of important information, determining what is important, being able to reason, reacting quickly and correctly to new situations, understanding visual images, processing and manipulating symbols, being creative and imaginative, and using heuristics (practical rules derived from experience) are some of the characteristics of intelligent behavior to be optimized in the execution of any functional activity in a company. More than that, they are also characteristics through which it is possible to observe a clear correlation between knowledge, learning, and creativity, as Stair (1998) points out.

Hence, from the above inputs, it emerges the theoretical construct of Creativity, and the above argumentation also allows to highpoint its relevance. Indeed, Creativity refers to novelty products and useful ideas from a person or group of individuals working together (Amabile, 1988; Madjar et al., 2002; Shalley et al., 2000; Zhou & Shalley, 2003). Following the same line of thinking, but exposing it in other words, considering the definition proposed by Plucker et al. (2004: 90) it is possible to ascertain that such theoretical construct refers to the "interaction among aptitude, process, and environment", which at its very core consists of a progressive dynamic over time lead by an individual or group, or in other words, by the core organizational actors, by whose efforts "a perceptible product" is produced, which assumes as "both novel and useful" within a given social context, to serve certain predefined needs.

So, extending on the above argument, it is arguable that Creativity, at its core, consists of the capacity to engender and produce outcomes which assume as both novel, different and relevant (Runco & Jaeger, 2012), henceforth, it connects to the novelty, significance, originality, usefulness, appropriateness, and unusualness of the production (Besemer & Treffinger, 1981; Ford, 2000; Runco et al., 2005; Runco & Okuda, 1988). In short, this argument culminates in the premise that Creativity precedes Innovation. Creativity develops all along with a spectrum, which means, it varies in terms of the magnitude of its expression and impacts: indeed, its different graduations have conventionally been categorized in literature via a framework which discriminates four ordered classes, comprising the "Four C Model of Creativity" (Kaufman & Beghetto, 2009). Presenting such categories in descending order, we then have the spectrum formed by the "Big- C", the "Pro-C", the "Little-c", and the "Mini-C".

"Big-C" Creativity consists of clear-cut, eminent creative contributions. Simonton's works on creative genius (e.g., 1977, 1991, 1994, 1997, 1998, 2004) are an example of studying "Big-C" Creativity, which the author has traditionally categorized to respect to typical creators whose works have lasted centuries, this way analyzing the first, best, and last contributions of scientists from a variety of scientific disciplines. A very useful and interesting example would be Simonton's extensive research on the relationship between age and achievement (for a review see Simonton, 1997), hence suggesting that creative output at the "Big-C" level begins around the '20s, which later ascends to an optimum level at some point near the age of 40 years old, and then gradually approaches back to zero in output levels.

"Pro-C" refers to professional-level accomplishments which are performed by a given domain's practitioner, and in this sense, they assume significant enough to contribute to this domain's growth and improvement, but the issue here is that they also assume not eminent within it. Following this explanation, and to better demonstrate it via practical examples, this type of Creativity (clearly also labeled as "Professional Creativity") can, for instance, take the form of an engineer finding new and cost-effective solutions to a problem (Cropley, 2015), or as well a flight attendant finding a creative way to deliver security instructions, to set a relaxed environment for passengers while gaining their attention (Waples & Friedrich, 2011).

The other predominant approach to Creativity (and in here we mention the "little-c" type, following the "Four C" Model categorization above exposed) sheds light on everyday activities, being, thus, practiced by virtually every individual throughout his/her routine, taking place in the simple activities around all spheres of life. So, it refers to those creative actions in which any nonexpert agent may indeed participate and add value, each day (Richards et al., 1988). Providing practical examples to better understand these manifestations of Creativity, we can see that it can take the form of discovering and implementing a new way to decorate a room, or even mixing cuisines to create a new meal. Also noteworthy to mention on this topic, such approaches have been mainly based on layperson perceptions of Creativity, which crystallize into theories that tend to de-emphasize analytical abilities – usually associated with IQ tests – to rather highpoint unconventionality, inquisitiveness, imagination, and freedom as core vectors to this Creativity manifestation (Sternberg, 1985).

Finally, "mini-c" refers to the refined and personally meaningful interpretations of situations, experiences, actions, and events – so, it assumes a more casuistic level of analysis, and is aimed mainly to express one's own emotions (Beghetto & Kaufman, 2007). These manifestations might not be public or tangible, yet they are meaningful to the individual creator and reflect the creation of (his/her) new and refined ideas and knowledge (Cotter et al., 2018).

To summarize and to set forward the synthesis postulated by Kaufman & Beghetto (2009), the "little-c" category has been useful for addressing common misconceptions about creativity (Plucker et al., 2004). For instance, too much of a focus on "Big-C" leads to the idea that only certain people can be creative, the only creativity that matters is that of the "Big-C" kind, or that creativity involves negative forms of deviance (e.g., drug use, mental illness). Moreover, the category of "little-c" helps underscore the important (and, at times, essential) role that creativity plays in everyday life (Richards, 2007) and points to the importance of identifying and nurturing creativity in everyday settings such as schools and classrooms (Beghetto & Plucker, 2006), the workplace (Agars et al., 2005; Bakker et al., 2006), and the home and social settings (Baer & Kaufman, 2005; Cropley, 2006).

Once again based on the argumentation postulated by Kaufman & Beghetto (2009), we see the Four C Model as representing a developmental trajectory of Creativity in a person's life, an argument that does not imply that creators necessarily pass through each category (or "stage of creativity") on their journey to become an eminent creator. Instead, the model offers a useful, flexible and evolving framework to conceptualize and classify different (nevertheless, interconnected) levels of creative expression, and also assumes particularly interesting as it points to potential paths of creative maturation.

One example of the above exposed indeed consists on Amabile et al.'s (1996) componential model of Creativity, in which it is argued that three variables are required so that manifestations of Creativity can occur, namely 1) domain-relevant skills, 2) creativity-relevant skills, and 3) task motivation. Following this line of reasoning, expanding the argument exposed in the last paragraph, the inputs of Ostrower (1993) assume relevance, as the author advocates that Creativity indeed is not possible to be understood as the exclusive property of a very few chosen ones, but instead consists of a potential inherent to the condition of being human, so, it does not consist of an asset exclusive to a given elite. From this, it results that Creativity is not possible to be apprehended as an isolated object, subsequently, it is indeed critical to avoid any simplification when trying to understand this theoretical construct, given that it consists of an interconnected and complex element within a widest social, economic, political, and cultural context, placed with specific configurations that hinder its free flow.

Deriving from the exposition all along with this chapter, it is to this moment possible to conclude that there is nowadays a consensual and wide-ranging recognition that Creativity consists of a key factor for Innovation and, subsequently, sustainable competitive advantage fuelling the long-term success of organizations. And the fact is that such premise has gained momentum due to the consolidated impacts of the new era brought about by Globalization, by which competition in the business environment and the rapid pace of change have strongly and constantly pressured organizations to engage in the design and implementation of strategies to make better use of their resources – here we would include the Creativity of their human resources, as a core organizational asset.

However, an important remark shall be done in regards to the above-mentioned. The fact is that in the current times there is a predominant conception of Creativity which postulates that this organizational asset consists of a phenomenon of mere intrapsychic nature, this to say, it is centered on the individual and is, hence, dependent on agent factors such as the person's personality traits, personal values, thinking styles, experiences, needs, and wishes, and motivation.

As a corollary, an anthropocentric perspective of Creativity stands out from such approaches, nevertheless, the impacts of Globalization and its crystallization on the current era have been assumedly questioning such theoretical perspectives on Creativity. This line of thinking has been set forward and further explained by Csikszentmihalyi (1998), who energetically reacted against the anthropocentric view on Creativity to rather advocate the relevance of the contextual environment – and its correspondent cultural, historical, social, political, economic, and financial forces – in the analysis around the different configurations of Creativity manifestations. In this sense, the author incisively reminds the need to evolve from the Ptolemaic theories around the construct of Creativity (as above explained, in which the person as an individual agent assumes as the vital center of every explanation, factor, and configuration), to rather design and develop a renewed model more inclined to a Copernican understanding of this construct, in which a refined level of analysis could be adopted and improved to conceive the person as an active, transformative and evolving part of a system of mutual influences and information. The key element on this idea roots on the premise that the two mentioned perspectives are not mutually exclusive: on the contrary, they emerge as dynamic evolutions, progressive improvements on the theoretical analysis, which climaxed to systematize that organizational environment factors influence creativity positively and negatively, but it is also needed to take in consideration the personal elements favorable and unfavorable to its expression. Exposed in other words, this argument advocates that the individual's behavior in the work environment, or also outside of it, is profoundly influenced by socio-cultural factors, which contribute considerably to the emergence, recognition, and cultivation of creativity, or, on the contrary, to its repression.

Based on the research developed by Guilford (1950), we can systematize the personal characteristics that, when manifested in an organization in a work context, negatively foster the expression of creativity: inflexibility, lack of motivation, lack of skills or experience, and lack of social preparation. On the other hand, opposite characteristics to the above have been observed in professionals who excel in creative production, such as a cast of personality traits, self-motivation, special cognitive skills, expertise in the area, and group skills.

As the origin of innovation lies in the creative ideas of individuals, creativity has received increasing attention, being pointed out as a critical human ability that must be channeled and strengthened in favor of the organization's success. Several authors (Alencar, 1995; Amabile, 1995; Bruno-Faria & Alencar, 1996; Locke & Kirkpatrick, 1995) have pointed out the characteristics of an organization that relate to creativity. Given the decisive role of creativity for the success of organizations in their process of dealing with change, uncertainty, instability, competition, and systematically promoting innovation, leaders must be attentive to the conditions prevailing in their work environment, undoing possible barriers to creativity, and maximizing opportunities for its expression. However, what has been observed is that many organizations tend to ignore the potential for competence, responsibility, and productivity, encouraging dependence and passivity rather than initiative and creativity. The basic needs for recognition, support, and security that should be met in the work environment are often ignored, affecting work motivation, and contributing to keeping the potential for creativity dormant.

In this logic, and enhanced by the current global paradigm, the literature review emphasizes that the individual creation and learning processes, similarly to the organizational process, imply a personal reinvention, i.e., they are associated with changes in mental models, cognitive maps, and behaviors, as well as with the search for great challenges and resolutions of internal tensions. Moreover, it is also found that individuals, in their creative and learning processes, depend on great intrinsic motivation, as well as on interaction with others, the combination of multiple perspectives and experiences, and personal trial and error. This approach indicates that knowledge management necessarily involves the systemic coordination of efforts at various levels: organizational and individual; strategic and operational; formal and informal norms. Underlying this is the recognition that human capital, formed by individual and organizational values and norms, as well as by the competencies, skills, and attitudes of each employee, is the "driving force" behind the generation of knowledge and value in organizations. This requires a constant stimulation of intrinsic motivation, the establishment of personal contacts, the analysis of different perspectives, openness to effective communication, and learning through individual experience, trial, and error.

In sum, we can infer some general strategic recommendations from this reasoning, which we present as a set of rules or indicators of success in knowledge management projects. We allude to the argumentation of Davenport et al. (1998): it should not be the project of an individual, but of the organization; develop a definition of knowledge in the company; emphasize knowledge as stock and flow; knowledge is inside and outside people's minds; create environments for the knowledge market to flourish (encouraged by trust and recognition); give equal importance to knowledge in different formats; contextualize knowledge; encourage learning and creativity; focus on the past/present as well as the future; recognize the importance of experimentation; give equal importance to the human and technological interface; seek ways to evaluate the knowledge initiatives undertaken, both by quantitative and qualitative measurement.

INNOVATION

Innovation assumes great relevance, both empirically and theoretically, because it is the cornerstone of economic development (Schumpeter, 1942). There is a clear relationship between business strategy, innovation, and organizational performance, based on the premise that innovation – consisting of the use of a new product, service, or method in business practice immediately after its discovery – influences the economic success and market share of the company/business in increasingly competitive global markets (Zahra, 1993).

There are several proposals for the definition of innovation. According to Chen et al. (2009), it refers to the introduction, in the productive system, of a new combination of factors essential to production. For Cardinal et al. (2001), it consists of a process that encompasses the technical, physical, and knowledge-based activities that are central to the development of production routines. Harkema (2003) proposes a knowledge process that aims at the creation, sharing, and assimilation of new knowledge via the development of commercially viable solutions, encompassing products and services, thus linking to change - radial or incremental. In the same vein, Gloet & Terziovski (2004) suggest that it concerns the implementation of discoveries and inventions, condensed into a process that creates new products/systems/processes. It is then consensual that this concept emphasizes the commercial application of new knowledge as a driver of growth and productivity in organizations and should be conceived holistically because it develops in a multidimensional process (Guellec & van Pottelsberghe, 2004). Along these lines, Lundvall & Nielsen (2007) systematize two reasons for conceiving innovation as a crucial outcome of knowledge production: on the one hand, innovation represents something new, adding aspects to existing knowledge; on the other, it represents the knowledge that has proven its relevance to the market economy, crystallized in an invention introduced into a given market.

It is also relevant to mention that innovation is a process of "creative destruction," since it opens the way to new markets and creates bases for the establishment of new businesses, while it closes existing markets, companies, and jobs (Schumpeter, 1942: 82-85). Arising from this logic, he highlights the moral depreciation of intellectual capital as another impact of innovation. Therefore, the change from a linear to an interactive model regarding the interconnection between innovation and knowledge aims to connect innovation and competence development based on an interactive and interdependent learning process in which all involved increase their competencies while engaging in the innovation process (Lundvall & Nielsen, 2007).

According to Drucker (1985), the concept of innovation refers to the tool par excellence of entrepreneurs, therefore, to how they explore change as a vehicle toward an opportunity for a new business or service. In the same vein, for Hartley (2014) the term consists of the successful development, implementation, and use of new or structurally improved products, processes, services, or organizational structures. Sarkar (2007), on the other hand, conceives the term as the exploration of new ideas that find market acceptance, often incorporating new technologies, processes, and even best practices. This argumentation, however, simultaneously merges and culminates in the ideas of Jacobs & Snijders (2008), when they suggest that innovation is the creation and implementation of something new, providing added value. Therefore, on a broader scale, it refers to the implementation of a new or significantly improved good, service, process, marketing method, or organizational structure.

According to Nicolau & Paranhos (2006), it is first necessary to highlight two vital aspects in the concept of innovation. On the one hand, it is important to retain that innovation is comprehensive, encompassing the activities of scientific research and technological development, as well as those of production and marketing; on the other, it should be noted that innovation occurs in a characteristic technological context, defined by the concepts of paradigm, trajectory, technological regime, and national innovation system.

In this sense, innovation management will consist of structuring a concrete innovation process, with a beginning (inputs), middle (processing), and end (outputs and generating results). In short, innovation management consists in establishing means and methods to generate value, making ideas concrete. This would derive into the premise which postulates that, given the concept of Innovation as laid out above, for the most part, innovations result from a conscious, constant, and intentional search for innovation opportunities. Therefore, we can see at the co-core of the concept of Innovation Management is a holistic analysis, corresponding to the joint and concerted management of the different actors and structures involved in the process: first, people as agents of innovation (who develop the ideas and apply them), and organizations and other attached structures, as operational structures that provide the conditions to effectively develop such innovations and capitalize on them.

Nevertheless, it is necessary to consider the opportunities for innovation as critical factors that may positively or negatively impact the innovation potential of given agents or organizations. It is, nonetheless, noteworthy to alter that these several sources or opportunities for innovation may occur at the same time, fostering an enriching context to boost innovation. In fact, and alluding to Drucker (2004), these are distinguished according to whether they occur within companies or industries, or whether, on the contrary, they are external to the company, this to say if they consist of aspects composing the surrounding social and intellectual environment.

In association with the concept of Innovation comes, therefore, the concept of Technology. Nevertheless, it should be noted that both are conceptually different, although often juxtaposition between them is observed. Well, the term technology, of Greek origin, is formed by *tekne* (referring to art, technique, or craft) and by *logos* (referring to the body of knowledge), therefore, it is used to define the knowledge that allows manufacturing objects and modifying the environment to meet human needs. In other words, technology is the set of tools, methods, and techniques that allow the practical utilization of scientific knowledge.

Therefore, the concepts of innovation and technology are interrelated, and they are not exhausted in themselves or in time. That is, both concepts imply, in their essence, an evolutionary dynamic, of constant and strategic transformations, of analysis and re-analysis, and of extension in time, in face of the challenges that the spatial and temporal contexts (whether in their economic, financial, cultural, political, environmental, social configurations, etc.) will pose to human beings. Therefore, nothing is ever finished in matters of invention, creativity, and innovation: discussions on these topics are permanently open debates.

Only in this way can societies evolve, transform themselves, add value, seek more sustainable ways of living, increase their quality of life, enhance their development, and reduce asymmetries. Society has been strongly impacted by the advancement of technology, and as a corollary to the arguments made in the previous paragraphs, we conclude that, if we cannot guarantee the future, we can predict it (and, more than that, build it). There is a huge space in the market for innovative ideas, and these will be the best bet to face the growing challenges of a globalized, fast-paced world, marked by intense complex interdependencies.

Aimed at defining the concept, Schumpeter (1988) differentiated between the conceptual constructs of both *invention* and *innovation*. Specifically, an *invention* hence refers to an idea, construct, sketch, or model directed and crystalized into a new, refined, or improved artifact, product, service, process, or system. On the other hand, an *innovation*, in the economic sense, may only be complete when there is a commercial transaction involving an invention, subsequently, generating wealth. Therefore, the scope, the core, and the dynamics of both instructs are different. From this line of reasoning, it derives that such a theoretical evolution that arises from the differentiation of such concepts entails the in-depth study of notions, ideas, models, and managerial practices which are strongly focused on Knowledge, Creativity, and Innovation, and, as a consequence, considers an ambitious and lengthened universe that largely goes beyond the internal context of organizations, this way configuring into an environment fruitful to innovation as it becomes dynamic and openly structured, via the design, implementation, improvement, and development of new information networks (Santos et al., 2011).

This alert advocated by Schumpeter (1988), as above mentioned, leads to, on the one hand, comparing the concepts of innovation and knowledge management, nonetheless, on the other hand, it as well provides fruitful insights to interconnect them. Hence, and expanding this idea, it is suggested that knowledge management aims to support innovation, this to say, it is directed towards the creation of new ideas and the exploitation of the organization's power of thought, including the capture of insights and experience to make them available and usable when, where and by whom needed, consequently leveraging and facilitating the access to the knowledge, to the experience, and the know-how. In this way, it favors the achievement of increasing and continuous collaboration levels, hence positively affecting knowledge sharing and constant learning, this way reaching process improvements by promoting better quality in decision making, ensuring efficiency and effectiveness on the exploitation of intellectual assets. Deriving from such a line of reasoning, knowledge management, on its side, consists of a planned, articulated, and structured approach to managing the creation, sharing, improvement, development, transformation, and transmission of knowledge as a vital organizational asset, being all such efforts articulated via a dynamic process directed at increasing the capability, speed, and effectiveness of an organization's performance levels in delivering products/services for the benefit of its customers, this planned, structured and developed within a given business strategy. Hence, it is possible to ascertain that knowledge management occurs at three levels, which interconnect even though they assume as different and separate, with their idiosyncrasies. Namely, it is possible to distinguish the individual, team, and organizational levels. As a corollary of such an idea, it then results that knowledge management may be understood as a holistic solution that incorporates a variety of organizational perspectives, actors, policies, practices, and structures and vectors – people, processes, culture, and technology (du Plessis & Boon, 2004).

In sum, from this argument, we see that knowledge management is not only focused on innovation but creates an enabling environment for innovation to occur knowledge management provides a knowledge-driven organizational culture in which innovations can incubate. Although knowledge management systems per se do not possess the qualities necessary to provide a sustainable competitive advantage to organizations, their combination with other resources and competencies is key to developing and maintaining sustainable competitive advantage through product/process innovation, as they enable the conversion of learning capabilities and core competencies into a sustainable advantage, revitalizing organizational learning and resource development processes.

We can then highlight three main drivers of the application of knowledge management in innovation, according to Cavusgil et al. (2003). The first is to build and sustain a competitive advantage using knowledge and collaborative practices. Building and sustaining an innovation program has become increasingly complex in the current era due to changing customer needs, extensive competitive pressure, and rapid technological change, so organizations find it increasingly difficult to internalize innovations. Knowledge management can facilitate this, as acquiring knowledge and capabilities via collaboration is an effective and efficient way to innovate. The second driver highlights that knowledge management can be a mechanism to address the complexity of innovation, as it helps the management of new knowledge created through the innovation process, but also the management of existing knowledge as an input resource in the innovation process. The third factor of applying knowledge management to benefit the innovation process focuses on the integration of knowledge internal and external to the organization, which implies that it can be shared, transacted, developed, refined, and made available at the point of need. Knowledge integration through knowledge management platforms and processes should therefore facilitate reflection and dialogue to enable personal/organizational learning and innovation, otherwise, organizations could underutilize knowledge as an innovation resource. In sum, from this argument, we see that knowledge management is not only focused on innovation but creates an enabling environment for innovation to occur knowledge management provides a knowledge-driven organizational culture in which innovations can incubate. Although knowledge management systems per se do not possess the qualities necessary to provide a sustainable competitive advantage to organizations, their combination with other resources and competencies is key to developing and maintaining sustainable competitive advantage through product/process innovation, as they enable the conversion of learning capabilities and core competencies into a sustainable advantage, revitalizing organizational learning and resource development processes.

In this line of thinking, the distinction between the concepts of innovation and creativity becomes clear, and although the concepts are distinct from each other, each of them presenting idiosyncrasies, the truth is that, nevertheless, it is possible to establish commonalities between both, since innovation requires creativity at its core. Therefore, it follows that the concepts of creativity and innovation cannot be considered synonymous, although it is necessary to recognize the juxtaposition that is often observed between the two. The main difference between these conceptual constructions lies in the focus of each one.

On the one hand, creativity exists as an ability that unleashes the potential for the human mind to process creatively, the human brain can conceive ideas and form thoughts that are capable of escaping from previously conceived and established patterns, launching ideas, products, and actions that are innovative and that challenge the status quo. It follows that creativity is a pre-existing ability in the human mind, certainly with the potential to evolve through the type of stimuli received and that it can be channeled into action.

Innovation, on the other hand, is more linked to action, since it consists in proposing or creating changes in systems that are considered stable. Innovating, then, means offering a new action, function, ability, product, or improvement in a system that we already know, making significant improvements in it, increasing its value. In short, this argument culminates in the premise that creativity precedes innovation. Creativity is the moment when the agents of innovation gather the references, this time idealizing an idea that is later transfigured into a product, starting the sketches for its conception. However, if such a product as a final conception is not put into practice, it will never be more than a creative effort, because it will not constitute innovation since it has not passed to the materialization phase. Innovation, on the other hand, happens when the idea is put into practice, thus generating value. As a corollary to this argument, there is the assumption that creativity is not measurable, whereas innovation is. In short, we can establish that, conceptually, creativity and innovation are in different spheres, but they are forces that coexist and, when they intersect, produce better results in adding value to a given process, good, or product.

What is Possible to Learn about Creativity from the Organizations' Response to Covid-19?

Creativity in the workplace has been regarded as an essential asset to organizational innovation (Akinola et al., 2019), acknowledging that, at its very essence, it corresponds to the creation of new and useful ideas or solutions, this way able to add value to the organizational value chain (Fisher & Barrett, 2019). Indeed, such relevance has been stressed out in the current Covid-19 times, and for sure by both the challenges and the opportunities that such context addressed to the organizational structures. From this logic, it derives the argument which postulates that creating, implementing, developing, and improving the conditions which allow employees to foster creativity consists of a vital twofold opportunity and challenge for managers, as people leading actors in organizational structures.

The core assumption on the argumentation all along with this chapter acknowledges that the scope of the potential and conceivable problems, as well as the corresponding resources related to a task, process, or responsibility, intrinsically outlines the configuration of the knowledge landscape in which people, as the essential organizational actors, must navigate as they search for refined, updated, novel and useful solutions to problems – creative and innovative approaches to the organizational challenges, this to say, as they try to add value in the organizational value chain, to achieve and sustain the organizational competitive advantage (Cromwell et al., 2018). The simultaneous interest, complexity, and density in analyzing this problem relate to the fact that the Covid-19 pandemic indeed drastically, rapidly, and substantially expanded the scope of possible problems to solve and opportunities to manage in the organizational environment on the business arena, while also restricting the resources available to develop creative and innovative solutions to those problems, so that organizations could not only survive to the challenging times but also capitalize on its challenges and opportunities, reviewing their business models and values chains, updating and strengthening their competitive advantage sources.

Both theoretical and empirical research on creativity and innovation problematics has been proving two predominant problem-solving processes that can be employed and developed by people to address and steer through such an uncertain, complex, delicate, and heterogeneous context. Namely, here we refer to a) directed creativity, which occurs when people start to develop innovative solutions to an issue stepping from a clearly defined problem, this way addressing and navigating through uncertainty and probabilities to search for a solution and potential added value; and, on the other hand, b) emergent creativity, which happens when people start to address a challenge via an incompletely designed and developed solution, as so, also facing uncertainty to search for problems, not directly to search for a solution (Cromwell et al., 2018).

Nonetheless, the critical issue here resides in the fact that, at its essence, creativity consists of a problem-solving ongoing process that is fundamentally affected by the environmental configurations of the time and space context in which people, its core agents, develop and operate. Therefore, from this line of reasoning, it derives that important dimension of the environment cannot be neglected, this way, refined analytical models (both to conceive creativity theoretically and operationally) need to be conceived. This is rooted in the premise that such a myriad of external vectors can positively or negatively affect creativity, given that they include the range of problems, goals, objectives, and criteria that need to be addressed, managed, overcome, and/or satisfied; and the resources available to develop the creative solutions, to add value to the organizational structure (in here, we are namely referring to knowledge, materials, tools, people, finance, time, etc.), following the argumentation of Cromwell et al. (2018).

Following this line of thinking, the critical problem in such a complex and delicate context arises from the case that creating these favorable conditions to creativity and innovation can be particularly hard and difficult during times of stress, heterogeneity, and uncertainty, which we have assisted particularly during the Covid-19 pandemic, times in which employees indeed tend to assign a greater priority to tasks that are certain and controllable, as a psychosomatic defense mechanism, rather than devoting efforts to try to be creative (Luis et al., 2020). The core premise to bear in mind here postulates that the fluctuation of problems and resources over time influences how much uncertainty, stress, and heterogeneity people experience during the process – this to say, may hinder their creative disposal and potential. If at least theoretically, such a line of reasoning can be established, the truth is that the Covid-19 pandemic offers a particularly timely and geographical context in which to sharply examine these issues, based on the premise that, given the great uncertainty in the workplace caused by the twofold challenges and opportunities such a global problem addressed, identifying ways to facilitate knowledge creation and sharing, learning tools, and initiatives, creativity and innovation platforms under such both complex and interesting conditions configures as a serious and perilous topic for organizational leaders to manage.

So, following the reasoning in the above paragraph, we can start from the assumption that uncertainty frequently conducts to stress (Peters et al., 2017) and anxiety (Counsell et al., 2017; Tseng et al., 2018) due to the disruption of routines and lifestyles, which, on their very end, should lead to opportunities to consider and implement creative actions. Nonetheless, stress and anxiety felt on such contextual configurations may configure as psychosomatic symptoms clearly and growingly felt by people facing such situations, being these increasingly reported throughout the Covid-19 outbreak, generally in all geographical areas affected by the pandemic. However, a line of reasoning can also be done to advocate those times of uncertainty may also positively act as catalysts for creativity (Barbot & Reiter-Palmon, 2019), based on the assumption that it is most certainly both welcome and needed in such unprecedented circumstances (Kaufman & Beghetto, 2013). Such an idea is based on the premise that uncertainty motivates the consideration, development, and exploration of creative actions and innovative tools and paths (Binnewies et al., 2008). In that sense, it can be argued that creative action usually entails challenging one's old assumptions and trying new things, therefore, this construct can be conceived as a process and a means to make sense of and cope with heterogeneity, uncertainty, and complexity (Barbot & Reiter-Palmon, 2019; Ford, 2000). Furthering this line of thinking, indeed the core assumption is that, in organizational and entrepreneurial contexts, uncertainty and its concurrent management may be conceived to be at the root of creative and innovative processes and their corresponding endeavors (Fong et al., 2018; Blauth et al., 2014).

In many countries, the Covid-19 pandemic produced impactful consequences both at a personal level, as well as in organizational structures, given that it resulted in a period of lockdown that impacted individuals' lifestyles in the overall spectrum. Even though the majority of empirical research under this area focuses on the negative effects of lockdown measures, hence reporting negative consequences on people skills, experiences, and habitats under such circumstances, the work of Mercier et al. (2021) assumes both interesting and extremely relevant, as it focusses on one particular positive outcome that might have emerged from the Covid-19 induced lockdown. In specific, the authors empirically examined the positive effects such a delicate and complex context provoked into creativity. The breakthrough results reported on this work provide fruitful insights to ascertain the relevance of creativity as a core organizational asset in the complex business environment brought about by the widespread impacts of the Covid-19 pandemic, this way as well allowing an insightful reflection around what this pandemic allowed academics, practitioners, managers, leaders, and employees to learn in regards to organizational and individual creativity as a means to face stress and uncertainty at a global level. Nonetheless, it is important to clarify that by no means do the authors neglect the potential and commonly reported prejudicial effects of such contexts in the physical and mental health of many people, consequences to be widely observed, regardless of the period, socio-economic level of living conditions, geographical area, among other variables. Hence, and to this end, the above-mentioned work compared self-reported professional creativity ("Pro-C", as already theoretically detailed in a previous point of this chapter) and everyday creativity ("Little-C") at two different points in time, organized into a continuous spectrum: namely, before and during the lockdown. Designing and making use of a questionnaire-based study conducted on a French sample (N = 1266), the authors expected the research participants to report a natural tendency to report creative solutions and innovative action paths during lockdown (than before such a period), and such a pattern of attitudes and behaviors to be observed in both their professional and everyday spheres. The case was, nonetheless, that, regarding "Pro-C", no significant variations and discrepancies were observed between the two temporal points in comparison; but in what concerns "little-c", the truth was that a significant increase has been reported to have happened during the lockdown period. Furthermore, the results provided by this work emerge as extremely relevant to such discussions because they suggest that participants disposing of lower levels of baseline creativity (this having been observed before lockdown) profited from the pandemic situation, such a conclusion to be derived if taken into comparison those individuals disposing of higher levels of initial baseline creativity. Hence, this work provides an enriching and useful basis to start theoretical discussions on the possible positive outcomes of uncertain and stressful contexts, such as the Covid-19 pandemics, into the organizational dynamics, this way boosting the debate into a more updated and refined analysis.

As we can see, the theoretical background of Mercier et al.'s (2021) work relates a set of concepts, articulating a multidisciplinary contribution to the empirical research and academic debate under the Organizational Creativity topic. Most of all, its roots rely on the investigation of psychosomatic responses to situations in which individuals face stress, anxiety, and uncertainty, and how such responses may then derive into organizational positive or negative impacts. In detail, it is possible to acknowledge that such work provides an enriching investigation around uncertainty, anxiety, solitude, boredom, exhaustion, stress, coping mechanisms, frustration, self-reflection, and self-transformation.

Furthering such a line or reasoning, we can start by advocating that solitude, at least discussing it in theoretical terms, should present as a positive feeling in a given situation that then proves beneficial to develop creative and innovative solutions and action paths, given that it can particularly enable renewed perspectives around a myriad of fields in the daily life, these motivated by the freedom of spirit which turns possible to be experienced in such a context (Simonton, 2000). Apropos, Long & Averill (2003) advanced two ways in which solitude could enable increased levels of Creativity, namely: one by promoting and energizing the individual's imaginative involvement into a myriad of realities, this to say, by both enabling and leveraging people's imagination, daydream, and wonder, creating alternative scenarios to the present reality (Barabasz, 1991). Although different, logically relating to this emerges the other possible way in which solitude could boost creative behaviors in such delicate contexts, based on the premise that solitude may be conceived as interestingly able to facilitate self-reflection and contemplation (Koch, 1994), which, at their very core, are key to the generation of alternative and refined perspectives on life, which later derive into the adoption of new-fangled attitudes and behaviors (Long & Averill, 2003). Hence, it can be defended that solitude can positively nurture creativity via the adoption of different, alternative, and complementary selves, this both via and resulting in a self-transformation process.

Furthermore, solitude has frequently and commonly been associated with boredom, both theoretically and empirically (Farmer & Sundberg, 1986; Galanaki, 2004), a fact that derives from the resulting deprivation of activities, experiences, and habits, foremost interpersonal interactions, that notably characterize situations of solitude and social isolation. This fact assumes relevant to the discussion in the present chapter, taking into consideration that such an uncertain, stressful, anxious, solitary, boring, and frustrating context has been broadly experienced throughout state-imposed lockdowns all around the globe in the current times, as a result of the governmental responses to the impacts derived from the Covid-19 pandemic (Brodeur et al., 2020). The key issue here is that not disregarding that empirical research and theoretical debates have commonly been associating boredom (just as solitude) with a series of negative outcomes (the work of Vodanovich, 2003 provides a fruitful review on the topic), recent research has also increasingly been confirming that boredom could derive into beneficial impacts to creativity (Mann & Cadman, 2014), thus conceiving creativity as a personal response to boredom in the sense that it develops alternative approaches to situations so that the individual or the organization can explore new ways to conduct and overcome boring tasks and/or rigid processes, via an attempt to make it more interesting, stimulating and/or engaging (Toohey, 2011). This way, it may be arguable that Creativity highlights into contexts marked by uncertainty, stress, anxiety, solitude, boredom, and frustration, if we defend that it consists, at its very essence, into "an alerting phenomenon" which calls our attention that "all is not well and something must be done" (Gaylin, 1979: 129).

As a corollary of such a discussion, it then turns defendable to postulate that, analogous to uncertainty, these contextual configurations so characteristic from the Covid-19 time and environment should favor creativity in both personal and professional spheres.

The Twofold Need For and Challenges Brought About by Creativity in the Organizational Environment, in the Covid-19 Era

The key problem to think about Creativity as an organizational response in terms of risk management approaches in times of crisis, such as the current one marked by the impacts of the Covid-19 pandemic, resides in the fact that, whereas the pandemic has tested the agility, competitiveness, sustainability, and resilience of organizations, it as well has forced both academicals, practitioners, leaders, managers, and overall employees to design, develop and implement a deeper look at the assumptions underlying the conventional theoretical frameworks so far used to guide managerial decisions and organizational practices (George et al., 2020).

In this regard, a first twofold need for and challenge brought about by Creativity in the organizational environment, in the Covid-19 era, can be mentioned to be the revised scope and content of problem categorization, which then develops to also profoundly impact the design, implementation, and development of risk management approaches. To base such an idea, we advocate that we have been assisting to a transversal, deep, and rapid disruption of the conventional problems employees, managers, academics, practitioners, scientists, and experts have been embedded in for decades, as their daily routines also profoundly alter in this accelerated and uncertain rhythm. As a consequence, it can be arguable that variations, even though subtle at the beginning, will occur not only on the frequency of problems that are addressed, but foremost on their type, scope, and geographical area. So, such differences will not only be observed on the problems that individuals choose to solve, but also (and more notably) on how these, on their hand, will affect the broader progress of the business atmosphere, of the organizational environment, and also of Science.

A second twofold need for and challenge brought about by Creativity in the organizational environment, in the Covid-19 era, can be mentioned to be the "forced" guideline to move from working in the workplace, to be working from home. Indeed, and alluding to George et al. (2020), it can be arguable that this has been the most notorious and significant organization design shock that the pandemic created, given that such new working habits resulted in a clear rush to design ways to implement working conditions at home, whenever possible, and to rapidly adapt to remote collaboration and its inherent technological infrastructure, which often needed to be updated. This way, working habits have faced a clear and profound transformation, evolving from a conventional notion of remote work, so far understood as a mere benefit/feature of multinational firms and open-source communities, to later be understood as a daily-life habit, almost a *sine qua non* condition of the current pandemic times, due to the forced "all remote, all the time" moto.

Creativity, thus, emerged in such a scenario as a simultaneous need and solution. Nonetheless, the challenge was the core theoretical problem inherent to the concept of creativity, which can be assumed to nurture, be expressed, and materialize quicker and easier when via interpersonal contacts. Fostering this idea, it can be defendable that collaboration and communication drive innovation behavior, so, the jointness of such vectors roots at the core of the construct of creativity, at least when conceiving it in theoretical terms. Henceforth, a key problem could arise, because it can be discussible if creativity, both in its personal and collective origin and manifestation, may suffer as a consequence of the forced "working from home" practices, so, from the impacts of the Covid-19 outbreak. Hence, it would be reasonable to question if collaboration patterns would be hindered in such a scenario, which means if they could change in terms of co-producers of knowledge and agents of knowledge transfer. Nonetheless, a different line of thinking is also legitimate to be addressed, by which it is possible to argue that conformity pressures that induce groupthink may be less prone to be generated and to prevail in virtual groups, which, on their hand, have been created as a result of this virtual "working from home" renewed practice. Not disregarding any of the mentioned ideas, the truth is that on the debates around this topic, it appears to be consensual that the current virtual work that emerged as a twofold need and consequence of the effects of the Covid-19 outbreak in organizational structures and working habits might indeed nudge the individual habits and patterns of thinking, investigation, and experimentation, in the sense that the locus of brainstorm, research, and tentative alters from co-located offices to indeed fix into geographically distant colleagues, in a myriad of spatial locations. Even though the complexification of the analysis can proceed and go deeper, because even based on such consensual assumption, it is legitimate to question if, in such a complex, accelerated, stressful, and uncertain scenario the rapid iteration and prototyping, characteristic results from creativity, can be feasible online.

Hence, we can postulate that Creativity entails, at least theoretically, a meaningful and substantial effect on the resilience ability of organizations, employees, and entrepreneurs, given that it induces both personal and organizational/collective adaptability, hence, developing important impacts on the survival and sustainability potential of the organizations and their inherent business value chains; idea which results on the assumption that Creativity must be advanced and followed by Innovation so that organizations can indeed assume able to survive during the Covid 19 times (Dora, 2021).

The so-far possible synthesis of such a logic of argumentation allows to advocate for the need for constant debate, given that the antecedents of creativity in virtual teams and people working outside the conventional workplaces remain still fundamentally unexplored (Litchfield et al., 2015), such a situation due as well to the fact that Covid-19 outbreak has been categorized as an unprecedented global crisis, provoking widespread consequences in a myriad of daily-life scopes, regardless of the conventional national, cultural, social, and geographical boundaries. Indeed, as systematizing the insights provided in the above paragraph, is it is consensual in the extant literature that, so far, face to face interactions has been the prevailing approach to boost knowledge creation, sharing, and transfer, and also to leverage creativity and innovation levels in organizational structures, hence defending that such approach entails a positive role in the generation, implementation, and enhancement of group-level creativity, according to George et al. (2020) another fact not to be overlooked also consists on the idea that it remains unclear if and how can it indeed be replicated or bettered in the current online contexts, structures, tools, and practices.

At the same time, the growing concerns on the virtualization of work cannot be neglected, and are compelling organizations to growingly look for hybrid solutions so that they can articulate a joint dynamic between virtual and present work (in-home, in the workplace). As a specific example, and perhaps the most frequent concern of both organizational managers and employees in what concerns this issue, refers to how can organizations onboard their new members, because it can for sure be questioned how teams can be structured from the ground, specifically from a variety of members who may have never met in personal terms and the physical world.

Not only is the question of collaboration assuming relevance in this scenario, but also the one referring to technology. Indeed, and furthering the insights already mentioned in the previous paragraphs, it is as well interesting to debate around the joint action between creativity, innovation, and collaboration networks, specifically if those may be boosted or hindered by the virtualization dynamics forced by the Covid-19 impacts, due to the "working from home" guidelines as a response to reorganize organizational structures and its correspondent working habits. Therefore, this line of reasoning hypothesizes if the network structure will profoundly and decisively change with such a forced need for virtualization, or, on the contrary, if the growing dynamics of experimenting with technology components will become more (or less) radical, or even if both options will coexist, and if so, in which combination. The rationale for these questions (which indeed consist of avenues for future research, together with the topics addressed in the previous paragraphs) roots on the core premise which postulates that, in this new, uncertain, complex, delicate, and stressful context, indeed social interactions are rapidly, constantly and deeply being reshaped, and, on the other hand, if innovation is established and predicted to emerge, develop and capitalize on mutual knowledge among employees articulated via a collaboration dynamics, occupying various and different structural roles and positions within a network of experts and scientists experimenting different theoretical and analytical approaches, as well as process tools and technology components (Schillebeeckx et al., 2019), then organizations will be challenged with the opportunity to conceive, develop and implement renewed patterns of knowledge creation and sharing, creativity, innovation, teamwork, and collaboration.

The pandemic has also speeded the adoption of virtual solutions, digital technologies, and cloud applications to leverage the delivery of core businesses, in face of the twofold challenges and opportunities brought about by Covid-19, which hindered the survival potential and sustainability of business value-added chains of many organizations. This way, George et al. (2020) advocate towards an industry disruption, which will result in the emergence of updated network infrastructures, new technologies, and refined business models as a consequence of the pandemics in which regards to Creativity, Adaption, Agility, and Innovation in the organizational context. The problem, nonetheless, resides in the fact that the dominant theoretical and empirical approach to Creativity and Innovation is in the extant academic research and organizational management. This emphasized the physical infrastructure required for creative efforts to be materialized into innovative outcomes as the dominant analytical rational and theoretical model for over a century. Nonetheless, the issue nowadays resides in the fact that the vast majority of the innovation infrastructure cannot yet be virtualized, which then hinders the efficiency potential of the business value-chain of many organizations in the current pandemic scenario. Therefore, it turns clear that this pandemic has been raising questions on the discovery pattern, the experimentation process, and how innovation infrastructure, including labs, buildings, and social ecosystems, might change (George et al., 2020). So, at its very core, the pandemics have given momentum to the thoughtful debate around the conduct of Science. Hence, furthering this line of thinking, in the current Covid-19 scenario it turns acceptable to debate if this pandemic will result in a need to rapidly and profoundly alert academics, practitioners, managers, leaders, employees, scientists, and experts to redesign how knowledge is created, developed, transferred, and shared, so also, how experiments are designed, conducted, and evaluated. This to say, as a corollary, it is increasingly discussible if the Covid-19 impacts will push the frontiers of science, namely, if certain scientific or innovation domains will have the potential to assume as more relevant or urgent due to contextual configurations, for example, due to flows of investment capital and managerial attention, which indeed, on their hand, mark the environmental configuration that resulted from this pandemic impacts.

Relating with the insights exposed in the above-mentioned paragraph, we can call out for a third twofold need for and challenge brought about by Creativity in the organizational environment, in the Covid-19 era. This refers to the dilemma concerning the digitalization and correspondent modularization of work, which may hinder the enhancement of collaborative patterns of action, cooperative approaches, and teamwork outcomes. In this regard, it has been increasingly questioned if the need for the modularization of work, brought about by the rapid, increasing, and somewhat forced triumph of the electronic interaction, will result in boosting or hindering the conventional dynamics of face-to-face interaction and work coordination. Two reasonable lines of thinking are being developed in this regard, one defending that face-to-face interaction and work coordination indeed come to be vital in such context; nonetheless, a distinct one argues that asynchronous tacit coordination configures as a potentially viable alternative (Srikanth & Puranam, 2014). Moreover, it can as well be defendable that perhaps a more efficient alternative perspective entails a combination of both approaches, based on the idea that such a joint method will allow the accelerated and complex interdependent work – so characteristic from the business environment brought about by the Covid-19 impacts – would, thus, be able to be executed in a distributed context, relying on both coordination and observation as management core vectors, being that coordination would root not on direct virtual contact, such as on video calls ("seeing the face"), but instead, it would be based on mutual and constant observation of work ("seeing the work") via a consensually defined set of systemic procedures to ensure compatibility.

Related to this, another relevant point to discuss arises – monitoring performance, team spirit, and individual agents' motivation and overall feeling of well-being in the workplace and the job function. Indeed, the rationale to such a debate consists of the problem that, if monitoring is expensive, from this assumption it could derive the argument that basic organization design principles will, subsequently, tend to an essential move and transformation from behavioral to outcome-based control (Puranam, 2018). The key conclusion from such an idea postulates that the digitalization of work may make the latter solution ("outcome-based control") both more likely to happen and feasible, henceforth providing, beneficial consequences, allowing workers to be evaluated on results, rather than behaviors. Nonetheless, a critical alert needs to be addressed given that, at the same time, a reasonable line of thinking may be defended to advocate that such digital technologies can enable intrusive surveillance.

Finally, a fourth twofold need for and challenge brought about by Creativity in the organizational environment, in the Covid-19 era, can be mentioned to be the unequal distributional consequences of Innovation which results from the creative boost this context enhances. The core vector to discuss concerning this postulates that this pandemic has amplified and intensified the dilemma between the beneficial effects of the increased and improved digital access (which is defendable to beneficially impact the well-being and life-quality standards to enhance people life chances), and the potentially harmful effects on equity, regional development, social cohesion, and inclusion. Furthering such an idea, it can be debatable that the consequences of Covid-19 have sharply impacted the self-awareness in what concerns the partiality, heterogeneity, and inequity patterns of Creativity inducement and Innovation outcomes, through its disproportional influence on individuals, geographical scopes, and disenfranchised communities, as to say that time is needed (and with it, also future and deeper research on the topic) aimed at studying the distributional consequences of creativity and innovation, as well as how these then result into the adoption of new technologies, based on the premise that, as we examine the effects of the pandemic, it starts to be clear that not everyone, every location, every community, and every organization benefits in equal levels and standards from the twofold challenges and opportunities brought about by Creativity and Innovation (George et al., 2020).

So, the corollary of such discussion culminates on the argument that technology tools and network infrastructures, as well as team structures and dynamics, namely the implications of the widespread of technologies and the virtualization of work on a dynamic of collaboration, teamwork, and creativity of all organizational individual agents, will assume gradually important in what concerns to Organizational Creativity and Innovation research (George et al., 2020).

SOLUTIONS AND RECOMMENDATIONS

Even though this work configures as an interesting starting point to boost discussions on the topic, it as well presents limitations. It is based on theoretical work, resulting from literature review, therefore, the theoretical insights hereby postulated shall be faced with and assessed by empirical testing, to prove their relevance and robustness as analytical lines of thinking. Hence, the contributes provided in this chapter shall be furthered, developed to be inserted into empirical and operational models, so that the impacts of creativity on organizational performance can be more sharply, incisively, broadly, and rigorously examined.

Expanding the above mentioned, it then turns clear that more empirical research shall follow, reviewing and furthering the present work, taking into consideration that the contextual situation which gave it its reason, motivation, and problematics (Covid-19 pandemic) prevails in time, given that solutions are not yet being provided and society is not yet capitalizing on its potential, indeed only mechanisms to control public health and to implement safety measures are mainly being actioned nowadays. So, as the problem keeps existing through time, more research is needed to fully understand it and to better both design strategies, policies, and tools to manage and to overcome it in the present, but studies are also needed to fully understand it so that the future can be best anticipated and analogous challenges overcome with potentially higher success rates.

FUTURE RESEARCH DIRECTIONS

The Covid-19 pandemic brought new ways of structuring organizations, doing business, managing teams, setting human resources policies and practices, balancing both personal and professional life, etc., according to which the challenge has been set forward so that organizations redefine the structures, processes, and tools how they operate to generate, implement, develop and sustain secure, successful and rewarding careers. The critical issue is that the Covid-19 crisis configured as a disruption to society, a new and ongoing situation with an unknown period, therefore, high levels of uncertainty are associated with such a context, which result in risk management approaches increasingly difficult to design and implement. Hence, such a scenario turns it difficult to try to foresee, anticipate and plan strategies for the present and the future.

Following the line of Ratten (2020), this represents both an opportunity and a challenge to the society, both in personal agents and their daily-life habits, both to the business environment and to the organizational value chains and teams, and the responsible approach and the corresponding success will indeed depend on how businesses address and respond to such challenges (Alon et al., 2020). Such a context emerged as a clear result of the widespread impacts of Covid-19, a pandemic that brought unparalleled challenges all over the world, regardless of conventional time and geographical frontiers. Such a delicate and complex context resulted in the organizational need to timely review and update the pre-designed crisis management plans (Mikusová & Horváthová, 2019), hence aiming at updating and sustaining business value chains, by creating refined work conditions to protect workers within an unforeseen situation.

In such a scenario, it can be defendable that a possible solution can root on the increasing need for cooperation particularly in terms of creating, developing, and sharing knowledge (Kirk & Rifkin, 2020). Creativity, hence, assumes as a critical vector to analyze to generate innovative tools, processes, and approaches to overcome the challenges and capitalize on the opportunities that current and future organizations face in such an environment. In conclusion, the present chapter provides a more inclusive and integrative framework to understand the relevance of creativity as a core organizational asset, so that business value chains can remain competitive in the current pandemic times, acknowledging that it offers fruitful insights which enable the deeper theoretical understanding, in terms of the business, organizational, and entrepreneurship literature, of the micro, macro, and meso environmental effects deriving from the Covid-19 crisis, fact that highlights as relevant taken into consideration that, in the present times, the effects of Covid-19 on entrepreneurship remain largely unexplored.

Another important aspect to expose evidence that the current research around creativity and innovation theory (foremost in what concerns their expression and management at the organizational level) largely ignores the role of the inherent individual idiosyncrasies, from which it highlights the creative self-efficacy (the work of Shao et al., 2019 constitutes, nonetheless, a notable exception to this tendency). Also, research (wrongly) tends to be parsimonious on the analysis, given that it frequently ignores and marginalizes the potential impact of the organizational context as an important vector impacting the organizational actor's propensity towards creative and innovative patterns of action (Radomska & Wołczek, 2020; Shao et al., 2019). Developing this line of reasoning, acknowledging the benefits of creative self-efficacy infers that managers shall configure as active, dynamic, exemplary, and reliable agents/models to provide their subordinates with key experiences and examples in which teams (as a collective structure) and employees (as individual agents) can generate, develop and implement creative ideas under uncertain conditions (Tierney & Farmer, 2002). Future research shall dedicate to investigating this issue. Providing employees with positive feedback about their creative behaviors, and also rewarding such efforts, may assume a positive attitude from the management team to improve the employees' self-confidence in their creative abilities, hence later boosting the expression and materialization of such skill. To help them in this process, the potential positive effects of creativity training sessions can be defendable as a useful approach to advance the employees' creative self-efficacy.

It shall be stressed that future research shall be furthered to design and implement instruments and analytical models which prove able to robustly assess the adherence and correspondent impact of stringency measures, governmental political configurations, social and historical background, and cultural dimensions into the possible positive contextual motivators of creativity and innovation so that we can have increasing insight on these relationships, which turns able the rigorous investigation of cross-cultural samples. This would configure an interesting line of research, based on the argument that the current empirical evidence has been reporting data which suggests that national creativity and innovation parameters are interdependent, indeed interconnected with the reported perceptions of overall subjective well-being at the country level (Kapoor & Kaufman, 2020). To be more specific on such an idea, empirical research on this issue is frequently evidencing that Eastern countries are more prone to report lower levels on what concerns overall creativity, compared to the ones scored by Western nations (Xie & Paik, 2019). The rationale to explain such discrepancies roots in the different perceptions towards innovative and creative outputs, climaxing into the assumption that these vary cross-culturally. Research has proven that the mentioned discrepancies may find their fundament on the different configurations of Hofstede's cultural dimensions (Hofstede, 1984), as the work of Morris & Leung (2010) reports, advocating that, for example, Western countries value novelty, whereas Eastern ones emphasize usefulness.

Also noteworthy to highlight, subsequent research on the Organizational Creativity topic can also compare global creativity and innovation indices in a temporal spectrum, this to say, in the years before and after the outbreak of the Covid-19 pandemic, which would configure an interesting line of empirical research given that it would allow exploring temporal changes in these outcomes.

As well, it shall be considered, into future research avenues, that the reported "Four C" Model to understand creativity and its manifestations, both in a personal as well as at the individual level, presents a good starting point to develop research on the topic, nonetheless, further revisions and updates to it are welcome. Specifically, by this idea, we convey that such an analytical and conceptual model, by focusing too closely and intently on the traditional ("Little-C", "Big-C") distinctions of Creativity, offers the dual risk of, on the one hand, overlooking – or even neglecting – the creative potential of children; and on the other hand, of minimizing professional-level creative productions of expert creators, given that it focusses mainly on the two extremes of the creativity spectrum.

Another crucial vector to emphasize in this section refers to the critical gap in the existing literature on the Organizational Creativity topic, which involves the lack of application of Paradox Theory to the existing theoretical debates and empirical models (Pinto, 2019). Such a limitation also occurs in this chapter, and it is vital to be highlighted taking into consideration that the Paradox Theory assumes particularly suited to investigate creativity because it allows incorporating the impacts of vectors deriving from both the role of individual differences (such as creative self-efficacy) and from the role of the organizational context (such as high uncertainty) into the theoretical and empirical research models.

To better understand this logic, we need to consider two basic premises: at first, we need to acknowledge that creating, constructing, and developing creative ideas intrinsically involves both divergent and convergent thinking (Miron-Spektor & Erez, 2017); simultaneously, we need to ponder that both cognitive flexibility and persistence are imbedded into creative thinking paths and innovative practical solutions (Nijstad et al., 2010). Based on this, it is possible to prove the relevance of the Paradox Theory to such debates, given that it rigorously analyses the inherent contradictions and inconsistencies derived from competing demands, such as (and namely) those experienced when indeed engaging in creative tasks.

CONCLUSION

Research around the topic of creativity and innovation at the organizational level is becoming more prolific, subsequently assuming as a growing field that has evolved beyond fundamental categorizations and basic initial models. If progress in scientific research on this topic has been achieved over time, often interconnecting several research domains via a multidisciplinary investigation approach, aimed at reaching a broad and rigorous analysis, it can as well not be neglected that it is nowadays vital to dispose of a specific understanding and classification of what it means to be creative, following the alerts addressed by Kaufman & Beghetto (2009).

The thematic analysis from the literature in which this chapter has been based has provided useful insights so that debates can follow and be developed into more rigorous levels of analysis, hence able to cover the four main inter-related twofold opportunities and challenges that contexts of crisis bring to organizations (from which it stands out, as a practical example, the effects brought about by the Covid-19 pandemic), including namely (i) cost and finance-related challenges, (ii) disruption of activities, (iii) workforce subsistence, and (iv) existential difficulties.

The present chapter advocates that argue that the "Four C" Model to classify the different types of creativity and manifestations assume relevant to investigate the expression of this skill both in a personal as well as in a professional level, given that it turns possible to consider personal interests, wishes, demands, aspirations, experiences, and pursuits, interconnected with creative ability, all this via a joint vector developed and operationalized at an appropriate level of specificity. Furthering this line, such a model assumes as a virtuous starting point to understand the issue under investigation, as it presents a framework in which previous theories, conceptual constructs, and core notions on knowledge, creativity, and innovation can be situated, connected, developed, implemented, and reviewed. Equally noteworthy to mention regarding this, is the fact that such a conceptual model for Creativity assumes as well relevant because it highlights aspects of creativity that require further clarification and investigation.

The argumentation exposed all along with this chapter provides fruitful insights which postulate that contexts of crisis – such as the current one, derived from the global impacts of the Covid-19 pandemic, which crosses demographic, geographic, and organizational boundaries – can enhance and facilitate people's efforts toward creativity and innovation under certain conditions, this being observed both in personal as well as in professional terms. Hence, contexts of stress, uncertainty, anxiety, solitude, boredom, doubt, and frustration for sure impel to negative outcomes on mental and physical health, nevertheless, it can also be advocated that beneficial effects may derive from such contexts, which can boost the search for creative and innovative approaches and outcomes to the problems faced on such scenarios. Specifically, by the argumentation exposed all along with this chapter it has been possible to advocate that employees may generally perceive work uncertainty brought by the pandemic as a motor to leverage their creativity so that they can start envisaging renewed and innovative approaches to the twofold challenges and opportunities such a context of stress, anxiety, uncertainty and solitude brought about.

By expansion of this argument, interconnecting the personal and the organizational level, the argumentation on this chapter defended that organizational creativity and innovation can be facilitated and positively boosted in such times of global crisis, indeed as a response strategy to such a scenario, imprinting the organization to develop higher levels of resilience and to address a more efficient response in order not only to manage risk and to overcome the challenge but foremost to capitalize on the opportunities brought about by such contexts. This consists, then, of the additional value provided by the present chapter, given that it provides a useful framework of recommendations to help organizations to enhance both their resilience, sustainability, and responsiveness in the context of the Covid-19 crisis, focusing on creativity as a core organizational asset, which is the same as to say, focusing on people as the organizational core asset to design, develop, improve and sustain the business added-value chain, and, subsequently and inherently, the organizational competitive advantage. These recommendations namely include collaboration, openness, durability, leading by example, and constant learning; attitudes and behaviors to be taken directly by the part of the management team, being the leaders and the managers the role models and motors of the subsequent employees' actions, which we can defend that will, then, derive as a spill-over effect of the above-mentioned ones.

The fundament to this argumentation basis on the very core concept of leadership, if we acknowledge that it essentially corresponds to the process of engaging with others aiming to set and achieve shared goals, which, logically, assumes indispensable in organizations, moreover in times of crisis such as the current ones we face due to the widespread global impacts of Covid-19 (Hayes & Wooten, 2010). Covid-19 has proven that risk management organizational strategies, policies, practices, and tools both require and result in consensual, concerted, coordinated, and mutual responses amidst uncertainty, transformation, acceleration, anxiety, and stress by a myriad of different teams that can cross multiple functions, scopes, roles, and localizations. Continuing such a logic of reasoning, it can be defendable that effective teamwork both roots and results on accurate, open, honest, timely, constant, dynamic, and problem-solving communications, which shall inherently and simultaneously convey and root on shared goals, shared knowledge, transparency, genuine care, and mutual respect — requirements which call out for the absolute need of relational coordination to both design and implement efficient risk management strategies to such global crisis contexts.

Following the line of Nembhard et al. (2020), which corresponds to the above argumentation, we can also conclude that this chapter encourages organizational managers and leaders to implement and perform seven actions, as their core business and management approach to boost personal and organizational creativity, aiming at improving the overall organizational performance so that beneficial effects can derive to improve the organizational resilience, sustainability, and risk management approach. Namely, this chapter highpoints 1) managing people as the center of the business environment and the organizational structure, 2) managing operations via creative paths of action and innovative solutions, 3) base interpersonal interactions in open, direct, fair, constant, and honest communication flows, 4) design and adopt learning rather than a performance mindset, 5) adhere to teamwork, progressive learning and knowledge sharing as the vital ways of working, 6) embrace clear, humble, coherent and transformative leadership, and 7) create outside partnerships which are transversal to a myriad of market scopes and geographical locations.

The problem here grounds on the fact that addressing such problems requires a systems' thinking mindset, able to conceive the scenario as a virtuous twofold challenge and opportunity, hence, imprinting a holistic perspective on the analysis. This perspective would be the most efficient one to address creativity at the organizational level because it would prove beneficial to recognize the importance of both internal and external operations, this way highlighting problems and analyzing strategies, structures, processes, policies, tools, and practices, subsequently having a great foundation allowing the creation and development of teams that foster internally and externally to create, share, transfer and expand knowledge sources to develop ideas, boosting personal and organizational creativity aimed at improving the innovation levels in the organization. Critical for such creativity is organizational support for the quick implementation of iterative changes to respond to new problems that may arise, and which may indeed trump planned strategy. Hence, organizations shall direct efforts to design and adopt a learning mindset, proven to be the most efficient one to boost personal and collective creativity, given that it explores new ideas and practices, hence relieving, overcoming, and mining the pressure created by the unproductive performance, setting employees free from such constraints so that they can have the right resources to offer creative ideas and innovative courses of action, this to say, to think, to hypothesize, to brainstorm with different agents and to experiment, to generate, develop, implement and evaluate updated, new-flanged and effective solutions. On the contrary, a performance mindset strictly follows and optimizes established practices, via pre-designed tools to achieve prior and often rigid strategies and goals, which hinders the room for creativity to bloom.

Nonetheless, an alert shall be done in regards to the just exposed, stressing that this idea is possible to be defended at least in theoretical terms, thus, of course, the logic of reasoning advocated in this work needs to be empirically tested and validated.

REFERENCES

Abramowitz, M., & David, P. (1996). Technological change and the rise of intangible investments: The US economy's growth path in the Twentieth Century. In D. Foray & B.-Å. Lundvall (Eds.), Employment and growth in the knowledge-based economy. Paris: OECD.

Agars, M. D., Baer, J., & Kaufman, J. C. (2005). The many creativities of business and the APT Model of creativity. *The International Journal of Creativity & Problem Solving*, *15*(2), 133–141.

Akinola, M., Kapadia, C., Lu, J. G., & Mason, M. F. (2019). Incorporating physiology into creativity research and practice: The effects of bodily stress responses on creativity in organizations. *The Academy of Management Perspectives*, *33*(2), 163–184. doi:10.5465/amp.2017.0094

Alencar, E. M. (1995). *Criatividade* (2nd ed.). Editora da Universidade de Brasília.

Alencar, E. M. (1998). Promovendo um ambiente favorável à criatividade nas organizações. *Revista de Administração de Empresas, 38*(2), 18–25. doi:10.1590/S0034-75901998000200003

Alon, I., Farrell, M., & Li, S. (2020). Regime type and COVID-19 response. *FIIB Business Review, 9*(3), 152–160. doi:10.1177/2319714520928884

Amabile, T. M. (1988). A model of creativity and innovation in organizations. *Research in Organizational Behavior, 10*(1), 123–167.

Amabile, T. M. (1995). Discovering the unknowable, managing their unmanageable. In C. M. Ford & D. A. Gioia (Eds.), *Creative action in organizations*. Sage Publications. doi:10.4135/9781452243535.n7

Amabile, T. M., Conti, R., Coon, H., Lazenby, J., & Herron, M. (1996). Assessing the work environment for creativity. *Academy of Management Journal, 39*(5), 1154–1184.

Baer, J., & Kaufman, J. C. (2005). Bridging generality and specificity: The amusement park theoretical (APT) model of creativity. *Roeper Review, 27*(3), 158–163. doi:10.1080/02783190509554310

Bakker, H., Boersma, K., & Oreel, S. (2006). Creativity (ideas) management in industrial R&D organizations: A crea-political process model and an empirical illustration of Corus RD&T. *Creativity and Innovation Management, 15*(3), 296–309. doi:10.1111/j.1467-8691.2006.00397.x

Barabasz, M. (1991). Hypnotizability in bulimia. *International Journal of Eating Disorders, 10*(1), 117–120. doi:10.1002/1098-108X(199101)10:1<117::AID-EAT2260100113>3.0.CO;2-T

Barbot, B., & Reiter-Palmon, R. (2019). Creativity assessment: Pitfalls, solutions, and standards. *Psychology of Aesthetics, Creativity, and the Arts, 13*(2), 131–132. doi:10.1037/aca0000251

Barry, C. B., & Brown, S. J. (1985). Differential information and security market equilibrium. *Journal of Financial and Quantitative Analysis, 20*(4), 407–422. doi:10.2307/2330758

Beghetto, R. A., & Kaufman, J. C. (2007). The genesis of creative greatness: Mini-c and the expert performance approach. *High Ability Studies, 18*(1), 59–61. doi:10.1080/13598130701350668

Beghetto, R. A., & Plucker, J. A. (2006). The Relationship Among Schooling, Learning, and Creativity: "All Roads Lead to Creativity" or "You Can't Get There from Here"? In J. C. Kaufman & J. Baer (Eds.), *Creativity and reason in cognitive development* (pp. 316–332). Cambridge University Press. doi:10.1017/CBO9780511606915.019

Besemer, S. P., & Treffinger, D. J. (1981). Analysis of creative products: Review and synthesis. *The Journal of Creative Behavior, 15*(3), 158–178. doi:10.1002/j.2162-6057.1981.tb00287.x

Binnewies, C., Ohly, S., & Niessen, C. (2008). Age and creativity at work: The interplay between job resources, age and idea creativity. *Journal of Managerial Psychology, 23*(4), 438–457. doi:10.1108/02683940810869042

Blauth, M., Mauer, R., & Brettel, M. (2014). Fostering creativity in new product development through entrepreneurial decision making. *Creativity and Innovation Management, 23*(4), 495–509. doi:10.1111/caim.12094

Brodeur, A., Clark, A. E., Fleche, S., & Powdthavee, N. (2021). COVID-19, lockdowns and well-being: Evidence from Google Trends. *Journal of Public Economics, 193*, 104–346. doi:10.1016/j.jpubeco.2020.104346 PMID:33281237

Bruno-Faria, M. F., & Alencar, E. M. (1996). Estímulos e obstáculos à criatividade no ambiente de trabalho. *Revista ADM, 31*(2), 50–61.

Cardinal, L. B., Alessandri, T. M., & Turner, S. F. (2001). Knowledge codifiability, resources, and science-based innovation. *Journal of Knowledge Management, 5*(2), 195–204. doi:10.1108/13673270110393266

Cavusgil, S. T., Calantone, R. J., & Zhao, Y. (2003). Tacit knowledge transfer and firm innovation capability. *Journal of Business and Industrial Marketing, 18*(1), 6–21. doi:10.1108/08858620310458615

Chen, Y. S., Lin, M. J. J., & Chang, C. H. (2009). The positive effects of relationship learning and absorptive capacity on innovation performance and competitive advantage in industrial markets. *Industrial Marketing Management, 38*(2), 152–158. doi:10.1016/j.indmarman.2008.12.003

Cotter, K. N., Silvia, P. J., & Fayn, K. (2018). What does feeling like crying when listening to music feel like? *Psychology of Aesthetics, Creativity, and the Arts, 12*(2), 216–227. doi:10.1037/aca0000108

Counsell, A., Furtado, M., Iorio, C., Anand, L., Canzonieri, A., Fine, A., Fotinos, K., Epstein, I., & Katzman, M. A. (2017). Intolerance of uncertainty, social anxiety, and generalized anxiety: Differences by diagnosis and symptoms. *Psychiatry Research, 252*, 63–69. doi:10.1016/j.psychres.2017.02.046 PMID:28254577

Cromwell, J. R., Amabile, T. M., & Harvey, J.-F. (2018). An integrated model of dynamic problem solving within organizational constraints. In R. Reiter-Palmon, V. L. Kennel, & J. C. Kaufman (Eds.), *Individual creativity in the workplace* (pp. 53–81). Academic Press. doi:10.1016/B978-0-12-813238-8.00003-6

Cropley, A. (2006). Creativity: A social approach. *Roeper Review, 28*(3), 125–130. doi:10.1080/02783190609554351

Cropley, D. H. (2015). Promoting creativity and innovation in engineering education. *Psychology of Aesthetics, Creativity, and the Arts, 9*(2), 161–171. doi:10.1037/aca0000008

Csikszentmihalyi, N. M. (1988). Society, culture, and person: A systems view of creativity. In R. Sternberg (Ed.), *The nature of creativity*. Cambridge University Press.

Darroch, J., & McNaughton, R. (2002). Examining the link between knowledge management practices and types of innovation. *Journal of Intellectual Capital, 3*(3), 210–222. doi:10.1108/14691930210435570

Davenport, T. H., De Long, D. W., & Beers, M. C. (1998). Successful knowledge management projects. *Sloan Management Review*, (Winter), 43–57.

Dnishev, F., & Alzhanova, F. (2016). Globalization of technological development and opportunities for national innovation systems of developing countries. *The Journal of Asian Finance, Economics, and Business, 3*(4), 67–79. doi:10.13106/jafeb.2016.vol3.no4.67

Dora, Y. M. (2021). Adaptability, Creativity And Innovation To Be Survive. Study On Arts And Culture Tourism Bandung. *Turkish Journal of Computer and Mathematics Education, 12*(8), 973–979.

Drucker, P. F. (1985). The discipline of innovation. *Harvard Business Review*, *63*(3), 67–72. PMID:10272260

Drucker, P. F. (1993). *Sociedade pós-capitalista*. Pioneira.

Drucker, P. F. (2004). A disciplina da inovação. *Harvard Business Review*, *82*(8), 80–85.

du Plessis, M., & Boon, J. A. (2004). Knowledge management in eBusiness and customer relationship management: South African case study findings. *International Journal of Information Management*, *24*(1), 73–86. doi:10.1016/j.ijinfomgt.2003.10.002

Edvinsson, L. (1997). Developing intellectual capital at Skandia. *Long Range Planning*, *30*(3), 366–373. doi:10.1016/S0024-6301(97)90248-X

Farmer, R., & Sundberg, N. D. (1986). Boredom proneness - the development and correlates of a new scale. *Journal of Personality Assessment*, *50*(1), 4–17. doi:10.120715327752jpa5001_2 PMID:3723312

Fisher, C. M., & Barrett, F. J. (2019). The experience of improvising in organizations: A creative process perspective. *The Academy of Management Perspectives*, *33*(2), 148–162. doi:10.5465/amp.2017.0100

Fleury, M. T., & Oliveira, M. M., Jr. (Orgs.). (2001). Gestão estratégica do conhecimento. São Paulo: Atlas.

Fong, P. S., Men, C., Luo, J., & Jia, R. (2018). Knowledge hiding and team creativity: The contingent role of task interdependence. *Management Decision*, *56*(2), 329–343. doi:10.1108/MD-11-2016-0778

Ford, C. M. (2000). Creative developments in creativity theory. Academy of Management. *Academy of Management Review*, *25*(2), 284–285. doi:10.5465/amr.2000.26079331

Galanaki, E. (2004). Are children able to distinguish among the concepts of aloneness, loneliness, and solitude? *International Journal of Behavioral Development*, *28*(5), 435–443. doi:10.1080/01650250444000153

Gaylin, W. (1979). *Feelings: Our Vital Signs*. Harper and Row.

George, G., Lakhani, K. R., & Puranam, P. (2020). What has changed? The impact of Covid pandemic on the technology and innovation management research agenda. *Journal of Management Studies*, *57*(8), 1754–1758. doi:10.1111/joms.12634

Gloet, M., & Terziovski, M. (2004). Exploring the relationship between knowledge management practices and innovation performance. *Journal of Manufacturing Technology Management*, *15*(5), 402–409. doi:10.1108/17410380410540390

Guellec, D., & Van Pottelsberghe de la Potterie, B. (2004). From R&D to productivity growth: Do the institutional settings and the source of funds of R&D matter? *Oxford Bulletin of Economics and Statistics*, *66*(3), 353–378. doi:10.1111/j.1468-0084.2004.00083.x

Guilford, J. P. (1950). Creativity. *The American Psychologist*, *4*(9), 444–454. doi:10.1037/h0063487 PMID:14771441

Harkema, S. (2003). A complex adaptive perspective on learning within innovation projects. *The Learning Organization*, *10*(6), 340–346. doi:10.1108/09696470310497177

Hartley, J. (2014). New development: Eight and a half propositions to stimulate frugal innovation. *Public Money & Management*, *34*(3), 227–232. doi:10.1080/09540962.2014.908034

Hofstede, G. (1984). Cultural dimensions in management and planning. *Asia Pacific Journal of Management, 1*(2), 81–99. doi:10.1007/BF01733682

Jacobs, D., & Snijders, H. (2008). *Innovation Routine: How Managers Can Support Repeated Innovation // Stichting Management Studies.* Van Gorcum.

Kapoor, H., & Kaufman, J. C. (2020). Meaning-making through creativity during COVID-19. *Frontiers in Psychology, 11*, 1–9. doi:10.3389/fpsyg.2020.595990 PMID:33391115

Kaufman, J. C., & Beghetto, R. A. (2009). Beyond big and little: The four c model of creativity. *Review of General Psychology, 13*(1), 1–12. doi:10.1037/a0013688

Kaufman, J. C., & Beghetto, R. A. (2013). Do people recognize the four Cs? Examining layperson conceptions of creativity. *Psychology of Aesthetics, Creativity, and the Arts, 7*(3), 229–236. doi:10.1037/a0033295

Kaya, F. B., Sahin, G. G., & Gurson, P. (2010). Intellectual capital in organizations. *Problems and Perspectives in Management, 8*(1), 153–160.

Kirk, C. P., & Rifkin, L. S. (2020). I'll trade you diamonds for toilet paper: Consumer reacting, coping and adapting behaviors in the COVID-19 pandemic. *Journal of Business Research, 117*, 124–131. doi:10.1016/j.jbusres.2020.05.028 PMID:32834208

Klein, D. A. (1998). *A gestão estratégica do capital intelectual: recursos para a economia baseada em conhecimento*. Qualitymark Editora Ltda.

Koch, P. (1994). *Solitude: A philosophical encounter*. Open Court Publishing.

Koenig, M. E. (1996). Intellectual capital and knowledge management. *IFLA Journal, 22*(4), 299–301. doi:10.1177/034003529602200411

Litchfield, R. C., Gilson, L. L., & Gilson, P. W. (2015). Defining creative ideas: Toward a more nuanced approach. *Group & Organization Management, 40*(2), 238–265. doi:10.1177/1059601115574945

Locke, E. A., & Kirkpatrick, S. A. (1995). Promoting creativity in organizations. In C. M. Ford & D. A. Gioia (Eds.), *Creative action in organizations*. Sage Publications. doi:10.4135/9781452243535.n13

Long, C. R., & Averill, J. R. (2003). Solitude: An exploration of benefits of being alone. *Journal for the Theory of Social Behaviour, 33*(1), 21–44. doi:10.1111/1468-5914.00204

Luis, D. J., Ruth, C. J., & Zhuofan, Z. (2020). Does stress lead to creativity? The relationship between occupational stress and individual innovative behavior. *Studies in Business & Economics, 15*(1), 21–30. doi:10.2478be-2020-0003

Lundvall, B. Å., & Nielsen, P. (2007). Knowledge management and innovation performance. *International Journal of Manpower, 28*(3/4), 207–223. doi:10.1108/01437720710755218

Madjar, N., Oldham, G. R., & Pratt, M. G. (2002). There's no place like home? The contributions of work and nonwork creativity support to employees' creative performance. *Academy of Management Journal, 45*(4), 757–767.

Mann, S., & Cadman, R. (2014). Does being bored make us more creative? *Creativity Research Journal*, *26*(2), 165–173. doi:10.1080/10400419.2014.901073

Martins, J. M. (2010). *Gestão do Conhecimento, Criação e transferência de conhecimento*. Edições Sílabo.

Mercier, M., Vinchon, F., Pichot, N., Bonetto, E., Bonnardel, N., Girandola, F., & Lubart, T. (2021). COVID-19: A Boon or a Bane for Creativity? *Frontiers in Psychology*, *11*, 1–10. doi:10.3389/fpsyg.2020.601150 PMID:33536973

Mikusová, M., & Horváthová, P. (2019). Prepared for a crisis? Basic elements of crisis management in an organisation. *Economic Research-Ekonomska Istrazivanja*, *32*(1), 1844–1868. doi:10.1080/1331677X.2019.1640625

Miron-Spektor, E., & Erez, M. (2017). *Looking at creativity through a paradox lens*. Oxford University Press.

Morgan, G. (1995). *Imagens da organização*. Atlas.

Morris, M. W., & Leung, K. (2010). Creativity east and west: Perspectives and parallels. *Management and Organization Review*, *6*(3), 313–327. doi:10.1111/j.1740-8784.2010.00193.x

Nembhard, I. M., Burns, L. R., & Shortell, S. M. (2020). Responding to Covid-19: Lessons from management research. *NEJM Catalyst Innovations in Care Delivery*, *1*(2), 1–7.

Nicolau, J. A., & Paranhos, J. (2006). Notas sobre o conceito de inovação. *Textos de Economia*, *9*(1), 23–37.

Nijstad, B. A., De Dreu, C. K., Rietzschel, E. F., & Baas, M. (2010). The dual pathway to creativity model: Creative ideation as a function of flexibility and persistence. *European Review of Social Psychology*, *21*(1), 34–77. doi:10.1080/10463281003765323

Ostrower, F. (1993). *Criatividade e processos de criação* (9th ed.). Vozes.

Peters, A., McEwen, B. S., & Friston, K. (2017). Uncertainty and stress: Why it causes diseases and how it is mastered by the brain. *Progress in Neurobiology*, *156*, 164–188. doi:10.1016/j.pneurobio.2017.05.004 PMID:28576664

Pinto, J. (2019). Key to effective organizational performance management lies at the intersection of paradox theory and stakeholder theory. *International Journal of Management Reviews*, *21*(2), 185–208. doi:10.1111/ijmr.12199

Plucker, J. A., Beghetto, R. A., & Dow, G. T. (2004). Why isn't creativity more important to educational psychologists? Potentials, pitfalls, and future directions in creativity research. *Educational Psychologist*, *39*(2), 83–96. doi:10.120715326985ep3902_1

Puranam, P. (2018). *The microstructure of organizations*. Oxford University Press. doi:10.1093/oso/9780199672363.001.0001

Radomska, J., & Wołczek, P. (2020). Integrative perspective on ambidexterity, creativity and networking: Literature overview. *European Research Studies Journal*, *23*(3), 31–49. doi:10.35808/ersj/1623

Ratten, V. (2020). Coronavirus (covid-19) and entrepreneurship: Changing life and work landscape. *Journal of Small Business and Entrepreneurship, 32*(5), 503–516. doi:10.1080/08276331.2020.1790167

Richards, R. (2007). Everyday creativity and the arts. *World Futures, 63*(7), 500–525. doi:10.1080/02604020701572707

Richards, R., Kinney, D. K., Lunde, I., Benet, M., & Merzel, A. P. (1988). Creativity in manic-depressives, cyclothymes, their normal relatives, and control subjects. *Journal of Abnormal Psychology, 97*(3), 281–288. doi:10.1037/0021-843X.97.3.281 PMID:3192819

Runco, M. A., Illies, J. J., & Eisenman, R. (2005). Creativity, originality, and appropriateness: What do explicit instructions tell us about their relationships? *The Journal of Creative Behavior, 39*(2), 137–148. doi:10.1002/j.2162-6057.2005.tb01255.x

Runco, M. A., & Jaeger, G. J. (2012). The standard definition of creativity. *Creativity Research Journal, 24*(1), 92–96. doi:10.1080/10400419.2012.650092

Runco, M. A., & Okuda, S. M. (1988). Problem discovery, divergent thinking, and the creative process. *Journal of Youth and Adolescence, 17*(3), 211–220. doi:10.1007/BF01538162 PMID:24277636

Santos, A. B., Fazion, C. B., & De Meroe, G. P. (2011). Inovação: Um estudo sobre a evolução do conceito de Schumpeter. *Caderno de Administração Revista da Faculdade de Administração da FEA, 5*(1), 1–16.

Sarkar, S. (2007). *Innovation, market archetypes and outcome: An integrated framework*. Springer Science & Business Media.

Schillebeeckx, S. J., Lin, Y., & George, G. (2019). When do expert teams fail to create impactful inventions? *Journal of Management Studies, 56*(6), 1073–1104. doi:10.1111/joms.12447

Schumpeter, J. A. (1942). *Capitalism, Socialism and Democracy*. Harper & Row.

Schumpeter, J. A. (1988). *Capitalismo, sociedade e democracia*. Abril Cultural.

Shalley, C. E., & Gilson, L. L. (2004). What leaders need to know: A review of social and contextual factors that can foster or hinder creativity. *The Leadership Quarterly, 15*(1), 33–53. doi:10.1016/j.leaqua.2003.12.004

Shalley, C. E., Gilson, L. L., & Blum, T. C. (2000). Matching creativity requirements and the work environment: Effects on satisfaction and intentions to leave. *Academy of Management Journal, 43*(2), 215–223.

Shalley, C. E., Zhou, J., & Oldham, G. R. (2004). The effects of personal and contextual characteristics on creativity: Where should we go from here? *Journal of Management, 30*(6), 933–958. doi:10.1016/j.jm.2004.06.007

Shao, Y., Nijstad, B. A., & Täuber, S. (2019). Creativity under workload pressure and integrative complexity: The double-edged sword of paradoxical leadership. *Organizational Behavior and Human Decision Processes, 155*, 7–19. doi:10.1016/j.obhdp.2019.01.008

Shinyashiki, G. T., Trevizan, M. A., & Mendes, I. A. (2003). Sobre a criação e a gestão do conhecimento organizacional. *Revista Latino-Americana de Enfermagem*, *11*(4), 499–506. doi:10.1590/S0104-11692003000400013 PMID:14748169

Silva, S. L. (2002). Informação e competitividade: A contextualização da gestão do conhecimento nos processos organizacionais. *Ciência da Informação*, *31*(2), 142–151. doi:10.1590/S0100-19652002000200015

Simonton, D. K. (1977). Creative productivity, age, and stress: A biographical time-series analysis of 10 classical composers. *Journal of Personality and Social Psychology*, *35*(11), 791–804. doi:10.1037/0022-3514.35.11.791 PMID:915694

Simonton, D. K. (1991). Emergence and realization of genius: The lives and works of 120 classical composers. *Journal of Personality and Social Psychology*, *61*(5), 829–840. doi:10.1037/0022-3514.61.5.829

Simonton, D. K. (1994). Individual differences, developmental changes, and social context. *Behavioral and Brain Sciences*, *17*(3), 552–553. doi:10.1017/S0140525X00035925 PMID:23218117

Simonton, D. K. (1997). Creative productivity: A predictive and explanatory model of career trajectories and landmarks. *Psychological Review*, *104*(1), 66–89. doi:10.1037/0033-295X.104.1.66

Simonton, D. K. (1998). Fickle fashion versus immortal fame: Transhistorical assessments of creative products in the opera house. *Journal of Personality and Social Psychology*, *75*(1), 198–210. doi:10.1037/0022-3514.75.1.198

Simonton, D. K. (2000). Creativity: Cognitive, personal, developmental, and social aspects. *The American Psychologist*, *55*(1), 151–158. doi:10.1037/0003-066X.55.1.151 PMID:11392859

Simonton, D. K. (2004). Psychology's status as a scientific discipline: Its empirical placement within an implicit hierarchy of the sciences. *Review of General Psychology*, *8*(1), 59–67. doi:10.1037/1089-2680.8.1.59

Srikanth, K., & Puranam, P. (2014). The firm as a coordination system: Evidence from software services offshoring. *Organization Science*, *25*(4), 1253–1271. doi:10.1287/orsc.2013.0886

Stair, R. M. (1998). *Princípios de sistemas de informação: uma abordagem gerencial.* Rio de Janeiro: LTC Editors.

Sternberg, R. J. (1985). Implicit theories of intelligence, creativity, and wisdom. *Journal of Personality and Social Psychology*, *49*(3), 607–627. doi:10.1037/0022-3514.49.3.607

Sternberg, R. J., O'Hara, L. A., & Lubart, T. I. (1997). Creativity as investment. *California Management Review*, *40*(1), 8–21. doi:10.2307/41165919

Tierney, P., & Farmer, S. M. (2002). Creative self-efficacy: Its potential antecedents and relationship to creative performance. *Academy of Management Journal*, *45*(6), 1137–1148.

Toohey, P. (2011). *Boredom: A lively history. Yale.* Yale University Press.

Tseng, M. L., Lim, M. K., Wong, W. P., Chen, Y. C., & Zhan, Y. (2018). A framework for evaluating the performance of sustainable service supply chain management under uncertainty. *International Journal of Production Economics*, *195*, 359–372. doi:10.1016/j.ijpe.2016.09.002

Vodanovich, S. J. (2003). Psychometric measures of boredom: A review of the literature. *The Journal of Psychology, 137*(6), 569–595. doi:10.1080/00223980309600636 PMID:14992349

Waples, E. P., & Friedrich, T. L. (2011). Managing creative performance: Important strategies for leaders of creative efforts. *Advances in Developing Human Resources, 13*(3), 366–385. doi:10.1177/1523422311424713

Wigand, R., Picot, A., & Reichwald, R. (1997). Information, organization and management. John Wiley.

Winter, S. (1987). Knowledge and competence as strategic assets. In D. Teece (Ed.), *The competitive challenge: strategy for industrial innovation and renewal*. Ballinger Publishing Company.

Xie, G., & Paik, Y. (2019). Cultural differences in creativity and innovation: Are Asian employees truly less creative than western employees? *Asia Pacific Business Review, 25*(1), 123–147. doi:10.1080/13602381.2018.1535380

Zahra, S. A. (1993). Environment, corporate entrepreneurship, and financial performance: A taxonomic approach. *Journal of Business Venturing, 8*(4), 319–340. doi:10.1016/0883-9026(93)90003-N

Zhou, J., & Shalley, C. E. (2003). Research on employee creativity: A critical review and directions for future research. *Research in Personnel and Human Resources Management, 22*, 165–217. doi:10.1016/S0742-7301(03)22004-1

ADDITIONAL READING

Amabile, T. M., Schatzel, E. A., Moneta, G. B., & Kramer, S. J. (2004). Leader behaviors and the work environment for creativity: Perceived leader support. *The Leadership Quarterly, 15*(1), 5–32. doi:10.1016/j.leaqua.2003.12.003

Ancillo, A. L., del Val Núñez, M. T., & Gavrila, S. G. (2021). Workplace change within the COVID-19 context: A grounded theory approach. *Economic Research-Ekonomska Istraživanja, 34*(1), 2297–2316. doi:10.1080/1331677X.2020.1862689

Doyle, D. (2020). Virtual creativity in COVID times. *Virtual Creativity, 10*(2), 137–140. doi:10.1386/vcr_00028_2

Hayes, E. J., & Wooten, L. P. (2010). *Leading Under Pressure: From Surviving to Thriving Before, During, and After a Crisis*. Routledge Taylor & Francis.

Hildebrandt, M. (2021). Creativity and Resilience in Art Students During COVID-19. *Art Education, 74*(1), 17–18. doi:10.1080/00043125.2020.1825597

Qin, X., Yam, K. C., Chen, C., Li, W., & Dong, X. (2021). Talking about COVID-19 is positively associated with team cultural tightness: Implications for team deviance and creativity. *The Journal of Applied Psychology, 106*(4), 530–541. doi:10.1037/apl0000918 PMID:34014708

Siswanti, Y., & Muafi, M. (2020). Empowering leadership and individual creativity: The mediation role of psychological empowerment in facing covid-19 pandemic. *The Journal of Asian Finance. Economics and Business*, *7*(11), 809–816.

Thukral, E. (2021). COVID-19: Small and medium enterprises challenges and responses with creativity, innovation, and entrepreneurship. *Strategic Change*, *30*(2), 153–158. doi:10.1002/jsc.2399

Tierney, P., Farmer, S. M., & Graen, G. B. (1999). An examination of leadership and employee creativity: The relevance of traits and relationships. *Personnel Psychology*, *52*(3), 591–620. doi:10.1111/j.1744-6570.1999.tb00173.x

Tran, D.-T., Lee, L.-Y., Nguyen, P.-T., & Srititiratku, W. (2020). How Leader Characteristics and Leader Member Exchange Lead to Social Capital and Job Performance. *Journal of Asian Finance. Economics and Business*, *7*(1), 269–278.

Chapter 3
Thermodynamics and Economics Analogies

Nima Norouzi
https://orcid.org/0000-0002-2546-4288
Bournemouth University, UK

ABSTRACT

A Gibbs-Duhem relationship in economics, Slutsky conditions, and the relationship proposed by marginalist theory between exchange value and value in use was found from a phenomenological description and tested considering the macroeconomic equations. To simplify the calculations, it was considered that the economic systems are in equilibrium to make an analogy with the thermodynamics of equilibrium. In addition, fundamental considerations were made, such as considering that a consumer's wealth or a small country is constant for the existence of a measurable (quantifiable) utility function. However, according to the discussion, it could be seen that W. Saslow's theory must be developed in the field of non-equilibrium thermodynamics, which is why the economy is not a system that is in equilibrium, but rather, on the contrary, the economy is a dynamic system.

INTRODUCTION

There are many analogies between economic and physical systems in economics; in particular, Thermodynamics has implications. These analogies are similar to each other, especially in how they deal with the concept of entropy. The idea of applying the way of thinking of thermodynamics to economics is not new, in fact, in the thought of Karl Marx. For Marxist thought, energy is a fluid, just like money is (Baierlein, 2001).

Bejan et al. (2020) shows that the sudden end of economic expansion (movement, wealth) emerges as a natural, physical feature of the spreading movement, which has access to power (money), freedom to morph, and power storage (savings) for future movement on even greater areas. The movement is driven by power generation, which is interspaced with power savings on the same area. The theory is constructed systematically from the physical basis of economics concepts (money, savings, time, and bubbles) to a physics model that accounts for the time-dependent spreading of movement on an area.

DOI: 10.4018/978-1-6684-2339-4.ch003

Previous study has shown that physics accounts for the proportionality between the annual wealth (GDP) of a population and the annual consumption of fuel to generate power for that population. The present theory extends this view to the more realistic situation where every movement in society (wealth and fuel consumption) is time dependent.

Jakimowicz (2020) shows that non-extensive cross-entropy econometrics is a valuable complement to traditional econometrics as it explains phenomena based on power-law probability distribution and enables econometric model estimation for non-ergodic ill-behaved (troublesome) inverse problems. Furthermore, the entropy economics has accelerated the emergence of modern econophysics and complexity economics. These new directions of research have led to many interesting discoveries that usually contradict the claims of conventional economics. Econophysics has questioned the efficient market hypothesis, while complexity economics has shown that markets and economies function best near the edge of chaos. Quantum economics has already appeared on the horizon, which recognizes money as a fundamental measurement device in the economy. The development of these sciences may indicate the need to reformulate all mainstream economics from its foundations.

A research Rashkovskiy (2021) consider the thermodynamic approach to the description of economic systems and processes. The first and second laws of thermodynamics as applied to economic systems are derived and analyzed. It is shown that there is a deep analogy between the parameters of thermodynamic and economic systems (markets); in particular, each thermodynamic parameter can be associated with a certain economic parameter or indicator. The economic meaning of such primordial thermodynamic concepts as pressure, volume, internal energy, heat, etc. has been established. The thermostatistics of the market is considered. It is shown that, as in conventional thermostatistics, many market parameters, such as price of goods, quantity of goods, etc., as well as their fluctuations can be calculated formally using the partition function of an economic system.

In this Tsirlin & Gagarina (2020), we consider optimal trading processes in economic systems. The analysis is based on accounting for irreversibility factors using the wealth function concept. The existence of the welfare function is proved, the concept of capital dissipation is introduced as a measure of the irreversibility of processes in the microeconomic system, and the economic balances are recorded, including capital dissipation. Problems in the form of kinetic equations leading to given conditions of minimal dissipation are considered.

Many consider that this case corresponds to a way of thinking of dialectical materialism. Historically, the development of thermodynamics originated due to the industrial revolution. The Nobel Prize winner in economics Samuelson suggested that economics and thermodynamics had much in common (Bedi et al., 2007), but his research was not as deep as Roegen Georgescu, who made a detailed study of one of the main variables of state of thermodynamics: Entropy (Bryant, 1982).

In 1998, Wayne Saslow, professor at Texas A&M University, proposed an analogy between thermodynamics and economics; specifically, his theoretical development was based on the Consumer Theory (Bryant, 2007). This analogy raises a phenomenological description of thermodynamic and economic systems. It is mainly based on the assumption that an economic system is in equilibrium; it also shows how some state variables of Thermodynamics have their respective analogs in Economics(Chakrabarti et al., 2006).

There are two ways of working in Thermodynamics and the first is finding the Internal Energy function of a thermodynamic system, the second is finding the Entropy function of the system. However, obtaining a function Internal energy or entropy theoretically is impossible, except that basic assumptions are made to obtain a function. Still, in reality, these functions can be experimentally obtained for a thermodynamic system (Chamizo & Novales Cinca, 2015).

From this point of view, W. Saslow seeks to describe an economic system based on a function by analogy with Internal Energy; the problem, the analogy proposed by Saslow, is that it does not clearly define a temperature and entropy for an economic system. However, those who take the path of looking for a Thermodynamic formulation by entropy find that its analog in economics is the Production Function using the analogy between the Second Law of Thermodynamics and the Second Law of Economics. However, the temperature in economics still is not a very clear concept (Cooter & Ulen, 1988; De León Arias, 2010). It should be noted that the basic hypotheses of W. Saslow's theory of the analogy between thermodynamics and economics are:

1. An economic system in equilibrium. In practice, we know that economic systems are neither isolated nor in equilibrium but are dynamic systems.
2. The existence of a Utility Function (the analog to Internal Energy)
3. Parameters are Measurable.
4. The wealth of a consumer or small country is constant.

The aim of this paper is to simplify the calculations, it was considered that the economic systems are in equilibrium to make an analogy with the Thermodynamics of equilibrium.

THEORETICAL FOUNDATION

Economic Analogies

Examples of economic systems of interest are an individual consumer or a small country, each incorporated into a large economic system. Assume an individual consumer. A fundamental assumption in economics is that the consumer uses a utility function U to choose to buy one good. We will assume that U is given in a convenient set of units, such as 200 suns, and we also assume that U is also measurable. Utility in economics is a psychological concept; it does not have a numerically measurable value. This tells us the (subjective) satisfaction that was owning an object produces. We begin by discussing certain fundamental relationships in economics(Erlichson, 1984).
We first consider the measurability of the economic quantity known as wealth W.

$$W = \lambda.M + p.N(Economy) \tag{1}$$

Where λ and M represent the value and amount of money, p and N represent the vectors of prices and numbers of goods. Wealth is not only the material part, that is, the amount of goods that an individual can possess, but it can also be the monetary part, for example, M represents the amount of money that the individual can have in the bank and λ represents how much your money increases. W is conserved in a transaction; since it is the total energy E of a thermodynamic system, it is tempting to consider them as analogs. From the point of view of maximization, a natural analogy is between –E and W(Felder & Felder, 2015). Economists assume that the value of currencies and goods of an individual consumer is adding by the value of U that typically exceeds W. The excess is known as surplus, for which we introduce the notation Ψ.

$$\Psi = U - W(\textit{Economy}) \tag{2}$$

In a primitive or very scarce economy, there is no surplus, so $\Psi = 0$. In this case, all people perform the same economic function in the same efficiency, and there is no benefit of specialization and trade. The surplus Ψ cannot be negative; for typical economic systems $\Psi < 0$.
The Helmholtz free energy of a system with N identical particles is defined as:

$$F = -P.V + \mu.N(\textit{Thermodynamics}) \tag{3}$$

Where P is the pressure, V is the volume and is the chemical potential μ of the particles. The quantity in thermodynamics analogous to the price p is the chemical potentialμ. To have a clear idea, the chemical potential is understood as a tendency to diffusion(Greiner & Müller, 2010). Let us consider a country that produces a certain good, and if the price of that good in the world market is higher than the sale price in that country, the usual tendency is for the country to export that good. Therefore there is a process of diffusion of country goods abroad. On the contrary, if the price of the same good is lower abroad than in the country itself, there is a process of diffusion of said good to the country's interior (Georgescu-Roegen, 1971).
Energy E is related to F in terms of temperature T and entropy S.

$$T.S = E - F(\textit{Thermodynamics}) \tag{4}$$

A comparison of equations (2) and (4) suggests another analogy, that Ψ and TS. Because the surplus is zero for an undeveloped economy, we attempt the common language of economic temperature T with economic development.

Thermodynamics, Entropy, and Balance

Thermodynamics operates with the transformation of heat into mechanical work and dates from the work of Carnot (1824), who established the principle of limiting the quantity of work from the heat under given conditions. The reasoning by which Carnot established his principle is one of the most extraordinary triumphs of the deductive method (Mankiw, 2014).

Carnot's thought is the basis of the Second Law of Thermodynamics. It is extraordinary that it was established long before the experimental studies of James Joule (1845), who established the First Law of Thermodynamics, that heat is a form of energy.

According to James Joule, a system with a fixed number of particles can change its energy in two ways. Heat δQ can enter the system or Work δW that the system can do. Conservation of energy is written as:

$$dE = \delta Q + \delta W \tag{5}$$

From Carnot's studies, we can show that in equilibrium δQ = TdS, where T is the temperature and the entropy S is a function of the system's state. Besides the work done on the system of δW =-PdV, the conservation of energy takes the form (McCauley, 2009):

$$dE = TdS + PdV \tag{6}$$

Although the energy of a thermodynamic system is uniquely defined, its contained heat (the integral of δQ) and its contained work (the integral of –PdV) are not so (Pindyck et al., 2013). A system can go from one energy state to another through an infinite number of processes where the contributions of work and heat differ [8]. Finally, if the particles can enter or leave the system by an amount dN, there is an energy change μdN. The fundamental relationship of thermodynamics combines these energy changes to give:

$$dE = TdS + PdV + \mu dN \tag{7}$$

Consider two identical gas chambers at the same temperature, pressure, and chemical potential. In the connection of the chambers, the energy E, the volume V, and the number of particles N of the combined system will be the sum of the energies, volumes, and number of particles N of the combined system of particles for the individual systems. Such variables are called extensive. The entropy S is also an extensive variable. On the other hand, the temperature T, the pressure P, and the chemical potential of the combined system do not change. Such variables are called intensive. Another way to write equation (7) is to consider energy as an extensive variable and a function of the three variables S, V, and N(Rogers & Rempe, 2012):

$$E = E(S, V, N) \tag{8}$$

Knowledge of E (S, V, N) characterizes the thermodynamic state of the system. For this reason, E (S, V, N) is called the system state function. Since changes in energy E are characterized by changes in S, V, and N, we have from equation (8)

$$dE = \left(\frac{\delta E}{\delta S}\right)dS + \left(\frac{\delta E}{\delta V}\right)dV + \left(\frac{\delta E}{\delta N}\right)dN \tag{9}$$

A comparison of equations (7) and (9) leads to the identification:

$$T = \left(\frac{\delta E}{\delta S}\right)_{V,N};P = \left(\frac{\delta E}{\delta V}\right)_{S,N};\mu = \left(\frac{\delta E}{\delta N}\right)_{S,V} \tag{10}$$

Maxwell's relations are a consequence of the fact that the crisscrossing wineskin of the derivatives of E (S, V, N) does not matter. Then, from equations (8) and (10), we have

$$\frac{\delta^2 E}{\delta S \delta V} = \frac{\delta^2 E}{\delta V \delta S} or \left(\frac{\delta P}{\delta S}\right)_{V,N} = \left(\frac{\delta T}{\delta V}\right)_{S,N}$$

$$\frac{\delta^2 E}{\delta S \delta V} = \frac{\delta^2 E}{\delta V \delta S} or \left(\frac{\delta \mu}{\delta S}\right)_{V,N} = \left(\frac{\delta T}{\delta V}\right)_{S,V}$$

$$\frac{\delta^2 E}{\delta N \delta V} = \frac{\delta^2 E}{\delta V \delta S} or \left(\frac{\delta P}{\delta N}\right)_{S,V} = \left(\frac{\delta T}{\delta V}\right)_{S,N} \tag{11}$$

These relationships guarantee that the integrals over dE in S, V, and N are independent paths. Because energy is an extensive quantity, it satisfies:

$$E(\alpha S, \alpha V, \alpha N) = \alpha E(S, V, N) \tag{12}$$

For a parameter α> 0. Scaling the extensive quantities S, V and N scale the extensive quantity E, which depends on S, V, and N [05]. Observe that T, P, and μ are not changed because this scaling corresponds to systems that are in great times α. Differentiating the left-hand part of the equation (12) concerning α, and then using equations (12) and (10) gives.

$$\frac{dE(\alpha S,\alpha V,\alpha N)}{d\alpha} = \left(\frac{\delta E(S,V,N)}{\delta S}\right)S + \left(\frac{\delta E(S,V,N)}{\delta V}\right)V + \left(\frac{\delta E(S,V,N)}{\delta N}\right)N$$

$$\frac{dE(\alpha S,\alpha V,\alpha N)}{d\alpha} = TS.PV + \mu.N \tag{13}$$

Differentiating the right-hand part of the equation (12)

$$\frac{dE(\alpha S,\alpha V,\alpha N)}{d\alpha} = E(S,V,N) \tag{14}$$

Equating equations (13) and (14) we obtain the fundamental relation:

$$E(S, V, N) = TS - PV + \mu N \tag{15}$$

Subtracting equation (9) from the differential of equation (15) gives the Gibbs-Duhem relationship [5, 11, 21]

$$0 = Sdt - VdP + Nd\mu \tag{16}$$

The variation of the chemical potential is not independent of the variations of temperature and pressure, but the variation of anyone can be calculated as a function of the variations of the other two (Richmond et al., 2013). The Gibbs-Duhem relationship presents the relationship between the intensive parameters in the differential form (Riley et al., 1999). For some purposes, the set (T, V, N) instead of (S, V, N) is a more natural set of variables. In this case, the appropriate thermodynamic potential is the Helmholtz F free energy. Combining equations (3) and (15), we obtain

$$F = E - TS \tag{17}$$

Using equations (10) and equation (17), dF satisfies

$$dF = dE - TdS - SdT = -SdT - PdV + \mu dN \tag{18}$$

Note that dF = 0 for a system at fixed V and N in contact with a thermal reservoir that sets T. But, by setting T and N, the dW =-PdV work done on the system equals dF. This relationship is the origin of the terminology free energy. Equation (17) is the Legendre transformation of energy and allows us to go from a function with natural variables (S, V, N) to one with natural variables (T, V, N) (Saslow, 1999). The variables S and T are dual to each other, like V and P and N and μ. If a system has two possible states with the same T, V, and N, the state with the lowest free energy F (T, V, N) is thermodynamically stable. Using these natural variables, equation (3) for F can be written as:

$$F(T, V, N) = -PV + \mu N \tag{19}$$

Thermodynamics Interpretation of Economics

Knowledge of the state function as a function of the appropriate economic parameters characterizes the economic system. From the economic relationship previously introduced

$$\Psi = TS \tag{20}$$

And from equation (2)

$$U = TS + W = TS + \lambda M + pN \tag{21}$$

A comparison of equation (21) for U to equation (15) suggests that, from the point of view of its set of natural variables, we have

$$U = U(S, M, N) \tag{22}$$

The equation (22) is our fundamental assumption. The economic equivalent of equation (7) is

$$dU = TdS + \lambda dM + pdN \tag{23}$$

Where

$$T = \left(\frac{\delta U}{\delta S}\right)_{M,N} ; \lambda = \left(\frac{\delta U}{\delta M}\right)_{S,N} ; p = \left(\frac{\delta U}{\delta N}\right)_{S,M} \tag{24}$$

In the book The Wealth of Nations, Adam Smith distinguished between two measures of utility. One measure is the "exchange rate." In economics, this is conventional to identifying the value in exchange with the price. From equation (24), we take this measure to be the marginal utility times dU/dN at fixed S and N. We want to identify "value in use" with marginal utility per good dU / dN for another set of fixed variables. For the sake of simplicity, we will do this by taking M as fixed, but we cannot be explicit about the second variable that will have to be fixed, and we simply denote it as x. From equation (23) below, we have

$$\left(\frac{\delta U}{\delta N}\right)_{x,M} = T\left(\frac{\delta S}{\delta N}\right)_{x,M} + p \tag{25}$$

One of the great triumphs of the nineteenth century in "marginalist" economic theory is in the following argument: A consumer purchases goods subject to the condition that the relation between the value of any good in use to its price p accepts in a common value (Saslow, 1999). This next statement requires that to set the market values of goods, U is maximized for each good. Fixing the market value means that products 1 and 2 are traded in the market subject to the condition.

$$0 = p_1 dN_1 + p_2 dN_2 \tag{26}$$

Maximizing U requires that

$$0 = \left(\frac{\delta U}{\delta N_1}\right)_{x,M} dN_1 + \left(\frac{\delta U}{\delta N_2}\right)_{x,M} dN_2 \tag{27}$$

Combining equations (26) and (27) below gives

$$\frac{1}{p}\left(\frac{\delta U}{\delta N}\right)_{x,M} = constant\ For\ each\ good \tag{28}$$

Therefore, the relation in value in use of prices is constant.
Using equation (25); Equation (28) can be expressed as

$$\frac{1}{p}\left(\frac{\delta U}{\delta N}\right)_{x,M} = \frac{T}{p}\left(\frac{\delta S}{\delta N}\right)_{x,M} + 1 = constant \tag{29}$$

We represent with m the value in use (marginal utility for good, at x and M fixed)

$$M = \left(\frac{\delta U}{\delta N} \right)_{x,M} \tag{30}$$

Notice that m is specified in monetary units. From equations (28) and (29), the m/p relationship has the dimensions for all goods. The m/p relationship can be generalized to include the value of currencies, thus allowing the study of savings. Specifically, we define

$$m\lambda = \left(\frac{\delta U}{\delta N} \right)_{x,M} \tag{31}$$

If equations (21) be used when relating W and U, the analogy with the expansion associated with F leads to

$$dW = -SdT + \lambda dM + pdN \tag{32}$$

where

$$S = \left(\frac{\delta U}{\delta T} \right)_{M,N} ; \lambda = \left(\frac{\delta U}{\delta M} \right)_{T,N} ; p = \left(\frac{\delta U}{\delta N} \right)_{T,M} \tag{33}$$

From equation (32), we can write the functional dependence

$$W = W(T, M, N) \tag{34}$$

Considering the market as a reservoir r, we have from equation (32)

$$dW_r = -S_r dT_r + \lambda r_d Mr + pr_d Nr \tag{35}$$

Subject to the conditions $dT = dT_r = 0$, the conservation of money ($dM + dM_r = 0$), we find by adding equations (32) and (35) that

$$dW + dW_r = (\lambda - \lambda r)_d M + (p - pr)_d N \tag{36}$$

The right-hand part of equation (36) is zero for arbitrary variations of dM and dN only if the value of the money for the consumer is the same as the value of the money for the market $\lambda = \lambda r$. Similarly, for the value of a good, we have $p = pr$.

Note that equation (1) gives the differential

$$dW = \lambda dM + pdN + Md\lambda + Ndp \tag{37}$$

The consistency of equations (37) and (32) requires that

$$0 = SdT + Md\lambda + Ndp \tag{38}$$

Equation (38) is the analog of equation (16), the Gibbs-Duhem relationship. It implies that a decrease in the price of currency or goods (as when the state develops an economic increase) is accompanied by an increase in economic temperature(Saslow, 1999). This qualitative behavior is expected from conventional economic reasoning. Specifically, if a common factor increases all prices and currency values, then the system does not change. By equation (38), the temperature is nothing more than an increase in the common factor. We can write the change in the Marshallian surplus, Ψ = TS as

$$d = TdS + SdT = TdS - Md\lambda - Ndp \tag{39}$$

We interpret the term TS as the change in the economic value of leisure. We will call TdS the Veblenian surplus. The term –Ndp is the change in the consumer surplus of the goods. We will call –Ndp the Smithian surplus. Likewise, the term –Mdλ can be interpreted as a surplus in the currency due to the efficiency produced by specialization.

To see that –Ndp is a surplus, note that the cost of purchasing goods increasing (where the first goods are scarce and therefore expensive) is $\int_o^N pdN$, where p (N) is the price of the good nth, it decreases as N increases (rate is, (dp/dN) <0). However, when all are purchased at once, the real cost for the consumer is Np. The difference is

$$\int_o^N pdN - Np = -\int_{p(=o)}^{p(N)} Ndp \tag{40}$$

The difference is positive since the limits of the integration of equation (40) dp are negative. Hence –Np is the change in consumer surplus of the goods. The following "Maxwell relations" are an immediate consequence of the fact that the interlocking order of the derivatives of U = U (S, M, N) does not matter.

$$\frac{\delta^2 U}{\delta S \delta N} = \frac{\delta^2 U}{\delta S \delta N} or \left(\frac{\delta P}{\delta S}\right)_{M,N} = \left(\frac{\delta T}{\delta N}\right)_{S,M}$$

$$\frac{\delta^2 U}{\delta S \delta M} = \frac{\delta^2 U}{\delta M \delta S} or \left(\frac{\delta \lambda}{\delta S}\right)_{M,N} = \left(\frac{\delta T}{\delta M}\right)_{S,N}$$

$$\frac{\delta^2 U}{\delta S \delta N} = \frac{\delta^2 U}{\delta S \delta N} or \left(\frac{\delta p}{\delta N}\right)_{S,M} = \left(\frac{\delta \lambda}{\delta V}\right)_{S,N} \tag{41}$$

In economics, these relationships are known as Slutsky conditions(Saslow, 1999). This guarantees that the integrals over dU in S, M, and N are independent paths.

THERMOMETRY AND THERMODYNAMICS

We have so far assumed that economic temperature is a well-defined quantity. However, in thermodynamics, it took hundreds of years before qualitative thermometers based on the height of a liquid column were properly calibrated against an absolute temperature scale T. The ideal gas law determines absolute temperature relatively simple. When applicable to real gases at low-density N/V and high absolute temperature T, the ideal gas law, PV=NkBT, makes thermometry relatively easy (kB is Boltzmann's constant). Therefore, a measurement of P and N/V gives T.

It will be difficult to find an inexpensive analog for the ideal gas thermometer. One difference is that the velocity of the particles of an ideal gas has no upper limit and a lower limit limits its energy, whereas an economic agent has defined values for its wealth and utility. However, magnetic salts behave, over a limited range of temperatures, as if they only have a finite number of energy levels(Saslow, 1999).

Common thermometry involves reading the height of a column of fluid or the position of a pointer connected to a coil of wire. Both quantities depend on the coefficient of thermal expansion. Other forms of thermometry depend on other temperature-dependent variables. These can be calibrated against another carbon resistance thermometer; its electrical resistance is a measure of temperature. In the millikelvin range, calibrations are made with thermometers that use certain magnetic salts. At very low and very high temperatures, it isn't easy to perform any thermometry at all. Therefore, we consider the general problem of how we can calibrate a measurement of a quantity (for example, electrical resistance), which we will call τ, against an absolute thermodynamic temperature T.

Recall that T = 0 for all scales is of thermodynamic temperature. However, there is no absolute scale for temperature. Recall that T = 0 for all thermodynamic temperature scales. However, there is no absolute scale for temperature. By setting (roughly) 273 degrees to be the water point or using the triple point temperature of pure material, you determine the Kelvin temperature scale from TK. On another planet, the thermodynamic temperature scale would be different, but only by a scale factor. If the inhabitants of another planet called the water point 546 degrees freezing, we would know that their temperatures are twice that on the Kelvin scale(Sciubba, 2005).

The determination of the temperature scale. In general, the thermometric property τ depends on T, P, and N. For this thermometer, and we must determine the thermodynamic temperature T as a function of τ, V and N. W. Saslow's discussion of it; is an extension of Landau and Lifshitz, where he considers a thermometer that measures a quantity τ depending only on the temperature T: $\tau = \tau$ (T) (Sciubba, 2005).

We will summarize (42 to 48) in this way. Consider a measurable quantity, the heat gain dQ = TdS, and how it varies with a change in pressure at a fixed temperature T and N. It will be useful to use Maxwell's relationship [based on Gibbs free energy G (T, P, N) = E - TS + PV] to rearrange everything and integrate to get the temperature scale T as T = T (τ, P, N) in terms of measurable quantities. Using fixed V and varying N, we can obtain T (τ, V, μ) through some relationships that indicate that there is more than one way to obtain a temperature scale. The above discussion suggests at least eight ways to perform thermometry. In fixed T, we measure the dependence of dQ to both dV or dP, either N fixed or μfix, we measure the dependence of dQ to both dN or dμ either fixed V or fixed P. Two pairs of these eight measurements lead to τ as a function of the same fixed variables.

THERMOMETRY IN ECONOMICS

Having described some developments in the complexity of thermometry in physics, we now indicate how to perform thermometry in economics. For this, it is essential that we measure dQ = TdS, the Veblenian surplus, and that we have a quantity that can serve as a thermometer. By equation (23), we have

$$TdS = dU - \lambda dM - pdN \tag{49}$$

Therefore, if we can measure dU, λ, dM, p, and dN, we can measure dQ = TdS. Let's suppose that we have some economic indicator t that depends on T, λ, and N. To use the relation τ (T, λ, N), which will require a new thermal Slutsky condition, we obtained using the economic free energy analog by Gibbs. As

$$V(T, p, M) = U - TS - \lambda M \tag{50}$$

Where = pN is obtained using the equation (21) and is the monetary value of the goods. Its differential satisfies

$$dV = -SdT - \lambda dM + pdN \tag{51}$$

Which follows in the substitution of the equation (23) in the differential of the equation (50). Then

$$\left(\frac{\delta^2 V}{\delta\lambda\delta T}\right)_N = \left(\frac{\delta^2 V}{\delta T\delta\lambda}\right)_N \tag{52}$$

What leads to

$$\left(\frac{\delta S}{\delta\lambda}\right)_{T,N} = \left(\frac{\delta M}{\delta T}\right)_{\lambda,N} \tag{53}$$

Using equation (53) and dQ = TdS, we have

$$\left(\frac{\delta Q}{\delta\lambda}\right)_{T,N} = T\left(\frac{\delta S}{\delta\lambda}\right)_{T,N} = -T\left(\frac{\delta M}{\delta T}\right)_{\lambda,N} = -T\left(\frac{\delta M}{\delta\tau}\right)_{\lambda,N}\left(\frac{\delta\tau}{\delta T}\right)_{\lambda,N} \tag{54}$$

which can be rearranged to write

$$1/T\left(\frac{\delta T}{\delta\tau}\right)_{\lambda,N} = -\frac{\left(\frac{\delta M}{\delta\tau}\right)_{\lambda,N}}{\left(\frac{\delta Q}{\delta\lambda}\right)_{T,N}} = f(\tau,\lambda,N) \tag{55}$$

The left side of the equation (55) is at fixed λ and N, so the right-hand side must be written in terms of the variables τ, λ, and N, which explains the notation *f*(τ, λ, N). Then we can integrate from T_0 and the corresponding τ_0 to obtain

$$ln\left(\frac{T}{T_o}\right) = \int_{\tau_0}^{\tau} f(\tau,\lambda,N)\,d\tau; f(\tau,\lambda,N) = -\frac{\left(\left(\frac{\delta M}{\delta \tau}\right)_{\lambda,N}\right)}{\left(\frac{\delta Q}{\delta \lambda}\right)_{T,N}} \quad (56)$$

where dQ = TdS as usual. From equation (56), we can obtain, in principle, the temperature scale T (τ, λ, N) in terms of the measurable quantities τ, λ, N. We have to be able to measure the quantities that appear in the expression for f (τ, λ, N). Especially difficult to measure is the change in the Veblenian surplus, dQ = TdS. By the equation (24), this quantity is also the change in utility in fixed goods and money, and thus its measurability is the subject of debate among economists. The present work is based on the assumption of the capacity to measure utility. Other forms of inexpensive thermometry are possible, which are perhaps more practical to apply by analogy to thermometry in thermodynamics(Tsirlin et al., 2005).

THERMODYNAMICAL INTERPRETATION OF MACROECONOMICS

Both thermodynamics and macroeconomics operate with variables that define the macroscopic state of the system under study. The passage of the variables defined by W. Saslow in the Theory of the Consumer to Macroeconomics is followed phenomenologically. In the case of the value of money λ and the amount of money M, each country has an amount of International Reserves (which will be M), a deposit of foreign currency controlled by central banks and other authorities. Monetary assets are made up of various reserve currencies, especially dollars and euros. And the value of money will be the price of said currencies in the country of origin, known as the exchange rate (now λ).

To consider a macroeconomic analog of entropy, the definitions that encompass this concept were taken into account. In the first place, entropy is understood as a degree of disorder of a system, a concept that is almost currently being discarded and is in question. The second concept is associated with the spontaneous dispersion of energy or as a measure of irreversible processes. But the clearest concept of entropy is a magnitude that always increases as time progresses. Hence entropy is considered as an arrow of time (Tisza & Manning, 1957).

For this reason, a macroeconomic variable was sought, which increases over time, considering that there are no externalities that could change or alter its behavior. Said macroeconomic variable turns out to be Real GDP since it expresses the monetary value of producing goods and services for final demand in a country during a given period. For the analogy proposed by Saslow, the Surplus Ψ (understood as what remains), for a country, becomes the Trade Balance (exports minus imports). The Trade Balance is a part of GDP shown in the following expression (Tsirlin et al., 2005).

$$Y = C + I + GP + BC \quad (57)$$

Where Υ is Real GDP, C is Consumption, and I is Investment, GP is Public Expenditure, and BC is Trade Balance. With this, according to the Second Law of Thermodynamics and the Second Law of Economics (Tisza & Manning, 1957), we can provisionally define a Level of Economic Development (T) given by the following equation:

$$T = \frac{BC}{Y} \quad (58)$$

An indicator will give the analogy for the price called the Consumer Price Index (CPI), which is a dimensionless quantity defined as:

$$CPI = \frac{\text{Price of the basket of goods and services in the current year}}{\text{Basket price in the base year}\left(\text{forexample,1950}\right)} \quad (59)$$

In analogy, the number of goods would have to be associated with a magnitude that allows us to approach some macroeconomic variable, but the turn in this search is based on the fact that this new variable carries its magnitude of monetary units (for example, S). This new variable is defined as the Quantity of Capital Goods or simply Capital Stock.

The Utility function defined by W. Saslow is not the Economic Utility Function since this would lead to measuring something that is not measurable (degree of satisfaction in a person when buying a good). Worse, according to the definition that he raises in equations (1) and (4), this would rather represent the Accounting Income function. Considering that the utility for an accountant is different from the utility for an economist since the latter considers variables, such as the opportunity cost, among other variables.

Table 1. Summary of the state variables suggested between the Microeconomic and Macroeconomic Thermodynamic systems

Thermodynamics		Microeconomics		Macroeconomics	
State variables	**Symbol**	**State variables**	**Symbol**	**State variables**	**Symbol**
Chemical Potential	μ	Price of the good	p	CPI	p or IPC
Particles Number	N	Number of Goods	N	Quantity of Capital Goods	N
Gibbs potential	G	Cost of Goods	C	Cost of Capital Goods	C
Entropy	S	Entropy	S	Real GDP	¡
Temperature	T	Level of development	T	Level of development	T
Hot	Q	Surplus	Q	Balance of trade	BC
Pressure	P	Value of money	λ	Exchange rate	λ
Volume	V	Money Amount	M	International Reserves	M
Helmholtz potential	F	Wealth	W	Wealth	W
Internal energy	E	Utility	U	Utility	U

DISCUSSION

One of the many fundamental assumptions of Saslow's theory is the existence of a Utility function, which is supposed to be measurable (measurable), but what Saslow did not take into account was that there are two types of utility functions, one is the function economic utility, which is what it assumes in principle, but which in reality is not measurable, because psychological aspects are associated with it (Tisza & Manning, 1957). As might be expected, it isn't easy to measure something that is associated with psychological aspects. The other utility function is the one commonly used by accountants. This function is quantifiable and differs greatly from the economic utility function. One consideration, based on a simple analysis of table I, we see that by analogy, wealth is the negative of the Helmholtz potential, according to Saslow. Helmholtz's Least Potential principle establishes that: "The equilibrium value of any internal parameter without bonds of a system that is in diathermic contact with a heat source minimizes the Helmholtz potential at constant temperature (equal to that of the heat source)." That is, in any process, the minimum of Helmholtz Potential will always be produced. The principle is perfectly fulfilled in nature. If we put the negative before that minimum, what we will do is maximize said function. What Saslow intended in his analogy was for wealth to be maximized. If we remove that negative, what we will have is that wealth will also have to be minimal. We also consider that wealth comprises two parts: an intangible (generated, for example, bank accounts) and a tangible (generated by the number of assets that one owns). Until a few years ago, many economists believed that wealth, for example, in a country, is an amount that is always increasing. Green economists, also called Ecological Economists, have taken a radical turn in this type of thinking by stating that the wealth of a country is constant since natural and human wealth is also considered (Saslow, 1999). In this condition, all the wealth constituted by tangible goods is obtained through a transformation. Therefore, we consider that the most convenient thing should be to minimize the extraction of natural resources, as the analogy between Helmholtz's potential and wealth would establish. Despite assumptions that in principle baffle any economist, this formulation developed by W. Saslow finds that the macroscopic variables that define a thermodynamic state have their respective analogs in microeconomics and Macroeconomics. Saslow gets very interesting outputs in his theory, taken by analogy economics from the point of view of thermodynamics(Greiner & Müller, 2010). He clearly shows the Slutsky relations obtained in the same way that the Maxwell relations are obtained, and an economic Gibbs - Duhem relationship, which was unknown until then, was also estimated. Furthermore, he deduces that the relationship between exchange value and value in use is a constant, an idea proposed by marginalist theory. It should be noted that the values obtained in our analysis of the data collected by various sources have a large margin of error and relative error. We consider that this is largely due to the elementary assumption of Saslow's theory that economic systems are in a state of equilibrium. This theory clearly shows the failure to apply theories based on equilibrium to such a dynamic system as the economy. Dynamical systems, their error exponentially as time elapses (Bryant, 1982). In addition, the fall in the Trade Balance shows us a clear example of how unstable an economic system is. In recent years, the Trade Balance values turned out to be negative, since lately more is imported than exported. If a thermodynamic analogy were made, it is as if the heat had been added to a thermodynamic system. Although W. Saslow indeed developed in a general way, an expression based on the analysis of thermometry, to develop a thermometric scale in economics, a problem arises when considering that a value of some thermometric property is required for the which is considered an acceptable temperature. For instance, Germany already develops this indicator, and it is considered that the stable temperature for an economic system is 36.5 ° C and 41 ° C when the economy is already becoming unsustainable[9]. In addition, to have a thermometric scale, the system must reach an equilibrium which, as we pointed out previously, is not so easy to develop since economic systems are dynamic systems(Greiner & Müller, 2010).

CONCLUSION

From Tables 1, we can see that they exist if it is possible to obtain by analogy the relationships between the thermodynamic and economic variables, as well as the Slutsky relationships (equations (41), (42), (43)) by analogy with the Maxwell relations, a Gibbs - Duhem relation (equation (39)), Saslow's theory, allowed to mathematize the "marginalist" proposal (equation (28)), as well as finding expressions that can be used to define thermometric scales in economics, as an indicator of the state of the economy. Despite all these beautiful achievements made by W. Saslow's theory, it is not applicable due to its elementary assumption that economic systems are in equilibrium. A condition that is not true in reality. This theory could not apply to thermodynamic processes since, in thermodynamic processes, quasi-static processes are considered. In addition, we must remember that in society, external factors occur, call it crises, catastrophes, wars; that can shift the economy of a country completely; under these conditions, Saslow's theory is not applicable since the functions (Cost of Capital Goods, Amount of Intangible Wealth, Trade Balance) as a function of its independent variables they are dispersing, which increases the error over time. It is proposed to study this theory, but from the point of view of Non-Equilibrium Thermodynamics, hoping to have results closer to reality. A theory could even be developed from a Statistical Mechanical approach to make corrections to Saslow's theory under microeconomic considerations. It should be borne in mind that today much of macroeconomic analysis focuses on microeconomic analysis. Finally, it is concluded that W. Saslow's Theory of the Analogy between Thermodynamics and Economics is not applicable due to obtaining considerable errors and considering economic systems as equilibrium systems. Briefly, to simplify the calculations, it was considered that the economic systems are in equilibrium to make an analogy with the Thermodynamics of Equilibrium. In addition, fundamental considerations were made, such as considering that a consumer's wealth or a small country is constant for the existence of a measurable (quantifiable) utility function. However, according to the discussion, it could be seen that W. Saslow's theory must be developed in the field of Non-Equilibrium thermodynamics, which is why the economy is not a system that is in equilibrium but rather, on the contrary, the economy is a dynamic system.

REFERENCES

Baierlein, R. (2001). The elusive chemical potential. *American Journal of Physics*, *69*(4), 423–434. doi:10.1119/1.1336839

Bedi, T., Coudouel, A., & Simler, K. (Eds.). (2007). *More than a pretty picture: using poverty maps to design better policies and interventions*. World Bank Publications.

Bejan, A., Errera, M. R., & Gunes, U. (2020). Energy theory of periodic economic growth. *International Journal of Energy Research*, *44*(7), 5231–5242. doi:10.1002/er.5267

Bryant, J. (1982). A thermodynamic approach to economics. *Energy Economics*, *4*(1), 36–50. doi:10.1016/0140-9883(82)90042-1

Bryant, J. (2007). A thermodynamic theory of economics. *International Journal of Exergy*, *4*(3), 302–337. doi:10.1504/IJEX.2007.013396

Chakrabarti, B. K., Chakraborti, A., & Chatterjee, A. (Eds.). (2006). *Econophysics and sociophysics: trends and perspectives*. John Wiley & Sons. doi:10.1002/9783527610006

ChamizoA.Novales CincaA. (2015). *Econometric models of credit spreads*. Available at SSRN 2707644.

Cooter, R., & Ulen, T. (1988). *Law and economics*. Academic Press.

De León AriasA. (2010). Teaching the Solow Through the Building of Economic Growth Scenarios: An Application to China. Available at SSRN 1749699. doi:10.2139/ssrn.1749699

Erlichson, H. (1984). Internal energy in the first law of thermodynamics. *American Journal of Physics*, *52*(7), 623–625. doi:10.1119/1.13601

Felder, G. N., & Felder, K. M. (2015). *Mathematical methods in engineering and physics*. John Wiley & Sons.

Georgescu-Roegen, N. (1971). *The Entropy Law and the Economic Process*. Harvard University Press.

Greiner, W., & Müller, B. (2012). *Quantum mechanics: symmetries*. Springer Science & Business Media.

Jakimowicz, A. (2020). The role of entropy in the development of economics. *Entropy (Basel, Switzerland)*, *22*(4), 452. doi:10.3390/e22040452 PMID:33286226

Mankiw, N. G. (2014). *Principles of economics*. Cengage Learning.

McCauley, J. L. (2009). *Dynamics of markets: The new financial economics*. Cambridge University Press. doi:10.1017/CBO9780511805448

Pindyck, R. S., Rubinfeld, D. L., & Rabasco, E. (2013). *Microeconomia*. Pearson Italia.

Rashkovskiy, S. A. (2021). Thermodynamics of markets. *Physica A*, *567*, 125699. doi:10.1016/j.physa.2020.125699

Richmond, P., Mimkes, J., & Hutzler, S. (2013). *Econophysics and physical economics*. Oxford University Press. doi:10.1093/acprof:oso/9780199674701.001.0001

Riley, K. F., Hobson, M. P., & Bence, S. J. (1999). *Mathematical methods for physics and engineering*. Academic Press.

Rogers, D. M., & Rempe, S. B. (2012). Irreversible thermodynamics. *Journal of Physics: Conference Series*, *402*(1), 012014. doi:10.1088/1742-6596/402/1/012014

Saslow, W. M. (1999). An economic analogy to thermodynamics. *American Journal of Physics*, *67*(12), 1239–1247. doi:10.1119/1.19110

Sciubba, E. (2005). Exergo-economics: Thermodynamic foundation for a more rational resource use. *International Journal of Energy Research*, *29*(7), 613–636. doi:10.1002/er.1096

Tisza, L., & Manning, I. (1957). Fluctuations and irreversible thermodynamics. *Physical Review, 105*(6), 1695–1705. doi:10.1103/PhysRev.105.1695

Tsirlin, A., & Gagarina, L. (2020). Finite-time thermodynamics in economics. *Entropy (Basel, Switzerland), 22*(8), 891. doi:10.3390/e22080891 PMID:33286661

Tsirlin, A. M., Kazakov, V., Alimova, N. A., & Ahremenkov, A. A. (2005). Thermodynamic model of capital extraction in economic systems. *Interdisciplinary Description of Complex Systems: INDECS, 3*(1), 1–16.

ADDITIONAL READING

Chouinard, Y., Ellison, J., & Ridgeway, R. (2011). The sustainable economy. harvard. *Business Review (Federal Reserve Bank of Philadelphia), 89*(10), 52–62.

Daly, H. E. (2007). *Ecological economics and sustainable development.* Edward Elgar Publishing.

Munasinghe, M. (1993). *Environmental economics and sustainable development* (Vol. 3). World Bank Publications. doi:10.1596/0-8213-2352-0

Pearce, D. W., & Warford, J. J. (1993). *World without end: economics, environment, and sustainable development.* Oxford University Press.

Robertson, J. (1999). *The new economics of sustainable development. A Report for the European Commission.* Kogan Page.

Seyfang, G. (2009). *The new economics of sustainable consumption. Minería transnacional, narrativas del desarrollo y resistencias sociales.* Biblos.

Tisza, L., & Manning, I. (1957). Fluctuations and irreversible thermodynamics. *Physical Review, 105*(6), 1695–1705. doi:10.1103/PhysRev.105.1695

Tsirlin, A. M., Kazakov, V., Alimova, N. A., & Ahremenkov, A. A. (2005). Thermodynamic model of capital extraction in economic systems. *Interdisciplinary Description of Complex Systems: INDECS, 3*(1), 1–16.

KEY TERMS AND DEFINITIONS

Circularity: A circular economy (also referred to as "circularity") is an economic system that tackles global challenges like climate change, biodiversity loss, waste, and pollution. Most linear economy businesses take a natural resource and turn it into a product that is ultimately destined to become waste because it has been designed and made. This process is often summarised by "take, make, waste." By contrast, a circular economy uses reuse, sharing, repair, refurbishment, remanufacturing, and recycling to create a closed-loop system, minimize resource inputs, and create waste, pollution, and carbon emissions. The circular economy aims to keep products, materials, equipment, and infrastructure in use for longer, thus improving the productivity of these resources. Waste materials and energy should become input for other processes through waste valorization: either as a component or recovered resource for another industrial process or as regenerative resources for nature (e.g., compost). This regenerative approach contrasts with the traditional linear economy, which has a "take, make, dispose" production model.

Eco Commerce: Eco commerce is a business, investment, and technology-development model that employs market-based solutions to balancing the world's energy needs and environmental integrity. Through green trading and green finance, eco-commerce promotes the further development of "clean technologies" such as wind power, solar power, biomass, and hydropower.

Eco-Tariffs: An eco-tariff, also known as an environmental tariff, is a trade barrier erected to reduce pollution and improve the environment. These trade barriers may take the form of import or export taxes on products with a large carbon footprint or imported from countries with lax environmental regulations.

Emissions Trading: Emissions trading (also known as cap and trade, emissions trading scheme, or ETS) is a market-based approach to controlling pollution by providing economic incentives for reducing the emissions of pollutants.

Environmental Enterprise: An environmental enterprise is an environmentally friendly/compatible business. Specifically, an environmental enterprise is a business that produces value in the same manner which an ecosystem does, neither producing waste nor consuming unsustainable resources. In addition, an environmental enterprise rather finds alternative ways to produce one's products instead of taking advantage of animals for the sake of human profits. To be closer to being an environmentally friendly company, some environmental enterprises invest their money to develop or improve their technologies which are also environmentally friendly. In addition, environmental enterprises usually try to reduce global warming, so some companies use environmentally friendly materials to build their stores. They also set in environmentally friendly place regulations. All these efforts of the environmental enterprises can bring positive effects both for nature and people. The concept is rooted in the well-enumerated theories of natural capital, the eco-economy, and cradle-to-cradle design. Examples of environmental enterprises would be Seventh Generation, Inc., and Whole Foods.

Green Economy: A green economy is an economy that aims at reducing environmental risks and ecological scarcities and that aims for sustainable development without degrading the environment. It is closely related to ecological economics but has a more politically applied focus.

Green Politics: Green politics, or ecopolitics, is a political ideology that aims to foster an ecologically sustainable society often, but not always, rooted in environmentalism, nonviolence, social justice, and grassroots democracy. It began taking shape in the western world in the 1970s; since then, Green parties have developed and established themselves in many countries around the globe and have achieved some electoral success.

Low-Carbon Economy: A low-carbon economy (LCE) or decarbonized economy is an economy based on low-carbon power sources with minimal greenhouse gas (GHG) emissions into the atmosphere, specifically carbon dioxide. GHG emissions due to anthropogenic (human) activity are the dominant cause of observed climate change since the mid-20th century. Continued emission of greenhouse gases may cause long-lasting changes worldwide, increasing the likelihood of severe, pervasive, and irreversible effects for people and ecosystems.

Natural Resource Economics: Natural resource economics deals with the supply, demand, and allocation of the Earth's natural resources. One main objective of natural resource economics is to understand better the role of natural resources in the economy to develop more sustainable methods of managing those resources to ensure their future generations. Resource economists study interactions between economic and natural systems intending to develop a sustainable and efficient economy.

Sustainable Development: Sustainable development is an organizing principle for meeting human development goals while simultaneously sustaining the ability of natural systems to provide the natural resources and ecosystem services on which the economy and society depend. The desired result is a state of society where living conditions and resources are used to continue to meet human needs without undermining the integrity and stability of the natural system. Sustainable development can be defined as development that meets the needs of the present without compromising the ability of future generations to meet their own needs. Sustainability goals, such as the current UN-level Sustainable Development Goals, address the global challenges, including poverty, inequality, climate change, environmental degradation, peace, and justice.

Chapter 4
Quality Function Deployment in the Agribusiness Supply Chain in the Food Sector:
Has Its Potential Been Thoroughly Exploited?

Juliani Elis Duarte Gotardi
São Paulo State University, Brazil

Eduardo Guilherme Satolo
https://orcid.org/0000-0002-8176-2423
São Paulo State University, Brazil

Priscilla Ayleen Bustos Mac-Lean
São Paulo State University, Brazil

ABSTRACT

The chapter aimed to understand the phases of quality function deployment (QFD) in agribusiness supply chain in the food sector. A literature review (LR) was analyzed a time frame of the latest 10-year publications of 30 scientific articles. Thus, results pointed out articles related to QFD applications in the food supply chain in which a scenario composed mostly of occasional publications is found. Such applications vary in product improvement or supply chain improvement, quality analysis of existing products, development of new products that satisfy consumers' desires, and quality planning; however, the application of other theories and tools related to QFD is recurrent, and there are few articles about the implementation of all QFD phases. It also found a greater predominance of performing the steps to build the HoQ (house of quality) corresponding to the "what?" of the voice of the customer matrix, the "how?" of the voice of the expert matrix, the "what and how?" of relationship and technical benchmarking matrices to determine target values.

DOI: 10.4018/978-1-6684-2339-4.ch004

INTRODUCTION

Agribusiness constitutes a complex supply chain that extends from rural producers to consumers, so that its latent agents are assigned to the manufacture of inputs, production in agricultural businesses, technical assistance, transformation and processing by agro-industries, commercialization, distribution, transport, export and retail (Gasques, 2004).

Therefore, the quality of consumer goods to be produced by this sector is also predicated on ensuring its market competitiveness. However, the concept of quality is a variable, which can lead to several conflicts in the supply chain, given that producers, suppliers and consumers evaluate quality as they conceive it (Brunsø et al., 2002).

The concept of quality incorporates two dimensions: a visible quality that refers to the physical features of the product, usually assessed by engineers or experts; and a subjective quality that concerns the one perceived by consumers. Therefore, an intersection between these two dimensions should be the target in the search for quality which will be a competitive parameter for agribusiness only if producers translate consumer desires into physical features for their product (Grunert, 2005).

In this line of thought, to create and manufacture products that meet total quality requirements and are outside the standards of imitation or copying, a system of tools and benchmarking techniques has been created by Yoji Akao and Shigeru Mizuno, which was originally called as the quality function deployment (QFD) (Akao, 1997).

This method is a way to translate the needs of customers based on their requirements or specifications, which can be used to correct, improve, or develop a product or provide a service so that it is as close as possible to meet actual customer needs (Akao & Mazur 2003). It can be implemented in any links of a supply chain.

According to Chan & Wu (2002), the functional fields of QFD applications are: product development (new products or model change), quality management (production processes or quality control improvement), analysis of customer needs (development of products with enhanced quality to meet or overcome customer expectations), product design (assisting the company in designing product quality by analyzing customer needs), process planning (so that issues can be found and resolved early on product development) and decision making (management of processes directed towards the customer based on a market analysis to aid decision-making processes).

Therefore, this chapter aims to answer the following question: how are publications about the Quality Function Deployment (QFD) applied and configured in agribusiness regarding food supply chain?

For such a purpose, the literature review (LR) is going to be presented, at first highlighting the concept of quality function deployment (QFD) and its history, in a second step punctuating the QFD application steps presented by several authors. Finally, to group concepts and application, we will present articles published of the last ten years in the agribusiness area where QFD was applied, regarding food supply chain, using all or some of the steps described as a development of the chapter.

In view of this, it is also aimed to contribute to show the importance of application of QFD in food supply chains and expand the boundaries of knowledge due to its originality, thus encouraging other researchers directly linked to production of food of animal and plant origin, to contribute to the enrichment of this area of knowledge, since QFD applications are common in several sectors, but incipient in agribusiness.

QUALITY FUNCTION DEPLOYMENT (QFD)

The *quality function deployment* (QFD) was created in Japan in the late 1960s after the Second World War with the aim of making the products developed countrywide original and satisfactory. In this period, *statistical quality control* (SQC) was the main activity undertaken in the country, and thus the search for quality became the main objective of automobile, electronics, and software companies, which has been quickly adopted by manufacturing industries. These aspects encourage its application from the perspective of manufacturing products and providing services under the control of total quality (Akao, 1997; Chan & Wu, 2002).

In view of its positive aspects, QFD was adopted by North American companies to be used in the set of planning tools for manufacturing their products and providing services in 1980. As a result, QFD was rapidly disseminated among industries in other countries, which is a fact that currently leads to insightful literature publications on the subject (Chan & Wu, 2002).

Quality deployment proposed by the QFD is essentially based on analysis, planning and structured development of products based on finding customer needs, which are to be interpreted and translated by the research and development team, as well as a team of engineers or experts who develop the method that are going to incorporate the customer's desirable quality features into the product through a detailed technical assessment after product planning (Benner et al., 2003; Hauser & Clausing, 1988).

In this sense, the QFD implementation enabled to plan and develop products based on a previously structured analysis, a condition that allows a company's product research and development to identify the demands and needs of its customers based on a systematic assessment of the proposed product, together with an analysis of consumer needs or expectations (Benner et al., 2003a).

The QFD has been drawing the attention of automobile manufacturing industries for quite some time, but nowadays it is flexible enough to be applied in several industrial sectors, such as the food sector. Theory has been applied in this sector since 1987, and currently several scientific articles report its application advantages based on an identification of the voice of the customer (Cardoso et al., 2015; Hofmeister, 1991; Benner et al., 2003; Costa et al., 2001).

How to Apply the QFD

QFD application starts from developing the product planning matrices which are the steps for building the *house of quality* - HoQ, since its structural appearance is like that of a house. In it, each component or process of product operation is associated and correlated to achieve desired quality (Benner et al., 2003).

Its objective is to capture the customer's requirements (quality desired by the consumer), turn them into product requirements (technical characteristics for the product found by the engineering team), relate them to obtain relative and absolute values (the higher they are, the greater the chance of satisfying customer's requirements) and translate them into respective target values, i.e. the one that must be implemented in the product (Takai & Kalapurackal 2012).

HoQ is the first and foremost translation that occurs in the step of product planning, given that specifications and improvement goals will be implemented through it. Thus, an understanding of QFD is achieved through the development of HoQ, so that each construction step guides the following process (Govers, 1996; Benner et al., 2003).

As a result, the QFD can be defined as a systematic implementation of operations functions required for the product, which allows improvement and attaining quality through understanding specific details that can be implemented by taking each step of every matrix (Cauchick Miguel, 2005).

The four steps of the HoQ include product planning, product design, manufacturing planning or process planning, and production planning, which can be understood as: step 1. product planning: translating customer needs into attributes for product design, corresponding to technical measures; step 2. product design: translating important technical measures into product characteristics; step 3. process planning: translating important features of the product into process operations; and step 4. production planning: translating key process operations into production requirements (Chan & Wu, 2004).

Given the above, the deployment of HoQ steps, shown in Figure 1, determines how outputs of one step determine the inputs of the following.

The HoQ preparation then depends on devising the planning phases, which is incorporated into any of the versions of the QFD. Given this, HoQ has been the main focus of literature on the method, as it contains the main quality planning data that the company needs to foster its relationship with consumers and maintain its stability and competitiveness in the market (Benner et al., 2003a; Costa et al., 2001).

Figure 1. HoQ steps deployment
Source: The authors based on Govers (1996); Abreu (1997).

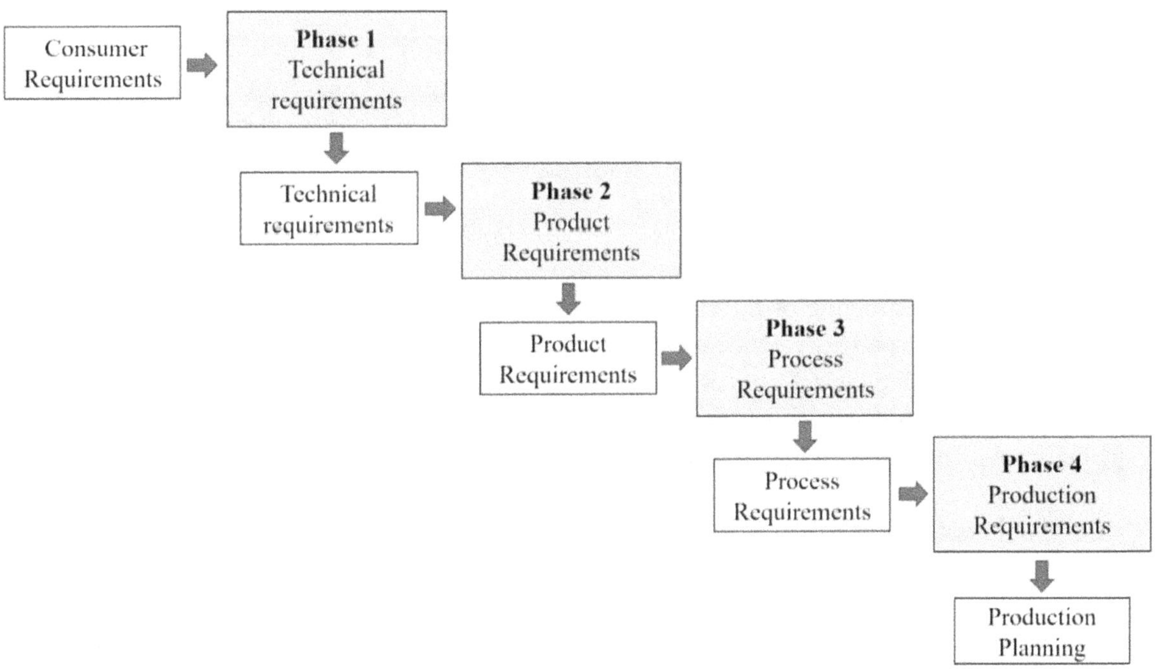

The basic structure of HoQ, shown in Figure 2, shows a step-by-step approach of its construction and how the guiding questions of the planning matrix lead to the deployment of structural steps.

Figure 2. HoQ structure
Source: The authors based on Benner et al. (2003); Costa et al. (2001).

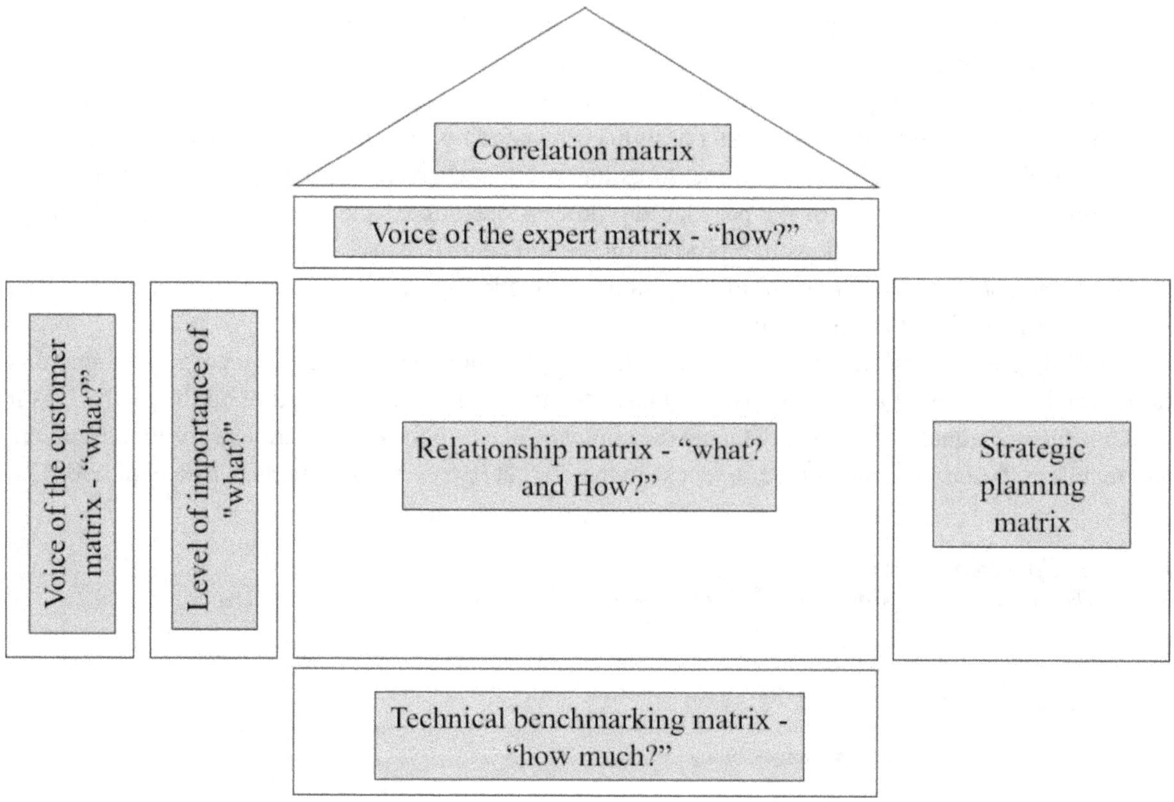

- Voice of the customer matrix - "what?"

Devising the basic structure of the HoQ starts by capturing the voice of the customer by asking "what?", which focuses on verifying the customer's requirements, i.e., what customers desire by listening to them and recording their desires and needs.

At this step customer demands are translated (Viaene & Januszewska 1999) that are usually obtained by qualitative market research (Hauser & Clausing 1988).

- Level of importance of "what?"

To identify whether the customer's preferences or needs are equally important, the next step is to measure the importance of "what?" from the considerations obtained by the voice of the customer.

The degree of importance is the quantitative aspect regarding the items described by consumers (Viaene & Januszewska 1999). Some companies use statistical techniques to determine the preference of customers in relation to existing products or those that will be launched in the market, or techniques of revealed preferences that capture the preference of consumers based on their actions or words (Hauser & Clausing 1988).

- Technical benchmarking matrix

This phase, on the right side of HoQ, is also known as the competition analysis matrix., and is causally linked to the expansion of the company's strategic vision, as it identifies opportunities for improviment for the studied product in relation to its competitors. As a result, technical benchmarking should be based on assessing scientific, qualitative, or quantitative surveys about customers, so that data can be classified or represented graphically in the matrix (De Pelsmaeker et al., 2014; Costa et al., 2001).

- Voice of the expert matrix - "how?"

At this step, the research and development team translate the customer's requirements into product requirements through an analysis.

According to Dolgun & Köksal (2018), this step is necessary, since it is common for customers to use vague, ambiguous, and imprecise terms to voice their concerns about their desired requirements.

This step filters the customer's requirements until technical features of the product are obtained, which will be used as parameters for the product planning process (Abreu, 1997).

- Relationship matrix - "what? and How?"

At the core of the HoQ is the relationship matrix. In it, the information between "what?" and "how?" are interrelate so as to identify how technical features affect quality characteristics (Viaene & Januszewska, 1999).

The idea is to show where decisions should be made on a positive or negative demonstration among related attributes (Hauser and Clausing 1988; Fernandes and Rebelato 2006).

- Technical benchmarking matrix - "how much?"

At the bottom of the HoQ is the value of technical priorities of the planning matrix, which correspond to the value of production requirements.

The "how much?" must be as measurable as possible, since the more measurable they are, the greater the number of opportunities for the research and development to analyze and improve the product will be (Benner et al., 2003).

- Correlation matrix

This matrix is located on the HoQ roof and is understood by means of a transversal analysis between correlations of technical features attributed by the voice of the expert.

The analysis is performed by using distinct symbols that encode positive, negative relationships and the intensity between elements through their correlation. This step allows observing and controlling interactions between the technical quality features of the product, along with the production process by projecting combined effects between technical quality features and highlighting possible opportunities for innovation, identifying incompatibilities between product characteristics and possible unwanted changes (Naspetti et al., 2015; Viaene & Januszewska, 1999).

APPLICATION OF QFD IN AGRIBUSINESS

The methodologies selected to construct this chapter are mainly based on the article published by Conforto et al. (2011) who had listed the stages of an LR through a roadmap. It has three methodological stages (input, processing, and output) together with the development of others that aim to guide the present methodology.

Thus, the methodological roadmap drawn by this research is according to the following stages:

Input

This stage corresponds to the selection of data collection procedures, which are going to be developed from support procedures to identify the problem and reach the research objective.

To achieve the objective of this research, which is to identify scientific publications about studies on *quality function deployment* (QFD) applied to agribusiness and food industries, four databases were selected: *Web of Science, Scopus (Elsevier), Wiley Online Library and Taylor and Francis.*

Subsequently, keywords were assessed to develop the strings that were cited isloatedly and in English at first, thus comprising the following: quality function deployment, agribusiness and food.

Next, strings were assessed by using the keywords presented above, together with Boolean operators "OR", "AND", quotes and parentheses, to cross-check the keywords from which the following combination was obtained *('Quality Function Deployment' OR 'QFD') AND (Agribusiness OR Foods).*

Processing

This stage consists in the manner through which information is going be generated and processed so that it is used in the output stage, corresponding to a summary of results.

Therefore, in the processing stage of this research, it was opted for restricted access to selected databases by using free search entries of isolated and strings, as referenced in the previous stage. The string choice was consolidated as: *('Quality Function Deployment' OR 'QFD') AND (Agribusiness OR Foods)*, due to presenting better results.

Thus, while consolidating the string choice, searches without filters (free searches) were carried out on the selected databases for identifying preliminary results.

Then, inclusion filters were selected based on the time frame proposed by the research objective, along with two other specifications. Therefore, the filters used in this search are as follows: No filter - no search filter; Filter 1- published in the latest ten years (2011 - 2020); Filter 2 - scientific articles only. Thus, the results in Table 1 were achieved.

Table 1. Results of the application of search filters on selected databases

Databases	Without Filter	Filter 1	Filter 2
Scopus	212	85	57
Taylor and Francis	323	186	186
Web of Science	65	54	39
Wiley Online Library	327	163	82
Total			364

Then, starting from a total of 364 articles, which results from using filter 2 together with the sum of results between databases, the sorting stage of articles by title and abstract begins to select accepted articles for data analysis. In this stage, it was adopted an inclusion criterion, which were: meeting the research objective, applying to agribusiness and food industries, and exclusion criteria, i.e., duplicate articles and those having technical or irrelevant subjects to the present research objective.

Therefore, due to the application of inclusion and exclusion criteria, the pre-selected articles totalled 31 articles. However, after reading and thoroughly analyzing articles, 29 were selected due to having verification priority, which consists of: 15 articles from the *Scopus* database; 11 articles from the *Web of Science* database, 2 articles from the *Wiley Online Library* database and 1 from *Taylor and Francis* database.

Output – Search Results

The results of selected articles are highlighted in Table 2. It is worth mentioning that this sort of investigation was restricted only to articles published in journals.

Table 2. Articles selected through a LR

Nº	Author	Year	Product type applied
1	Rosado Júnior et al.	2011	Bulls
2	Bevilacqua et al.	2012	Olive oil
3	Naspetti et al.	2012	Organic sheep cheese
4	Park et al.	2012	Korean beef barbecue
5	Kuo et al.	2012	Black bean
6	Kristianto et al.	2012	Flour
7	Cardoso et al.	2014	Organic fruit jelly
8	Alba-Elías et al.	2014	Food additive for pickled vegetables
9	Pelsmaeker et al.	2014	Stuffed chocolates
10	Dill et al.	2014	Pork loin supply chain
11	Naspetti et al.	2015	Poultry meat in the organic sector
12	Hsia, Liao & Huang	2016	Rice
13	Mardar et al.	2016	Grain-based product
14	Suryaningrat	2016	Cassava processed for cake
15	Ekawati & Bazarado	2016	Processed carrot
16	Elleuch et al.	2016	Supply chain for agrifood products
17	Sayadi et al.	2017	Agricultural practices in olive growing
18	Dolgun & Köksal	2017	Yogurt
19	Jambrak et al.	2017	Apple juice and nectars
20	Djekic et al.	2018	Chicken meat supply chain
21	Schiavo et al.	2018	Frozen chicken meat supply chain
22	Ahmadabadi et al.	2018	Sesame
23	Liu et al.	2018	Supply chain of fresh products (vegetables, meat, and others.)
24	Kaletnik et al.	2019	Nuts value chain
25	Vanany et al.	2019	Halal chicken meat
26	Wang et al.	2019	granola bar
27	Djekic et al.	2019	Chain of fruits (apple)
28	Cappelli et al.	2019	Value chain of extra virgin olive oil
29	Dias Júnior et al.	2020	Barbecue Charcoal

The discussion of how authors have performed the QFD phases and their main research results have been presented in increasing chronological order, next to the Discussion section as follows.

DISCUSSION

A discussion has been additionally presented in increasing chronological order about how authors have performed the QFD phases and their main research results, in which:

Rosado Júnior et al. (2011) used the QFD to analyze the quality required for bulls to enhance the method by Ribeiro; Echeveste; Danilevicz (2000) and propose a new one aimed to analyze and relate products and services together. The authors collected the voice of the customer through a structured questionnaire with discursive questions asked to customers and technicians to elaborate the "what?" and "how?" phases and draw a relationship between "what? and how?" from the HoQ. Subsequently, they developed a technical benchmarking matrix to analyze the difficulties in achieving the required quality, which have been found through a competitive analysis. An adaptation of the method provided a joint analysis of products and services, thus confirming the hypothesis that there is a cross-relationship between them and enabling further studies on the product. The authors developed a HoQ by focusing on finding the relationship matrix results without applying and analyzing the correlation matrix.

Bevilacqua et al. (2012) employed a QFD model proposed by Bech et al. (1997) which promotes an integration between sensory and market analyses of food products. The authors also included a joint evaluation by using the fuzzy logic and QFD to analyze the quality of different types of olive oil. The authors captured the voice of the customer through a structured questionnaire applied randomly to perform the "what?" and how?" steps. Therefore, they assign the "what? and how?" relationship matrix to the fuzzy logic with the purpose of capturing the elasticity of imprecise requirements from customers. The use of mathematical tools, such as the fuzzy logic, proved to be applicable to QFD. And a fuzzy suitability index was calculated for each type of oil analyzed according to a equation written by Facchinetti; Ghiselli Muzioli; Ricci (1998) to compare and classify them. The research focused on collecting the main items of data from HoQ without using the technical benchmarking and correlation matrices.

Naspetti et al. (2012) collected the voice of the customer in Italy in Udine, Potenza and Ancona to improve the quality of organic sheep cheeses, but only consumers of organic products were assessed by employing the likert scale so as to determine the level of importance of "what?". The research demonstrated how consumer choices affect the production phase and agricultural strategies, and vice versa, thus highlighting the opportunity to connect producers and customers towards continuous product improvement. The authors have not adapted the QFD or used other methods or tools, nor developed analyzes on the technical benchmarking and correlation matrices, which reveals the need for further studies to analyze these parameters.

Park et al. (2012) employed the voice of customer at random involving experts and consumers of Korean restaurants located in Manhattan and New York to improve the quality of Korean beef barbecue. The authors comprehensively assessed the HoQ by developing and employing all phases of the technical benchmarking matrix. In addition, they pointed out that it is essential to integrate planning phases for devising product improvement strategies. It was suggested five main strategies for improving Korean bulgogi barbecue by considering relative importance, technical difficulties, and a correlation of engineering characteristics. Furthermore, other methods or tools were not employed in their research, and an adaptation of the method was not performed either.

Kuo et al. (2012) integrated the Kano model into the QFD, which is a quality management tool that seeks to improve processes, products or services, whose main basis is also to analyze customer needs. The authors interviewed consumers over the age of 20 who are on a healthy diet in central Taiwan, Nantou County, Changhua County and Taichung City. The Kano model focused on the "what?" to identify the relationship between quality attributes of healthy black bean foods together with quality factors generated by the Kano model. Then, the "how?" was deployed and it was carried out an analysis of the relationship between "what? and how?", as well as a competitive analysis of products (technical benchmarking matrix). The statistical package for the social sciences 16.0 (SPSS) software was used to analyze quantitative data and measure research reliability, which proved to be satisfactory, as the results serve as reference for companies to allocate their resources efficiently, establish product development strategies, improve quality according to consumer satisfaction, reduce time in product development and increase the success rate of their products in the market.

Kristianto et al. (2012) employed the QFD to assess and improve the effectiveness of a total quality management (TQM) program adopted by a wheat flour milling company. TQM is a strategy that seeks to improve quality, productivity, and competitiveness, which integrates all organizational functions to meet the needs of customers. In this sense, data on the voice of customers of the flour processing company has been collected, in which an integrated approach of the method assisted in designing a more competitive product predicated on the fact that the quality program should extend to all operational levels of the company, as well as on what resources should be aligned to meet customer needs. It was focused on the results of the "what? and how?" to obtain the technical benchmarking matrix without considering considerations by the strategic benchmarking and correlation matrices.

Cardoso et al. (2014) integrated the supply chain information (SCIM) model suggested by Benner et al. (2003b) into the QFD who had proposed to insert supply chain information into the initial "what?" matrix to develop a new organic fruit jelly product. Thus, the research captured the requirements of customers and experts on the production process of fruit jelly and the supply chain in terms of production by linking its main elements to the required quality. The used model comprises the voice of the customer, quality planning, production planning and a correlation between raw materials and quality features. It was considered that an adaptation of the initial matrix, which includes information about the agro-industrial supply chain, was fundamental to ensure levels of quality required by consumers of organic products, along with considerations about possible changes that the food may undergo during the production process, i.e. natural product variations. However, the strategic planning and correlation matrices were not used in their analysis.

Alba-Elías et al. (2014) used the QFD to improve the production process of vegetable preserves together with the creation of a regulator in the form of tablets for using food additives[1]. The authors used the method created by Hauser and Clausing (1988) and used the HoQ to compile a part deployment matrix, which was used to assemble an regulator. The part deployment matrix was initiated by assigning the HoQ's most important "whats?" to the "hows?" to obtain a relationship matrix of "what? and how?" and the technical benchmarking matrix. Subsequently, the correlation matrix was assesed, and it was found that there is no need to apply and evaluate the technical benchmarking matrix in the research. The method ensured that the designed device met all the top requirements of customers, thus increasing productivity and safety of the regulator.

De Pelsmaeker et al., (2014) employed the model proposed by Costa; Dekker; Jongen (2001) to improve the quality of stuffed chocolate, which consists in a combination of sensory analysis (appearance, aroma, texture and taste) with instrumental analysis and production features at the HoQ. This model is used to translate consumer requirements into sensory attributes measurable by a descriptive sensory analysis, in which the voice of the customer is highly detailed from sensory and instrumental analysis. Sensory measures aim to establish target values for the most favorite chocolates as input to the HoQ, and thus a quantitative descriptive analysis was used together with a sensory panel composed of eight trained evaluators. The authors deployed all steps of the HoQ and provided precise details on the design of matrices. The research suggested the correction of fat bloom to optimize the production of chocolates, in addition to mentioning that the HoQ is a good tool due to demonstrating the relationship between product features and those of the process, therefore new sensory studies have been proposed to highlight that the method requires technique, knowledge and skill from food company personnel.

Dill et al. (2014) captured the voice of the customer through a structured questionnaire sent by email aiming to improve pork loin supply chain, whose data was collected by the software Sphinx Léxica version 5.0 from a random sample of customers. The authors used the HoQ to analyze the relationship between consumer needs and the processes that must be improved in the supply chain, which required analyzing supply chain sectors (production, transport, processing, and distribution). Results were achieved by understanding the technical benchmarking matrix data, mainly by focusing on identifying that food safety is an attribute which is quite relevant from the customer's perspective, along with improvements in the product image and different stages of the supply chain processes, such as: animal handling, genetic selection and feeding are associated with the end product's quality, as well as animal welfare practices for transport and slaughter, i.e. all aspects related to obtaining a more succulent and less fibrous product . Furthermore, the research has neither exclusively assessed the technical benchmarking matrix, nor the correlation matrix.

Naspetti et al. (2015) used the voice of the customers of buyers from three regions of Italy using the standpoint of experts to obtain the voice of the expert and analyzing the correlation matrix to improve poultry meat. Two measurable visual sensory attributes (meat color and presence of fat) were included in the research analysis to emphasize the understanding of results through the HoQ relationship, technical benchmarking and correlation matrices by deploying and describing each phase of the research analysis. The results revealed that to meet customer requirements, the poultry meat sector must focus on more extensive production systems, such as those applied in integrated or organic systems, operating specifically along the supply chain as for the type of poultry housings, animal genotype, animal welfare and stocking density, in addition to collecting more accurate data on product labeling. Moreover, the possibility of integrating sensory aspects to the method is a good alternative to further investigate the relationship between the processing stages and poultry meat quality.

In a study conducted by Hsia et al. (2015), the manager, engineers and workers of a milling plant were asked to describe and discuss the eight stages of rice processing to elaborate the voice of the customer regarding "what?" aiming to improve it. Therefore, the "what?" and "how?" were translated and applied in a relationship matrix to obtain a technical benchmarking matrix. No other matrices were used (technical benchmarking and correlation matrices), but other tools and techniques were implemented to achieve objectives, starting from a risk assessment and analysis to determine the three stages of greatest risk in rice milling processes (drying, cooling and color classification). Then, the cause-effect diagram was used and eight risk factors that lead to a decrease in the quality of the three stages have been identified. In addition, the eight risk factors have been analyzed by the HoQ relationship matrix, which showed that the quality of ground rice can increase as a result of an operational improvement of the four main factors, namely: grain color analysis, refrigeration system for storage, installation of ventilation and improvement of the subsystem of air pressure nozzles of the grinding machine, which has been achieved by using the Russian method *theoria resheneyva isobretatelskehuh zadach* (TRIZ), i.e. a method of systematic innovation for problem solving.

Mardar et al. (2016) captured the voice of the customer through a questionnaire structured in accordance with the Kano model for customer satisfaction, which has been divided into three requirements corresponding to the product (essential product requirements, satisfaction features and attractiveness requirements) for creating a new grain-based product. The voice of the customer was interpreted by the technical team with the aid of two quality management instruments, i.e., the affinity and tree diagrams, whose results were used to compile the relationship matrix afterwards. All the phases of the method have been deployed and the technical benchmarking matrix that assessed the competitiveness of popular grain biscuit brands was highlighted, as well as the correlation matrix that pointed out the difficulties in correlating the product's technical features, therefore an empty sign was used for such cases. The research showed that the creation of a new grain-based product that improves customer health depends on the type of raw material being used, and on the rich food additive and energy value of the finished product. As for the method applied to new grain-based products, it proved to be satisfactory at determining customer requirements, as well as for prioritization purposes based on customer preferences through the translation of detailed technical features which facilitated the elaboration of a competitive product.

Suryaningrat (2016) empoyed the QFD method to assess customer expectations and develop a sustainable supply chain for processed cassava products, which is cassava cake in this case. The first step to prepare the HoQ (voice of the customer) was obtained by means of a structured questionnaire addressed to loyal consumers, product retailers and the production unit itself (employer and employees). Then, it was discussed the translation of the voice of the expert and the assessment of data from the relationship matrix to compile a technical benchmarking matrix. Analyzes of the correlation matrix or the right side of the HoQ have not been addressed in his research, which exposes the competitive analysis. The study showed that nine items affect product performance, which are related to ingredients used, production method and packaging process. As regards the supply chain, production execution and quality control stand out, as raw material processing activities tightly integrates with all items translated by the voice of the expert, which must be in accordance with production planning, control and inventory, as well as performing an evaluation for raw material acquisition.

Ekawati & Bazarado (2016) after using the voice of the customer, identified and designed two new food products based on carrots, i.e. candied carrot and carrot syrup, which were applied to the QFD, especially for product design. The authors developed four HoQ matrices with two different objectives for each product, one for product evaluation (phase 1) and another for packaging evaluation (phase 2), with separate evaluations for each product. The results of phase 1 were the basis for determining technical requirements of phase 2, which has been subsequently used for packaging. The product design specifications about "how?" were obtained through brainstorming sessions in the form of discussions and interviews involving agricultural or food technology experts. Furthermore, the authors deployed all steps of the HoQ and pointed out that product design development, elaborated through the QFD, resulted in critical elements for producing carrot products and their packaging.

Elleuch et al. (2016) addressed the problem of mitigating vulnerabilities by improving resilience within an agribusiness by carrying out a specific analysis of its supply and production chain to improve the agribusiness supply chain. The QFD was used as a structuring tool to find vulnerabilities and potential resilience capacities of the supply chain, which has been performed by assessing the relationship and technical benchmarking matrices, without the deployment of other HoQ matrices. However, conducting the research is complex, since it comprises four phases prior to the method application, namely: 1. identification of vulnerability factors in the supply chain through interviews invoving experts and application of results together using the Ishikawa diagram; 2. assessment of the severity of vulnerability factors by means of a binary comparison of the analytic hierarchy process (AHP); 3. identification of potential resilience capabilities through literature review; 4. application of the QFD method with the purpose of minimizing vulnerability factors through mitigation solutions called resilience capabilities. The result was deemed pertinent by the company's operational team and fuzzy methods were used to meet the research uncertainties through a binary comparison of vulnerabilities regarding the assessment of resilience in relation to vulnerabilities.

Sayadi et al. (2017) captured the voice of the customer through survey questionnaires to improve agricultural practices of olive cultivation. These questionnaires have been analyzed by the National Institute of Statistics of Spain in which a large quantitative study was developed to analyze habits, behaviors, and attitudes of customers as regards olive oil consumption. The demand for quality aspects included broad requirements, such as chemical and sensory aspects, marketing (price, place of purchase and type of bottle), certification and quality assurance, social aspects, and environmental issues. In addition, an analysis was carried out involving experts in the sector (olive growers, researchers, technicians and oil mills runners, quality laboratories, etc.) to obtain technical features and understand the relationship and correlation matrices. The research has not addressed the right side of the HoQ (strategic planning) since it was not necessary for achieving the research objective. The target values shown by the HoQ pointed out that the search for olive oil quality starts from agricultural operations to the extraction process (transport, storage, processing, and olive oil storage) that must be combined and integrated to preserve the qualitative features of olive oil.

Dolgun & Köksal (2018) used the QFD for improving the quality of Yogurt and thus the method was coupled with the strengths of Kansei Kougaku (engineering of the senses) to support processes and define goals between the customers' requirements and the product's technical features, specially for the sensorial, project and statistical modeling analyses of hedonic and technical data coupled with the analytic hierarchy process (AHP) so as to better understand the results. The voice of the customer was captured rigorously (only regular product customers) by collecting demographic information and patterns of purchase/consumption, along with the sensory analysis of the studied product and three more competing samples for a sensory comparison. Data was translated by the marketing, manufacturing, research and development, and quality control team of the yogurt company. The research focused on obtaining data from the technical benchmarking matrix and its evaluation, which were coupled with other support tools without the application of a correlation matrix. The ordinal scale assessments proposed by the HoQ were converted into proportional scales and applied to AHP. According to the authors, peer comparisons require less cognitive effort than direct scoring methods, and the approach based on the results of statistical analysis, technical and prediction models significantly increased the effectiveness of decision-making processes if compared to previous attempts of adding information from the HoQ.

Jambrak et al. (2018) analyzed the quality features of apple juice and nectars treated with high powered ultrasound, in which sensory and physical attributes have been evaluated by the HoQ based on the customers' requirements. The voice of the customer was randomly captured, and general demographic information was obtained about interviewees, as well as their impressions about tasting. Seven basic quality sensory features about apple juice and nectars have been classified and translated into ten technical characteristics by the research and development. The HoQ was adapted for employing sensory characteristics and it was usd the required quality, technical quality characteristics, relationship matrix and target values obtained by technical benchmarking observed for the treated sample. The correlation matrix and the technical benchmarking matrix were not used. Treated samples have been evaluated by the software Statgraphics Centurion for statistical understanding and used to compile a semantic differential chart showing the final quality index to assimilate results. It was found that processing using high powered ultrasound assists in preserving the drink and maintains its main sensory properties, such as taste and scent.

Djekic et al. (2018) analyzed the chicken meat supply chain and thus focused their efforts on the analysis of the voice of the customer, voice of the expert, relationship, and technical benchmarking matrices, but without the application of other matrices of the HoQ. The voice of the customer was captured through five actors in the chicken meat supply chain: customers, retailers, meat processors, slaughterhouses, and farms. The authors built five HoQ, one for each actor in the supply chain, which have been developed with the aid of the Delphi method that provides consensus and synthesizes the opinions of experts. The results were analyzed from the customers' standpoint to the supplier at different stages of the chicken meat supply chain and therefore the HoQ was employed from the consumer to the farm. The Mann-Whitney test was carried out at statistical significance level using Microsoft Excel 2010 and SPSS Statistics 17.0 to determine the differences between customers and suppliers in relation to the ranking of variables along with quality aspects for interactions. The method provided an identification of quality features among the five actors in the chicken meat supply chain by emphazing animal welfare requirements in the first link of the supply chain. The research allowed improving cohesion and enhancing the supply chain, in addition to the possibility of formulating policies to improve the supply chain's competitive strategies.

Schiavo et al. (2018) analyzed the quality of frozen chicken meat supply chain by consolidating the voice of the customer based on considerations of two groups: those who buy chilled chicken meat to be resold (retail, supermarkets, butchers and steakhouses) and those that add value to products by preparing chicken for end consumers (restaurants and fast food), which have been translated by the HoQ in the "how?" phase and the next phase of the relationship matrix was subsequently employed to prioritize the phases of frozen chicken meat production process using an exponential scale of 1-3-9, i.e. the most popular scale according to Takai and Kalapurackal (2012). By evaluating the relationship and technical benchmarking matrices, it was identified that there are differences in the priorities of retailers and restaurants, but that both present congruence to requirements such as food safety, in which time and temperature are crucial variables in relation to trasport and distribution.

Ahmadabadi et al.(2018) coupled the QFD with the design of experiments (DOE) to identify a new sesame product, which is generally used to recognize and compare the effects of quality features during a process, along with the BAH structure (Booz, Allen and Hamilton) for the development of new food products. 384 people were interviewed to survey the voice of the customer and record new customer needs, in which it was obtained sesame chocolate. The new product was examined and translated into technical features towards the advancement of "how?", and then they elaborate the relationship and technical benchmarking matrices to assess the ingredients and plan the sesame chocolate production process. The next phase of the research used Taguchi's method of experimental design (TMED) to improve the quality of products through statistical concepts in which ten phases of procedures based on its method have been implemented. The study confirms that the QFD and DOE applied to a new product can meet customer demands, while at the same time meeting the requirements of product nutritionists and technologists. However, the actors have not used the correlation or technical benchmarking matrices.

Liu et al. (2018) proposed a new decision-making method invoving multiple statistical criteria for selecting green produce suppliers of fresh produce to improve its supply chain process, which consists of two stages: 1. selection criteria establishment; 2. presentation of the selection method. In the first stage, the HoQ was used to compile two relationship matrices to assess the interrelationships between customer requirements, company strategies and selection criteria, which have been considered separately in the two matrices. However, one of the main limitations of QFD was the application of absolute priority to determine target values among customers' requirements. Thus, to overcome such an issue, the analytic hierarchy process (AHP) and analytic network process (ANP) were integrated into the QFD with the purpose of selecting suppliers and calculating priorities between customer requirements and company strategies, together with subjective criteria weights. In the second stage, Shannon's entropy method (used to measure the average degree of uncertainty regarding sources of information) was used to obtain the objective criteria weights, thus generating qualitative assessment and multi-objective optimization by reason analysis, in addition to the fuzzy MULTIMOORA method which was used to classify the supplier. The method proposed by the authors considered the interrelationships between selection criteria for green produce suppliers and the research focused on obtaining the technical benchmarking matrix from the HoQ without developing other HoQ matrices.

Kaletnik et al. (2019) considered economic and mathematical methods, and the theory of constraints for the development of provisions for new products, in addition to the QFD methodology for creating innovative products. Healthy, organic, and unrefined vegetable oils have been identified, which is walnut oil. It was performed interviews involving experts and conducted experimental laboratory studies for the production and implementation of requirements for the developed innovative products. The HoQ was used to link customer desires to the company's marketing complex. The features met the requirements of the gradual process of quality function deplotyment in the following order: voice of the customer, voice of the expert (through the company's marketing complex for the characteristics of components) that is typical of the process, and production planning features, which resulted in the weighting coefficient (target value) of the QFD. The study offered a model for the value chain and nut processing, in addition to new production technologies for designing new products, and the integrations of technical benchmarking and correction matrices were not considered.

Vanany et al. (2019) proposed a multiphase QFD model to identify key processes and prioritize programs to improve the production of halal chicken meat. The matrix of the first step of the HoQ was compiled according to the Halal Assurance System (HAS) requirements, in which two halal auditors represented the voice of the customer which was translated by experts from the Assessment Institute for Food, Drug and Cosmetic, The Indonesian Council of Ulama (AIFDC - ICU) as regards technical control features for the product. Thereafter, the authors described the three phases of the QFD and their interrelationships using the proposed multiphase model for halal food control, which refers to: 1st. prioritizing the production process based on HAS attributes; 2nd, prioritizing the critical halal factors in terms of quality; and 3rd. the priority of improvement programs is assessed by and the degree of criticality of halal foods is determined. Thus, the relationship and technical benchmarking matrices were devised to establish priorities and assessing the requirements inserted in the HoQ, without an interpretation of technical benchmarking and correction matrices. The target values proposed by the technical benchmarking matrix assisted the chicken processing plant in minimizing possible problems, thus identifying the main improvement programs.

Wang et al. (2019) proposed a seven-phase QFD approach, which emerged by modifying the nine-phase model proposed by Chan and Wu (2005) coupled with the sensory model proposed by Bech et al. (1997), in addition to performing a competitive analysis so as to find the final level of importance of customers' requirements, together with the fuzzy logic applied in the results of interviews, in which it was replaced a precise value of relative technical ratings with classification priorities. The model presented by the authors identified the customers' requirements, evaluated the level of importance of the competitive analysis results (technical benchmarking matrix), identified the sensory attributes of the product found and presented how strong the relationship between "what? and how?" is through its technical features. Therefore, the relative technical ratings (priority ratings according to the product development team) have been determined based on the importance of "what?" along with the strength of relationships between "what? and how?". The correlation matrix was not integrated with the proposed model, and results show that relative technical ratings reflect the customer's needs and the relative importance they attached to them.

Djekic et al. (2019) addressed the four main actors regarding apple supply chains (consumers, retailers, cold stores, and orchards) with the purpose of understanding quality requirements for agents in fruit supply chains. The authors analyzed the points of view of customers and suppliers at different links of the apple supply chain (from orchards to consumers) by applying the QFD of consumers to the orchard through the elaboration of three HoQ matrices that related: orchards and cold storage; cold storage and retail; retail and consumer. The development of HoQ focused on the elaboration of the relationship matrix to obtain target values of the technical benchmarking matrix. The voice of the customer was captured through the participation of four main actors involved in apple production, which were translated into quality features and applied to the relationship matrix to verify integrations. Integrations were assessed by scientists from the faculty of agriculture (Institute of Food Technology and Biochemistry and Horticulture Institute) who have classified relationships using the Delphi method (used by experts to reach a consensus) with the purpose of stimulating and synthesizing opinions between integrations. The results suggest that the perceptions of quality are different among actors in the apple supply chain, ranging from internal quality parameters, such as juiciness, to the production system employed in orchards, which can be confirmed by comparing the results in two directions - from orchards to consumers (from bottom to top) and from consumer to orchards (from top to bottom).

Cappelli et al. (2019) utilized the QFD to improve the quality of the extra virgin olive oil (EVOO) production process. The research was applied to companies that produce high-quality EVOO and the voice of the customer was captured by an experienced group of people involved in the process of extracting olive oil through brainstorming, in which they raised the critical phases capable of influencing product quality performance and identified the olive washing process as the most critical, as it presents great potential for microbial load when poorly washed. Thus, HoQ was applied to the olive washing machine to identify the main aspects that must be improved in it. The identification took place through the experts' evaluation of the relationship and technical benchmarking matrices, which revealed the following as target value: processing time reduction, washing system improvement, water consumption reduction, and worker safety increase. The results obtained have been translated into technical characteristics for the washing machine, which were subsequently introduced into a prototype that was developed and tested by eight experts in the engineering sector. According to the results, the application of the proposed approach made it possible to identify the critical issues of EVOO and led to a complete reinvention of the olive washing system.

Dias Júnior et al. (2020) presented the second phase of a study conducted by Dias Júnior et al. (2015), in which it was aimed to broadly identify the quality requirements of barbecue charcoal. In this research, the authors integrated the Kano model to relate product requirements to consumer satisfaction and the cause-effect or Ishikawa (fishbone) diagram has been used to ascertain the main factors that contribute to the quality of charcoal, i.e., from production to making the product available for customers. The voice of the customer was collected through questionnaires through a technical-qualitative approach to deepen the analysis of customers' perspective on the product. The "what?" was translated by a team of producers, distributors, traders, technical experts, and others interested in matters related to charcoal supply chain. In addition, all HoQ matrices have been applied and deployed by experts. The projected quality results (target values) revealed the main quality features of barbecue charcoal in an extensive and decreasing order (ease of ignition, rapid formation of embers and flames, rapid preparation of barbecues and affordable price to customers) and highlighted that the factors related to the raw material (wood), labor and methodology (carbonization) indicated by the cause-effect diagram exert more decisive effects on the quality of barbecue charcoal.

The research results comprise a discussion of 30 selected articles, in which the type of product used, purpose of application, illustration of the HoQ steps and a description of application steps of HoQ has been analyzed and summarized in Table 3.

Table 3 (a). Overall analysis of 30 articles selected through the literature review

Nº	Purpose of application					
	Analysis of the required quality applied	**Analysis of the quality of an existing product**	**Analysis of the required quality of an existing product**	**Development of a new product**	**New product design development**	**Improving the quality of an existing product**
1	x					
2		x				
3						x
4						x
5						x
6						x
7				x		
8						x
9						x
10						x
11						x
12						x
13				x		
14						x
15					x	
16						x
17						x
18						x
19		x				
20		x				
21		x				
22				x		
23						x
24				x		
25						x
26						x
27			x			
28						x
29		x				

Source: The authors.

Table 3 (b). Overall analysis of 30 articles selected through the literature review

Description of the QFD application phases					
Voice of the customer matrix - "what?"	Voice of the expert matrix - "how?"	Relationship matrix - "what? and How?"	Strategic planning matrix	Technical benchmarking matrix - "how much?"	Correlation matrix
x	x	x	x	x	
x	x	x		x	x
x	x	x		x	
x	x	x	x	x	x
x	x	x	x	x	
x	x	x		x	
x	x	x		x	
x	x	x		x	x
x	x	x	x	x	x
x	x	x		x	
x	x	x		x	x
x	x	x		x	
x	x	x	x	x	x
x	x	x		x	
x	x	x	x	x	x
x	x	x		x	
x	x	x		x	x
x	x	x	x	x	
x	x	x		x	
x	x	x		x	
x	x	x		x	
x	x	x		x	
x	x	x		x	
x	x	x		x	
x	x	x		x	
x	x	x	x	x	
x	x	x		x	
x	x	x		x	
x	x	x	x	x	x

CONCLUSION

Out of the 29 scientific articles analyzed, it was found that publications related to the QFD characterize a scenario of predominantly one-off publications (28 publications), except for the case of charcoal, which presents a second stage of the study.

Regarding the variations in method application, it was observed that these are configured as: product improvement or supply chain improvement (17 publications), quality analysis applied to existing products (5 publications), development of new products that satisfy customer requirements (4 publications) and quality planning (1 publication).

As for the application of other theories and tools linked to the method, it was found that it is recurrent (18 publications), since authors increasingly seek to improve the applications of QFD in the food industry to restrict the method to its research objective and obtain more secure data.

Moreover, it was found that few articles are aligned with the application of all steps of the HoQ (5 publications), corresponding to studies carried out by Park et al. (2012); De Pelsmaeker et al., (2014); Mardar et al. (2016); Ekawati & Bazarado (2016); and Dias Júnior et al. (2020).

The matrices with greater application predominance are: the "what?" voice of the customer matrix, the "how?" voice of the expert matrix, the "what? and how?" relationship matrix and the technical benchmarking matrix for obtaining target values (29 publications).

As for the divergent points, these are related to a lack of application of the technical benchmarking matrix (9 publications) and correlation matrix (9 publications), and there is no simultaneous application of these two phases in 24 other publications.

This demonstrates that the QFD has been partially used in studies identified for the agribusiness supply chain in the food sector, thus its full potential is wasted for analyzing quality in product development.

Thus, it is relevant to conduct studies on the agribusiness supply chain in the food sector aimed to extend knowledge about it and minimize conflicts between agents in the supply chain regarding variation in the perception of quality, since it is currently driven by all actors involved in agribusinesses (Grunert, 2005).

ACKNOWLEDGMENT

National Council for Scientific and Technological – CNPq [grant number 424722/2018-6].

REFERENCES

Abreu, F. de S. (1997). QFD - desdobramento da função qualidade - estruturando a satisfação do cliente. *Revista de Administração de Empresas*, *37*(2), 47–55. doi:10.1590/S0034-75901997000200005

Ahmadabadi, H. Z., Zamzam, F., Meybodi, F. R., Streimikiene, D., & Mardanim, A. (2018). Developent of a new sesame product using QFD and DOE methods: A case study of sesame product in yazd. *Montenegrin Journal of Economics*, *14*(1), 27–44. doi:10.14254/1800-5845/2018.14-1.2

Akao, Y. (1997). Past, Present, and Future. *International Symposium on QFD '97 – Linköping*, (2), 1–12.

Akao, Y., & Mazur, G. (2003). The leading edge in QFD: Past, present and future. *International Journal of Quality & Reliability Management*, *20*(1), 20–35. doi:10.1108/02656710310453791

Alba, F. E., de La Nieta, A. G. S., Casas, A. G. S., González, A. M., Cendõn, J. A., & Limas, M. C. (2014). A new device for dosing additives in the food industry using quality function deployment. *Journal of Food Process Engineering*, *37*(4), 387–395. doi:10.1111/jfpe.12094

Bech, A. C., Hansen, M., & Wienberg, L. (1997). Application of house of quality in translation of consumer needs into sensory attributes mea- surable by descriptive sensory analysis. *Food Quality and Preference, 8*(5-6), 329–348. doi:10.1016/S0950-3293(97)00062-1

Benner, M., Geerts, R. F. R., Linnemann, A. R., Jongen, W. M. F., Folstar, P., & Cnossen, H. J. (2003b). A chain information model for structured knowledge management: Towards effective and efficient food product improvement. *Trends in Food Science & Technology, 14*(11), 469–477. doi:10.1016/S0924-2244(03)00154-7

Benner, M., Linnemann, A. R., Jongen, W. M. F., & Folstar, P. A. (2003a). Quality Function Deployment (QFD) - Can it be used to develop food products? *Food Quality and Preference, 14*(4), 327–339. doi:10.1016/S0950-3293(02)00129-5

Bevilacqua, M., Ciarapica, F. E., & Marchetti, B. (2012). Development and test of a new fuzzy-QFD approach for characterizing customers rating of extra virgin olive oil. *Food Quality and Preference, 24*(1), 75–84. doi:10.1016/j.foodqual.2011.09.005

Brunsø, K., Fjord, T. A., & Grunert, K. G. (2002). Consumers' Food Choice and Quality Perception. *Aarhus School of Business, 77*, 1–60.

Cappelli, A., Parretti, C., Cini, E., & Citti, P. (2019). Development of a new washing machine in olive oil extraction plant: A first application of usability-based approach. *Journal of Agricultural Engineering, 50*(3), 134–142. doi:10.4081/jae.2019.949

Cardoso, J. de F., Casarotto Filho, N., & Cauchick Miguel, P. A. (2015). Application of Quality Function Deployment for the development of an organic product. *Food Quality and Preference, 40*, 180–190. doi:10.1016/j.foodqual.2014.09.012

Cauchick Miguel, P. A. (2005). Evidence of QFD best practices for product development: A multiple case study. *International Journal of Quality & Reliability Management, 22*(1), 72–82. doi:10.1108/02656710510573002

Chan, L. K., & Wu, M. L. (2002). Quality function deployment: A literature review. *European Journal of Operational Research, 143*(3), 463–497. doi:10.1016/S0377-2217(02)00178-9

Chan, L. K., & Wu, M. L. (2004). A systematic approach to quality function deployment with a full illustrative example. *Omega, 33*(2), 119–139. doi:10.1016/j.omega.2004.03.010

Conforto, E. C., Amaral, D. C., & Silva, S. L. (2011). Roadmap for systematic literature review: application in product development and project management (in portuguese). *Congresso Brasileiro de Gestão e Desenvolvimento de Produto,* (8), 1-12.

Costa, A. I. A., Dekker, M., & Jongen, W. M. F. (2001). Quality function deployment in the food industry: A review. *Trends in Food Science & Technology, 11*(9–10), 306–314.

De Pelsmaeker, S., Gellynck, X., Delbaere, C., Declercq, N., & Dewettinck, K. (2014). Consumer-driven product development and improvement combined with sensory analysis: A case-study for European filled chocolates. *Food Quality and Preference, 41*, 20–29.

Dias Júnior, A. F., Andrade, C. R., Milan, M., Brito, J. O., Andrade, A. M., & de Souza, N. D. (2020). Quality function deployment (QFD) reveals appropriate quality of charcoal used in barbecues. *Scientia Agrícola*, *77*(6), 1–11. doi:10.1590/1678-992x-2019-0021

Dill, M. D., Revillion, J. P. P., Barcellos, J. O. J., Dias, E. A., Mércio, T. Z., & de Oliveira, T. E. (2014). Procedural priorities of the pork loin supply chain. *Journal of Technology Management & Innovation*, *9*(1), 84–92. doi:10.4067/S0718-27242014000100007

Djekic, I., Radivojevic, D., & Milivojevic, J. (2019). Quality perception throughout the apple fruit chain. *Journal of Food Measurement and Characterization*, *13*(4), 3106–3118. doi:10.100711694-019-00233-1

Djekic, I., Skunca, D., Nastasijevic, I., Tomovic, V., & Tomasevic, I. (2018). Transformation of quality aspects throughout the chicken meat supply chain. *British Food Journal*, *120*(5), 1132–1150. doi:10.1108/BFJ-08-2017-0432

Dolgun, L. E., & Köksal, G. (2018). Effective use of quality function deployment and Kansei engineering for product planning with sensory customer requirements: A plain yogurt case. *Quality Engineering*, *30*(4), 569–582. doi:10.1080/08982112.2017.1366511

Ekawati, Y., & Bazarado, M. (2016). Designing food products based on carrots using the product design phase of Quality Function Deployment. *Journal of Engineering and Applied Sciences (Asian Research Publishing Network)*, *11*(5), 3109–3116.

Elleuch, H., Dafaoui, E., El Mhamedi, A., & Chabchou, H. (2016). A Quality Function Deployment approach for Production Resilience improvement in Supply Chain: Case of Agrifood Industry. *IFAC-PapersOnLine*, *49*(31), 125–130. doi:10.1016/j.ifacol.2016.12.173

Facchinetti, G., Ricci, R. G., & Muzzioli, S. (1998). Note on ranking fuzzy triangular numbers. *International Journal of Intelligent Systems*, *13*(7), 613–622. doi:10.1002/(SICI)1098-111X(199807)13:7<613::AID-INT2>3.0.CO;2-N

Fernandes, J. M. R., & Rebelato, M. G. (2006). Proposta de um método para integração entre QFD e FMEA. *Gestão & Produção*, *13*(2), 245–259. doi:10.1590/S0104-530X2006000200007

Gasques, J. G., de Rezende, G. C., Verde, C. M. V., Salerno, M. S., da Conceição, J. C. P. R., & Carvalho, J. C. de S. (2004). *Desempenho e Crescimento do Agronegócio no Brasil*. Instituto de Pesquisa Econômica Aplicada – IPEA.

Govers, C. P. M. (1996). What and how about quality function deployment (QFD). *International Journal of Production Economics*, *46–47*(95), 575–585. doi:10.1016/0925-5273(95)00113-1

Grunert, K. G. (2005). Food quality and safety: Consumer perception and demand. *European Review of Agriculture Economics*, *32*(3), 369–39. doi:10.1093/eurrag/jbi011

Hauser, J. R., & Clausing, D. (1988). The House of Quality. *Harvard Business Review*, *66*(5/6), 63–73.

Hofmeister, K. (1991). Quality Function Deployment: market success through customer driven products. New York. In E. Graf & I. S. Saguy (Eds.), *Food product development: from concept to the market place* (pp. 189–210). doi:10.1007/978-1-4613-0675-7_9

Hsia, T. C., Liao, R. C., & Huang, S. C. (2015). Enhancing the quality of rice milling by systematic innovation techniques. *International Journal of Systematic Innovation*, *3*(4), 27–36.

Kaletnik, H., Lutsiak, V., Melnichuk, O., Dovhan, Y., & Malicki, M. (2019). Organizational basis of the development of innovative functional food products by the Ukrainian enterprises of deep walnut processing. *Ukrainian Food Journal*, *2*(1), 1–148.

Kristianto, Y., Ajmal, M. M., & Sandhu, M. (2012). Adopting TQM approach to achieve customer satisfaction: A flour milling company case study. *The TQM Journal*, *24*(1), 29–46. doi:10.1108/17542731211191203

Kuo, C. M., You, S. H., & Lu, C. Y. (2012). Integration of the Kano and QFD model in health food development: Using black beans as examples. *Quality & Quantity*, *48*(1), 225–242. doi:10.100711135-012-9762-8

Liu, A., Xiao Y., Ji, X., Wang, K., Tsai, S. B., Lu, H., Cheng, J., Lai, X., & Wang, J. (2018). A novel two-stage integrated model for supplier selection of green fresh product. *Sustainability, 10*(7), 1–23.

Mardar, M., Zhygunov, D., & Znachek, R. (2016). QFD methodology to develop a new health-conducive grain product. *Eastern-European Journal of Enterprise Technologies*, *2*(11), 42–47. doi:10.15587/1729-4061.2016.65725

Naspetti, S., Alberti, F., & Solfanelli, F. (2012). Quality determinants in the organic cheese supply chain: A Quality Function Deployment approach. *New Medit, 11*(4), 61–64.

Naspetti, S., Alberti, F., & Solfanelli, F. (2015). Quality function deployment in the organic animal food sector: Application to poultry meat. *Italian Journal of Animal Science*, *14*(3), 544–550. doi:10.4081/ijas.2015.4050

Park, S. H., Ham, S., & Lee, M. A. (2012). How to improve the promotion of Korean beef barbecue, bulgogi, for international customers. An application of quality function deployment. *Appetite*, *59*(2), 324–332. doi:10.1016/j.appet.2012.05.008 PMID:22634190

Režek Jambrak, A., Šimunek, M., Grbeš, F., Mandura, A., & Djekic, I. (2018). Analysis of apple beverages treated with high-power ultrasound: A quality function deployment approach. *Journal of the Science of Food and Agriculture*, *98*(6), 2258–2266. doi:10.1002/jsfa.8714 PMID:28981162

Rosado Júnior, G. A., Lobato, J. F. P., Echeveste, M., Gerber, A. S., Yang, S., & Rossini, K. (2011). Identification of demand characteristics in the production of sires using a conceptual model of quality function deployment: A case study. *Revista Brasileira de Zootecnia*, *40*(1), 210–220. doi:10.1590/S1516-35982011000100029

Sayadi, S., Erraach, Y., & López, C. P. (2017). Translating consumer's olive-oil quality-attribute requirements into optimal olive-growing practices: A quality function deployment (QFD) approach. *British Food Journal*, *119*(1), 190–214. doi:10.1108/BFJ-05-2016-0228

Schiavo, G., Korzenowski, A. L., Batista, E. R. S., de Souza, D. L., & Scavarda, A. (2018). Customers' quality demands as directions to the cold chicken supply chain management. *Business Process Management Journal*, *24*(3), 771–785. doi:10.1108/BPMJ-11-2016-0224

Suryaningrat, I. B. (2016). Implementation of QFD in food supply chain management: A case of processed cassava product in Indonesia. *International Journal on Advanced Science, Engineering and Information Technology*, *6*(3), 302–305. doi:10.18517/ijaseit.6.3.713

Takai, S., & Kalapurackal, R. M. (2012). Sensitivity analysis of relative worth in quality function deployment matrices. *Concurrent Engineering, Research and Applications*, *20*(3), 195–202. doi:10.1177/1063293X12442411

Vanany, I., Maarif, G. A., & Soon, J. M. (2019). Application of multi-based Quality Function Deployment (QFD) model to improve halal meat industry. *Journal of Islamic Marketing Journal*, *10*(1), 97–124. doi:10.1108/JIMA-10-2017-0119

Viaene, J., & Januszewska, R. (1999). Quality function deployment in the chocolate industry. *Food Quality and Preference*, *10*(4–5), 377–385. doi:10.1016/S0950-3293(99)00007-5

Vinayak, K., & Kodali, R. (2013). Benchmarking the quality function deployment models. *Benchmarking*, *20*(6), 825–854. doi:10.1108/BIJ-07-2011-0052

Wang, T. Y., Hsiao, H. I., & Sung, W. C. (2019). Quality function deployment modified for the food industry: An example of a granola bar. *Food Science & Nutrition*, *7*(5), 1746–1753. doi:10.1002/fsn3.1014 PMID:31139387

ENDNOTE

1. Food additive means any substance that is not typically consumed as food itself, and that is not usually used as a typical food ingredient, either with or without nutritional value, whose addition is intended for the manufacture, processing, preparation, treatment, packaging, transport or storage of food products (Alba-Elías *et al.*, 2014).

Chapter 5
CSR Programs of Financial Institutions: Development-Oriented Issues or Just Greenwashing?

Mirela Panait
https://orcid.org/0000-0002-5158-753X
Petroleum-Gas University of Ploiesti, Romania & Institute of National Economy, Romanian Academy, Romania

Lukman Raimi
Universiti Brunei Darussalam, Brunei

Eglantina Hysa
https://orcid.org/0000-0002-3429-5738
Epoka University, Albania

Abiodun S. Isiaka
https://orcid.org/0000-0002-5508-784X
University of Regina, Canada

ABSTRACT

This chapter explores the contrary views on CSR activities of financial institutions by drawing attention to the purported chameleon behavior of banks in promoting various CSR programs, adopting equator principles in lending activity, conducting financial education campaigns to increase the degree of financial inclusion of the population versus the claim about deceptive promotional techniques, practicing abusive contractual clauses in order to maximize profits at the expense of consumers. The chapter is distinguished by the critical attitude towards the behavior of FTNCs which knows significant differences depending on the area of manifestation – in the country of origin or in the host countries, developing countries. In addition, these entities take advantage of international instruments set up such as the equator principles or non-financial reporting standards to create a positive image among stakeholders, although their behavior is not socially responsible.

DOI: 10.4018/978-1-6684-2339-4.ch005

INTRODUCTION

Financial Transnational Corporations (FTNCs) are the giants of the world economy; despite not having a productive activity, they have a major impact on the real life because all companies and consumers need different financial products and services. FTNCs ensure the financing of economic agents not only from the countries of origin but also from the host countries, thereby influencing transnational transactions at the global level. The business method of FTNCs have a worldwide impact, from Small and Medium Enterprises (SMEs) to financial consumers and investors. FTNCs have evolved from the sole goal of maximizing profit at any cost for shareholders' interest to satisfying various stakeholders. The effects of the financial crisis of 2008 felt by people across the globe, generated a rather hostile attitude towards FTNCs.

The uncontrolled feeding of the financial innovation process by banking institutions and the poor disposal of banking products by rating agencies contributed to the financial crisis in 2008. The devastating impact of toxic assets on the real economy was further highlighted during the global crisis. For these reasons, we are currently witnessing the metamorphosis of the international financial system, under the sign of sustainable development and corporate social responsibility (CSR). It has been noted that financial institutions have different behavior in host and home countries. The application of the double standard suggests a chameleon behavior that on the one hand promote social responsibility but, on the other hand, generate negative externalities especially in developing countries characterized by abusive behavior towards consumers coupled with poor financial education. Seeking to exploit the low level of financial education, banks with foreign capital practice use abusive contractual clauses and deceptive promotional techniques. Therefore, the presentation by financial institutions of detailed information on their websites or even the publication of sustainability reports is considered by some specialists as a greenwashing technique that tries to beautify the real situation and create a very good image among stakeholders.

In terms of CSR, banks are very involved and run various programs that focus initiatives such as the development of local communities and improvement of financial education. Such programs ensure an improvement of the image among consumers and enhances public opinion (Perez & del Bosque, 2009; Pérez & del Bosque, 2015; Ruiz, 2018; Palazzo et al., 2020; Pérez et al., 2020; Siano et al., 2020; Stancu et al. 2020). To shed more light on the discussions surrounding the motives of FTNCs, this work investigates the genuineness or greenwashing of CSR activities of the FTNCs, and how they promote the principles of sustainable development by adopting green standards such as the Equator principles (EPs).

In specific terms, the study sets out to discuss CSR activities of the FTNCs, and how they promote the principles of sustainable development by adopting green standards such as the Equator principles (EPs). EPs represents a CSR framework that is voluntary and serves as a basis of determining, assessing and managing environmental and social projects[1].

CSR is also very important in financial institutions mainly because of the role in capital allocation and infrastructural development (Osellame, 2013). For example, Canadian Bankers Association are also committed to sustainability initiatives. To demonstrate their commitment, the association have adopted sustainable banking through some of their operations in areas such as purchasing green power and reducing the amount of paper used[2]. Furthermore, the Bank Act (459.3) requires Canadian banks with more than one billion dollars to publish a statement describing the contribution to the Canadian economy and society (The Bank Act, 1991)[3]. This requirement further emphasizes the importance of sustainability initiatives within the banking industry.

Despite these efforts, there is a growing concern that banks may not be sincere when it comes to sustainability initiatives as financial institutions are generally profit oriented enterprises. The incentive to voluntarily adopt a CSR initiative-and its associated costs- is therefore baffling. Is it all green washing or do these banks really care? We find that Financial corporations from many countries implement and comply with the Equator Principles (EPs) as a means of demonstrating CSR initiatives. This is contrary to the criticism that banks adopt EPs to be seen as responsible. For example, Canadian banks that adopt EPs are regarded as an "Equator Principle Financial Institution". Such banks are allowed to use this title on their websites and in press releases (Robertson, 2011). The use of such title confers reputational advantages to adopting banks. In addition to this criticism, researchers have also expressed concerns that financial institutions are free-riding to gain reputational benefits and not necessarily complying with EPs requirements (Amalric, 2005; Osellame, 2013) mainly because banks have been noted to evaluate the cost versus benefit of implementing EPs, and will likely adopt EPs if the expected benefits outweigh the identified costs (Amalric, 2005). Exploring sustainability initiatives through EPs adoption further provides the much needed insight into the CSR activities and the associated motives of this very important sector.

This study therefore explores the contrary views on CSR activities of financial institutions by drawing attention to the purported chameleon behavior of banks in promoting various CSR programs, adopting Equator principles in lending activity, conducting financial education campaigns to increase the degree of financial inclusion of the population versus the claim about deceptive promotional techniques, practicing abusive contractual clauses in order to maximize profits at the expense of consumers. The present work is distinguished by the critical attitude towards the behavior of FTNCs which knows significant differences depending on the area of manifestation - in the country of origin or in the host countries, developing countries. In addition, these entities take advantage of international instruments set up such as the Equator Principles or non-financial reporting standards to create a positive image among stakeholders, although their behavior is not socially responsible.

This study is important for three main reasons. First, this article contributes to research on sustainability and sustainability initiatives. As organizations aim to be seen as socially responsible in order to gain reputation advantage, it is important for the public to be well informed regarding the motive behind such CSR activities. Second, this study contributes to the debate on the need for a framework to report CSR activities. A review of the Equator principles compliance suggests banks are genuine in the application of the principles. This review can motivate other financial institutions to voluntarily comply and also motivate other regulatory bodies to float similar initiatives. Finally, the findings are important to multiple stakeholders. Managers of organizations and those responsible for company image and reputation are more informed about activities of competitors. Regulators are better informed about compliance of financial companies to existing regulations and researchers are provided with opportunities for further research in areas such as the impact of digitalization in the financial services industry.

To present the complex activity carried out by financial institutions, the chapter has three sections. The first section presents the background information on the convergent and divergent actions of banks in promoting the principles of sustainable development. The authors focused on the CSR in the financial industry including a discussion on the greenwashing strategies of these firms and proposes the research questions in this study. The second section discusses the methods and analysis. In the next section the authors present the main directions in which financial institutions are involved in promoting the principles of sustainable development. In the second section, the authors present the applicability of Equator principles by financial institutions in the lending process, the involvement of banks in financial education programs and also the cases of greenwashing detected in different countries. The third section concludes with considerations regarding the divergent actions in which the credit institutions are involved as well as proposed directions for future research.

LITERATURE REVIEW

Corporate Social Responsibility in Financial Field

For several decades, the concern of different categories of stakeholders has been about the process of sustainable development. More and more often, researchers and practitioners talk about sustainability and CSR in terms of sustainable industry, sustainable agriculture, sustainable tourism, and social responsible investors. Sustainable development is a concept that has found its rightful place in all areas (Matei, 2013, Raimi et al., 2019; Gigauri & Damenia, 2020, Morina et al., 2020; Raimi, 2020). Increasing number of companies and public authorities are pursuing environmental, social and economic issues in their activities, and business models being remodeled under the sign of CSR. Under pressure from stakeholders, more and more companies are getting involved in CSR programs focused on environmental protection, developing local communities, respecting employees' rights or fighting corruption. The interest of the companies for carrying out the CSR actions has to do with the field of activity, the size of the company but can depending on other economic and social variables (Mukherjee, 2013).

The three P's: People, Planet and Profit or "three pillars of sustainability" are elements that need to be considered in order to assess the performance of different organizations in a broader context. "Triple Bottom Line" involves the incorporation of economic, environmental and social aspects in the activity of any company or public institution. Sustainable development is the most important goal of humanity, but its complexity involves the implementation of the principles of sustainability in all areas and the mobilization of stakeholder groups, whose activity does not necessarily have a negative impact on the environment or resource consumption. In fact, sustainable development and social responsibility are two concepts that could shape the activity of each public or private entity, because any economic agent must not only limit its negative impact on the environment, but must also intensify its positive effects on the society in which it acts through different tools established to protect the environment, contribute to the development of the local community, promote ethical business behavior, and fight against corruption (Ciutacu et al., 2005; Hysa, 2011; Matei, 2013; García-Sánchez et al., 2019; Pérez et al., 2020; Stancu et al. 2020;Gigauri, 2021; Voica & Stancu, 2021). According to the study of Rehman et al. (2020), government financed research and development can help the private sector investment in research and development, to quickly restore the economy after one crisis happens. Concerning the impact of university research and development intensity on private research and development intensity, European economies are superior in terms of university-industry linkages for boosting technological innovation (Rehman et al., 2020).

At the international level, the concentration of efforts by international organizations such as the United Nations (UN), Organisation for Economic Co-operation and Development (OECD), International Monetary Fund, World Bank are remarkable. These institutions seek to shape the sustainable and responsible behavior of various actors in the world economy and to provide multiple economic operators instruments and tools through which they can meet the objectives of sustainable development (Matei, 2013; Osellame, 2013; Raimi et al., 2015; Hysa et al, 2020; Panait et al., 2020; Raimi and Isiaka, 2021). As a first step, companies and institutions in different fields of activity became aware of the contribution they can make to sustainable development and then identified the means by which they can do so taking into account the specifics of the activity. Internationally, the activity of institutions such as the United Nations, which launched the principles of corporate social responsibility (Global Compact Principles), principles that were later adapted in collaboration with other organizations for other entities or fields of activity (for investments responsible for agriculture, for portfolio investors, etc.).

Given the specific categories of stakeholders, social responsibility and corporate governance actions have specific characteristics in this field. The robustness of the international financial system was called into question after the bank failures that broke out in 2008, and experts realized that risk management is a particularly important issue for financial institutions given the specific categories of stakeholders, namely depositors, life insurance policyholders, pension scheme beneficiaries, savings account holders, portfolio investors. The bankruptcy of a bank or its financial difficulties can generate devastating effects on their customers who have deposited their savings, but also on other institutions given the interdependence relations specific to the sector. The manifestation of systemic risk, the domino effect and the decline of investor confidence are realities demonstrated by the international financial crisis. Analyzes by European authorities on the work of financial institutions have highlighted the misunderstanding of risks by employees involved in the risk management process and their lack of training on risky products (COM / 2010/284).

Since 2001, the European Commission has noted the involvement of financial institutions in promoting CSR among their clients by using social and environmental criteria in evaluating investment projects that are to be financed by bank credit (COM 366/2001). Building on the policies developed by the International Bank for Reconstruction and Development and the International Finance Corporation on assessing investment projects not only financially but also socially and environmentally, the world's largest commercial banks have taken the initiative to launch the Equator Principles in 2001 (the principles were revised in 2006 and 2013). These Principles aim to determine, assess and manage the environmental and social risks related to investment projects financed (minimum USD 50 million) by banks. Adherence to the principles also involves the annual publication of reports presenting synthetic data on projects financed under these principles. In Romania, the analysis of CSR activities carried out by the most important banks in the banking system demonstrates their involvement, especially in the social field, a special importance being given to financial education programs.

Another area of interest of financial institutions is impact investments. Impact investments create a positive social or environmental impact in addition to the profit generated, being oriented on the basis of the economic pyramid (bottom of pyramid) - they aim either to improve the living conditions of poor or vulnerable people or to obtain positive effects on the environment. The financing instruments for impact investments are the traditional ones such as stocks, bonds, but there are also innovative products such as Social Impact Bonds already launched in the UK, even the evaluation of financial performance is done traditionally. Measuring social and environmental impact is much more difficult to achieve, but there are some concerns in this regard: Global Impact Investing Rating System (GIIRS) or Impact Reporting and Investment Standards. Therefore, the vulnerabilities generated by the international financial crisis are managed with the help of new financial instruments and strategies, noting the involvement of international bodies such as the World Bank or national public authorities.

"Because CSR activities are driven by industry-specific factors and the societal environment in which organizations operate" (Singh et al., 2020, p 1), financial institutions are focused on financial inclusion programs. Credit institutions are increasingly interested in attracting people with low incomes, disabilities or from disadvantaged areas to the banking system, both to increase the number of customers and implicitly the profit, but also to offer them the possibility to carry out specific operations in a certain safe and fast way in order to meet current needs but also to achieve long-term goals. In this way, through the financial inclusion of unserved people, the sustainable development of the company is achieved through a better management of financial resources by consumers, by attracting their savings in the formal financial system, by a better substantiation of the decisions they take on long term and expanding the financing possibilities they have for satisfying personal needs or for starting a business (Maei, 2013; Vasile et al., 2021).

Financial inclusion assumes that "all working-age adults have effective access to the following financial services provided by formal institutions: credit, savings (defined broadly to include transaction accounts), payments, insurance, and investments." (GFPI 2016, p. 6). Financial inclusion involves combined efforts by credit institutions, companies by conducting financial education programs for consumers and employees so that they understand and be able to use increasingly complex financial instruments, to substantiate the decisions as accurately as possible, savings and investment plans given the impact of these decisions on personal and professional life. Financial inclusion has acquired new values with the digitalization of economic activities. This new type of financial inclusion (digital financial inclusion) uses digital technology in the design and delivery of specific financial products. The integration of digitalization in the process of financial inclusion has beneficial effects by attracting in the banking circuit people in disadvantaged areas (where there are no banking agencies) or people with disabilities who travel more difficult. The positive externalities generated by financial inclusion are broad and complex, in addition to increasing the degree of social inclusion and the standard of living for consumers, it can be seen as complementary to anti-money laundering and counter-terrorist financing, as it enhances transparency (De Koker & Jentzsch, 2013; Grigorescu et al., 2017; Raimi et al., 2021; Singh et al., 2021; Vasilc ct al., 2021).

Figure 1. Dimensions of financial institutions activities on international market
Source: *authors based on selected scholarly papers*

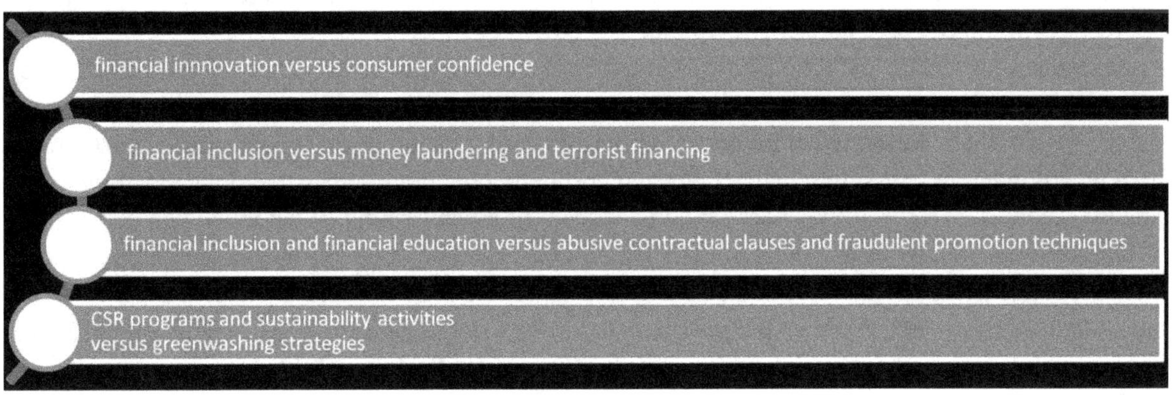

The international financial crisis has demonstrated the negative implications of the financial innovation process (figure no 1). Out of a desire to face the increasingly fierce competition on the financial market, banks and stock exchanges have competed in launching new financial products, in some cases very sophisticated and complex, being difficult to assess even by rating agencies. The financialization of the economy, the loss of the link between the real and the symbolic economy have dramatic consequences that have led to weakening investor confidence in financial market institutions, including financial supervisors who demonstrated, during the crisis, the limits of their competencies (Matei, 2013; Voica, 2017; Andrei et al., 2018; Panait et al., 2018).

Given the specifics of the activity carried out by banks, the first of the research questions posed in this chapter could be formulated.

(RQ1) Are FTNCs involved in promoting the principles of sustainable development and CSR through several instruments?

Greenwashing Strategies in Financial Field

As good as the philosophy of CSR is, both in theory and practice, a number of industry reports and scholarly articles explained that the underlining motives behind the green behaviours of corporate firms are being undermined by unethical and deviant behaviour of companies (Vollero et al., 2016; Volgger & Huang, 2019). In the global financial industry, the phenomenon of green banking (otherwise called social or responsible banking) is being promoted by the banks as climate-sensitive sectors to reduce global carbon emission and in full compliance with their economic, environmental and social obligations (Park & Kim, 2020). Consequently, it has become fashionable for banks and other financial institutions to make false claims in advertising in order to attract green shareholders and socially responsible investors that attaches importance to Environmental, Social and Governance (ESG) initiatives when making investment consideration (Financial Times, 2019; Financial Times, 2021). This unethical practice is called greenwashing or window dressing CSR (Volgger & Huang, 2019). The scholars noted that the green claims of many corporate businesses apparently contradict best practices in CSR and international sustainability guidelines of the (OECD), the United Nations and local sustainability reporting documents. This unethical behaviour is worrisome, hence firms manifesting this tendency have been referred to using several terminologies such as ‘the helpful hypocrites’, ‘the rogue corporation’, ‘the scrooge posing as Mother Teresa’ and hypocritical organisations within the CSR literature (Scheidler et al., 2019; Babu, De Roeck & Raineri, 2020; Glozer and Morsing, 2020).

A number of financial institutions in the United States, and Europe fit the descriptions above, as they deliberately and falsely cover-up their unethical behaviours by flagging social responsibility programmes, and projects in their operating environment. There are several examples of the phenomenon of greenwashing and window-dressing social actions and corporate donations across the globe. To progress in the discussion, it is important to understand the concept of greenwashing. What then is greenwashing?

Different definitions have emerged, but they all converged on the fact that, greenwashing occurs when corporate firms intentionally make false claims regarding their environmental practices and social causes to the customers and other multiple stakeholders in order to gain corporate reputation, influence purchase intentions, and enhance consumer loyalty (Delmas & Burbano, 2011; Lyon & Montgomery, 2015). In other words, greenwashing is a phenomenon where corporate firms flagging CSR have not genuinely matched their claims on environmental performance, product and service quality, social issues and other green causes with actual corporate activities. It is a case of corporate firms telling the truth, telling lies or half-lies about their green behaviours with respect to global warming, clean energy, sustainable production, pollution, deforestation, species extinction, and resource depletion and other and social initiatives (Walker & Wan, 2012; de Jong, Huluba & Beldad, 2020).

The motives behind the practice of greenwashing by the financial institutions include building positive reputation through green posturing, gaining strong environmental credentials, attracting customer loyalty, attracting government attention and patronage, boosting public perception as eco-friendly financial institutions, benefitting from the emerging green products, forging partnership green investors/green savers, gaining tax waivers from tax authorities, and gaining long-term financial performance associated with green banking (Oni, 2016; Financial Times, 2019; Park & Kim, 2020; Financial Times, 2021).

With regards to types of greenwashing, a number of extant literatures categorised greenwashing using fascinating taxonomies. There are good and bad greenwashing; and there are authentic and misleading greenwashing depending on the severity of the greenwashing. For the purpose of guiding the customers and other critical stakeholders in identifying products of corporate firms with misleading environmental claims, TerraChoice (2009) categorised greenwashing into seven sins, namely: (a) sin of the hidden trade-off, (b) sin of no proof, (c) sin of vagueness, (d) sin of worshiping false labels, (e) sin of irrelevance,, (f) sin of lesser of two evils, and (g) sin of fibbing.

According to UL (2021), the sin of the hidden trade-off refers to an environmental claim that is based on a narrow set of product attributes without any link to environmental issues such as zero-tolerance for greenhouse gas emissions and management of chlorine and other bleaching chemicals. The sin of no proof on the other hand refers to spurious environmental claim that is not substantiated by accessible certification and/or verifiable information from reliable third-party authorities such production of facial tissues and toilet tissue products from recycled contents without providing evidence. Moreover, the sin of vagueness describes an environmental claim that has vague meaning and poorly defined essence that could mislead the consumer. Whereas, the sin of worshiping false labels refers to a misleading claim about products through the words, labels and images that convey the impression of third-party endorsement, but proved to be false, deceitful and lacking endorsement. Furthermore, the sin of irrelevance refers to environmental claim that apparently are truthful, but the claims are common knowledge in the public domain, hence unhelpful for the consumers seeking environmentally preferable products. Sin of lesser of two evils however refers to a claim that is factual within a specific product category, but distract the consumer from the greater environmental impacts and risks posed by other whole product category. Finally, the sin of fibbing refers to environmental claims that are apparently false, and is designed to mislead.

The phenomenon of greenwashing is a pervasive corporate practice and come in different forms (Atkinson and Kim, 2014). Truth in Advertising (2020a) identified corporate firms accused of greenwashing to include:

- Blueland (claimed 100% recyclable' cleaning products);
- Quorn Foods (carbon footprint claims);
- Tide purclean ('plant-based' laundry detergent);
- Windex and other household cleaners (claimed non-toxic');
- Volkswagen/BMW/Chevy/Ford/Mercedes-Benz (claimed clean diesel autos);
- GreenPan (claimed green cookware);
- Nestle (claimed sustainably sourced cocoa beans);
- Nest Labs (claimed programmable thermostats);
- Kauai Coffee (compostable coffee pods);
- Charmin Freshmates (claimed flushable wipes);
- Rainforest Alliance (Chiquita bananas, coffee, tea, etc.);
- Reynolds American (Natural American Spirit cigarettes);

- AJM Packaging Corporation (claimed paper plates);
- LEI Electronics (claimed carbon neutral batteries); and

SeaWorld (claimed killer whale shows). The environmental massages and sustainability claims of these fifteen (15) corporate firms sound attractive and had continued to elicit patronage and loyalty from the buying consumers, but the claims were found to be half-truth or complete lies about the environmental performance of their respective products.

Given the scale of the greenwashing phenomenon in the world economy, the authors asked the second research question.

(RQ2) Are FTNCs involved in greenwashing actions to improve image among stakeholders?

This article aims to fill the gap detected in the literature because the identified scientific papers focus either on the involvement of FTNCs in promoting sustainable development or on building a false positive image among stakeholders although the financial institution abuses the asymmetry of market information and poor financial education of consumers especially in developing countries.

METHODS AND ANALYSIS

The unit of analysis of this study is organisational level with a cross-continental focus. Furthermore, the study adopts a qualitative research method, while relying on a desk research technique. To gain deeper insight into the CSR activities of FTNCs, the authors sourced the required data from scholarly articles, SDD - GRI Database, reports on greenwashing, and other online resources. The data gathered from diverse sources were objectively appraised and analysed using a critical review of the literature (CLR). For objectivity in the selection of literature, the study searched for relevant papers on ScienceDirect, EbscoHost, JSTOR, Google Scholar, and ProQuest. From over 200 scholarly articles, texts, and working papers previewed, the authors selected 80 articles using the purposive sampling technique. The research methodology follows a three-stage protocol for selection and analysis of articles.

Stage 1: Data sourcing - This stage focuses on sourcing relevant articles that better provides clarity and direction to the intent of the research. The articles selected through the purposive sampling technique followed the selection criteria below:

- The selected articles discuss corporate social responsibility (CSR) in relation to sustainable development;
- The articles chosen focus on CSR activities of financial institutions particularly the Financial Transnational Corporations (FTNCs);
- The articles discuss the phenomenon of greenwashing of the FTNCs;
- The articles selected explicate the principles and measures for mitigating greenwashing of the FTNCs;
- The selected article cut across major continents.

Stage 2: Data development and conversion – At this stage, the insightful information from the selected articles were compiled and integrated t in readiness for analysis data.

Stage 3: Data analysis – The information extracted on CSR programs promoting sustainable development and greenwashing were appraised and analysed using a content analysis. This methodological approach is supported by (Jepson, 2009; Williams & Shepherd, 2017).

The selected articles focused on two major aspects that guide the activity of FTNCs and that demonstrate the chameleon behavior of these entities (1) involvement in promoting sustainable development by using specific tools such as CSR programs and (2) greenwashing actions through which they try to mask unethical behavior towards various stakeholders by practicing abusive contractual clauses or using deceptive promotional techniques So, the selected scholarly works were content analysed and contextualised to answer the question of whether the CSR programs of the financial institutions are targeted at development-oriented issues or just greenwashing palliatives (figure 2).

Figure 2. The main areas of interest for published articles
Source: *Authors based on selected scholarly papers*

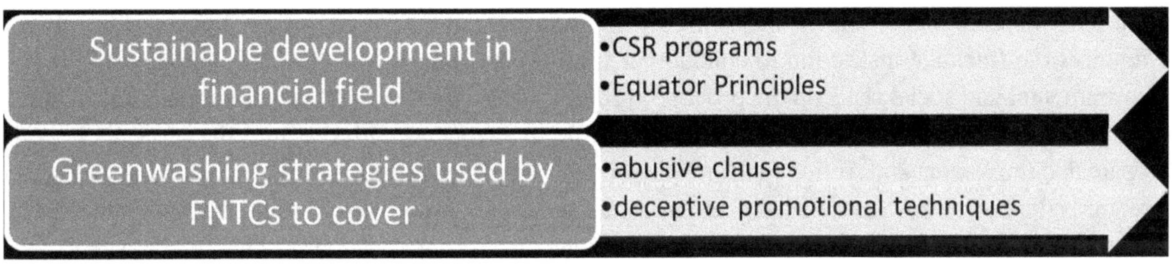

CSR AND SUSTAINABLE DEVELOPMENT IN THE FINANCIAL SECTOR: AN APPLICATION OF THE EQUATOR PRINCIPLES

This section utilizes the Equator Principles (EPs) to illustrate the role of Corporate Social Responsibility (CSR) in the financial sector. Specifically, we examine the motive and the steps that financial institutions are taking to implement and comply with these principles. The EPs represent a risk management framework to determine, assess, and manage environmental and social risks projects within the institution. Set in motion on June 2003 by 10 international banks, these principles signify the commitment of these banks to ensuring that the projects they support are developed in a socially responsible manner (Scholtens & Dam, 2007). At the time of writing, 116 financial institutions in 37 countries have adopted the EPs[4]. The institutions within the countries are called Equator Principles Financial Institutions (EPFIs). A list of these institutions is presented in Appendix 1 (Equator Principles, 2020).

CSR is important in financial institutions mainly because of their role in capital allocation and infrastructural development (Osellame, 2013). There are however concerns that banks may not be sincere when it comes to sustainability initiatives as financial institutions are generally profit-oriented enterprises. For instance, the use of "Equator Principle Financial Institution" title on websites and in press releases confer reputational advantages to such institutions (Robertson, 2011). Researchers have also expressed concerns that financial institutions are free-riding to gain reputational benefits and not necessarily complying with EPs requirements (Amalric, 2005; Osellame, 2013). Another concern is that banks have been noted to evaluate the cost versus benefit of implementing EPs, and will likely adopt EPs if the expected benefits outweigh the identified costs (Amalric, 2005), therefore, adoption of EPs may not be entirely altruistic. In view of the importance of CSR to financial institutions and the wide adoption of EPs, this section provides insight into the importance of CSR within the framework of EPs. The findings of which will enlighten the public and perhaps encourage more financial institutions to adopt these principles.

The EPs represent a voluntary code of conduct in which financial institutions that sign on agrees to comply with the stipulated requirements in financing projects. The principles serve as a benchmark for financial institutions to manage social and environmental issues encountered in project financing (Lawrence, 2009). EPs apply to five financial products when supporting a new project with total capital cost of $10 million or more.[5] These products are Project Finance Advisory Service, Project Finance, Project-Related Corporate Loans, Bridge Loans, and Project-Related Refinance, and Project-Related Acquisition Finance. The latest version, EP4 was released on November 18, 2019 and effective October 1, 2020.

The EPs consists of 10 guiding principles. *Principle 1* requires each EPFI to categorize each project. The project types are classified according to three categories. Category A represents projects with potentially significant adverse social and environmental impacts that are diverse, irreversible, or unprecedented. Category B represents projects with potentially limited adverse social or environmental impacts while Category C represents projects with minimal social or environmental impacts. *Principle 2* requires the financial institution to conduct an appropriate process to address the impact of relevant environmental and social risks on the proposed project. *Principle 3* requires the assessment to consider the relevant host country's laws and regulations while *principles 4 and 5* require the project's host to prepare and implement and Action Plan and demonstrate an effective stakeholder engagement. *Principle 6* requires the borrower to establish a grievance mechanism to ensure consultation, disclosure and community engagement. *Principle 7* provide the guidelines for an independent review of the assessment process while *principle 8* stipulates the inclusion of the EP's requirements into loan covenants. *Principle 9* requires monitoring and reporting of the EPFIs and *principle 10* requires the member bank to report its EPs implementation on an annual basis (Equator Principles, 2020).

Theories of CSR propose that firms will engage in CSR if they anticipate benefits from such activities (Bagnoli & Watts, 2003; McWilliams & Siegel, 2001). From an economic perspective, banks will likely adopt EPs if the perceived benefits outweigh the expected costs. There are however concerns that EPs may not represent a CSR model due to free rider issues and adverse selection (Schepers, 2011). Scholars have suggested that the ability of EPs to serve in a CSR capacity depends on how they are enforced and the compliance by the reporting entities. The legitimacy of EPs has also been questioned due to gaps in the governance structure and the lack of transparency and accountability which has left the processes and content to the discretion of the banks (Schepers, 2011). Irrespective of the shortcomings, EPs represent a major step by financial institutions to be socially responsible, and compliance to the principles will greatly benefit the society.

Similar to other industry self-regulations, there are concerns about EPFIs underlying motive considering financial institutions are mostly profit-oriented. Industry self-regulation is prone to problems such as implementation and compliance with the regulation (Garvin, 1983). Studies on voluntary disclosure suggests there are incentives for firms to provide credible disclosures. Credible voluntary disclosure has been linked to reduced cost of capital (Botosan, 2006) and improved reputation (Friedman & Miles, 2001). In line with the belief that sustainability initiatives add value, reduces risk, and results in improved business outcome (Lawrence, 2009), the authors also expect that EPs will be relevant and add value to adopting financial institutions. Scholars have suggested that signing into the principles will enhance firm value (Amalric, 2005), strengthen corporate reputation (Wright & Rwabizambuga, 2006) or signal their superior environmental and social responsibility (Scholtens & Dam, 2007). Amalric (2005) formulated three hypotheses to explain the motives behind EPs. The first hypothesis suggests that EPs create a level playing ground in the industry for firms with different reputation risks. A reason for this is that more exposed banks often have to adopt higher standards of corporate behaviour to manage their reputation risks. The second motive suggests these principles serve as a screen for projects with social and environmental risks while the third hypothesis predicts EPs adopt the principles to counter critics that large development projects do not support sustainable development.

Other studies on motives of EPs adoption also aligns with Amalric (2005). Contreras, Bos, & Kleimeier (2019) highlight the role of peer pressure by the finding that banks collaborate (i.e. syndicated loans) with adopters rather than non-adopters. The authors also find that external pressure through public campaign influence the adoption of EPs. These findings imply that adoption of EPs by banks will confer legitimacy benefits to the banks which will enhance the reputation and credibility of the principles (Contreras et al., 2019). While some studies find benefits such as improved CSR rating (Scholtens & Dam, 2007), there is no evidence of improved financial performance (Chih, Chih, & Chen, 2010; Scholtens & Dam, 2007) suggesting that profit maximizing may not be an incentive to adopt EPs. This is corroborated by the findings that EPFIs have higher liquidity compared to non-EPFIs possibly because of the higher risks that accompanies EPs adoption and the additional monitoring and regulations (Chen, Huang, & Lin, 2018). Scholtens & Dam (2007) also support the notion that adopting EPs is not mere window dressing because the EPFIs incur real costs, and that the benefits (reduced risk) of signing up outweighs the cost. Scholars also find that EPFIs have changed their organizational structure, policies and procedures to reflect their adoptions of the EPs (Meyerstein, 2015).

The major concerns with EPs are the lack of transparency of the disclosures possibly to protect the clients (Kass & McCarroll, 2006) and the lack of compliance with the codes due to the absence of a regulatory body (Schepers, 2011). Although EPs have recorded success in aligning banks' strategic motivation, the lack of transparency results in the inability to evaluate the environmental benefit of the EPs (Macve & Chen, 2010). The author's opinion is that these concerns are not limited to EPs and are not significant enough to negate the benefits that having an industry-wide CSR framework confers on the society.

Therefore, the authors of this study agree with scholars that there are some genuine motives behind the adoption of EPs, the major concerns relate to transparency and monitoring of compliance to the reporting system in order to provide more transparency and credibility to the framework. Thus, focusing specifically on the Equator principles, findings support the first research objective that financial institutions (of which FTNCs are inclusive) promote principles of sustainable development and CSR through specific instruments.

FINANCIAL INDUSTRY AND GREENWASHING: DEFINITIONS, MOTIVES, CASES AND EFFECTS

With specific reference to the financial industry, the incidences of greenwashing are tabulated and discussed analytically in Tables 1 to 5 hereunder.

Table 1. Sunmark Credit Union and allegation of falsehood in overdraft fee

Case Name (Date)	Dadm Kidz Cuts II LLC et al v. Sunmark Credit Union 21-cv-178, N.D.N.Y. (Feb. 2021)
Product/Service	Financial services in checking account
Greenwashing Allegations	The litigant filed a class-action lawsuit against Sunmark Credit Union for falsely misrepresenting when customers will be charged overdraft fees. There was no clarity on the financial services for customer using checking accounts.
Status of the case in court	The case is pending.

Source: Truth in Advertising (2020b)

Table 2. Patelco Credit Union's overdraft fees

Case Name (Date)	Wicks et al v. Patelco Credit Union 20-cv-4586, N.D. Cal. (May 2020)
Product/Service	Checking accounts
Greenwashing Allegations	The litigant filed a class-action lawsuit against PATELCO Credit Union of falsehood and misrepresenting when consumers will be charged overdraft fees
Status of the case in court	The case is pending in the state court

Source: Truth in Advertising (2020c)

Table 3. Overdraft fees at Pathfinder Bank

Case Name (Date)	Claflin et al v. Pathfinder Bank 21-cv-380, N.D.N.Y. (Apr. 2021)
Product/Service	Checking accounts
Greenwashing Allegations	The litigant filed a class-action lawsuit against Pathfinder Bank of falsehood and misrepresenting when consumers will be charged overdraft fees
Status of the case in court	The case is pending in the state court

Source: Truth in Advertising (2020d)

Table 4. HSBC Suspected of Greenwashing (Europe)

Resolution (Date)	117 individual filers including Amundi, Sarasin & Partners and Man Group, 15 major financial institutions are just some of the major investors supported a resolution against HSBC
Product/Service	Financial services and investment banking
Greenwashing Allegations	HSBC claimed to be environmentally friendly, committed to strategic transition to a low-carbon economy and poised to investing $100 billion into sustainable finance by 2030. This was observed to be a misinformation, and the investors consequently passed a resolution against HSBC accusing it of funding fossil fuel and coal projects to the tune of $1.8 billion despite its pledge to go carbon neutral.

Source: Williams (2021)

Table 5. Bank of England accused of Greenwashing (United Kingdom)

Resolution (Date)	In 2020, Environmental activists, Protesters, Fossil divestment groups such Fossil Free London and Extinction Rebellion Hammersmith & Fulham passed a resolution and queried hypocrisy of Governor Andrew Bailey.
Product/Service	Regulators, Financial services and investment banking
Greenwashing Allegations	Bank of England claimed to be committed to a low-carbon economy and poised to preserving the planet for generations to come. But, the banks provided bailouts to the UK's companies affected by Covid-19 under the Corporate Covid Financing Facility (CCFF), without social or environmental strings attached to the support interventions. This was observed to be a misinformation, and the investors consequently passed a resolution against HSBC accusing it of funding fossil fuel and coal projects to the tune of $1.8 billion despite its pledge to go carbon neutral.

Source: Dewhirst (2020)

The negative effects of greenwashing are colossal, and call for serious remediation actions. The literature identified a number of negative effects of greenwashing on the shoppers/consumers, investors and other critical stakeholders in the society. The effects include proliferation of widespread misinformation on environmental compliance in sustainability reports and CSR reporting; false labeling of products has encouraged the consumption of trans-fat, sugar, sodium and saturated fat by the consumers; fake advertisements have encouraged proliferation of unsustainable products; false sustainability reports deceived investors into investing in harmful corporations posing as eco-friendly entities, underserved awards have been given by the rating agencies to underserved corporations; rogue corporations deploying greenwashing antics have gained underserved customer loyalty for activities that are environmentally-unfriendly, and irresponsible corporations have achieved financial performance through misleading green communication (Gatti, Seele & Rademacher, 2019). For the companies flagging CSR as greenwashing, the effect includes litigations, financial costs of litigations, reputational risks, sanctions from courts, divestment by investors, damaging protests by the environmentalists, loss of customer loyalty and negative financial bottom-line (Oni, 2016; Financial Times, 2021).

Greenwashing deserves punishments and sanctions because the alleged corporate firms have reaped the benefits of a green positioning flagging environmental responsiveness without behaving accordingly (de Jong et al., 2020). Unfortunately, there is no punishment or legal sanction on corporate firms for using greenwashing to mislead, deceive and misinform the public. Considering the damage that the unethical practice of greenwashing has done to the wellness of the people and the society, a number of Non-Governmental Organizations (NGO) such as Friends of the Earth Sweden and Truth in Advertising and others have become watchdogs and underdogs checkmating the incidence of greenwashing. The NGOs impose sanctions, court actions, boycotts and reputational damage to corporate firms found to be culpable (Dzafic & Petersson, 2016; Truth in Advertising, 2020a). Annually, the Friends of the Earth Sweden present the Swedish Greenwash Award to notorious corporate firm that spread false environmental claims and misleading messages on green behaviours (Dzafic & Petersson, 2016).

With increased awareness by the investors and the consumers, corporate firms found to have greenwashed have suffered financial losses through withdrawal of investments by socially responsible investors/shareholders, sanctions by governments/regulatory authorities, withholding of social licenses by the host community, switching of loyalty by the consumers and fines by the courts (TerraChoice, 2009; Truth in Advertising, 2020a).

To curb this dangerous development, the European Union introduced new rules with wider series of green finance regulations, broad-based classification of investment products and tough disclosure requirements. These new rules were promulgated to prevent banks and from engaging in greenwashing (Financial Times, 2021). Similarly, the financial sector regulatory authorities in Nigeria, China, and Bangladesh have respectively developed mandatory guidelines called Nigerian Sustainability Banking Principles (NSBP), Chinese Green Credit Guidelines (GCG) and the Bangladeshi Environmental Risk Management (ERM) that guide the financial institution as well as curb excesses in annual, sustainability and corporate social responsibility reports (Oni, 2016).

An important policy prescription for stemming the tide of greenwashing is for the governments and its regulatory authorities, environmental activists, socially responsible investors and green consumers in all countries to deploy a combination of voluntary and mandatory compliance regulations that would guide, limits and control the manners of communicating and disclosures of CSR philosophies, accomplishments, programmes and interventions in order to avert the growing trends of greenwashing that overstretches CSR messages and claims.

FUTURE RESEARCH DIRECTIONS

Given the complexity of the phenomenon analyzed and the challenges posed by the COVID crisis, the future studies will focus on the impact of digitalization on the phenomenon of financial inclusion and financial education. Digitization in the financial sector increases the degree of financial inclusion, especially in countries where the physical infrastructure is less developed. The development of digital infrastructure must be accompanied by adequate financial education programs so that consumers know how to properly use mobile devices, electronic platforms but also understand the risks and limitations of new products and services. Financial institutions need to take steps to better manage risk. Risk management is a key element of corporate governance and can restore investor confidence in the financial system. Therefore, a key role is played by the board of directors and employees, which must not only aim to increase the financial performance of institutions in the short term, but also to increase their stability and investor confidence in these entities in the long term. Therefore, consolidating a healthy culture in terms of risk management (COM / 2010/284; Noja et al., 2021) and increasing the social involvement of financial institutions are essential. Restoring investor confidence in the financial system is a difficult process that requires sustained action by financial institutions: Credit institutions need to run financial education programs for their clients so that they understand the risks involved in operating in the financial market. Financial education programs are essential because the lack of financial culture has serious long-term consequences on the well-being of citizens and, implicitly, of society in general. The actions of financial institutions are also supported by public authorities, in some countries, there are national financial education strategies, and public institutions such as the ministry of education, the national bank or the financial supervisory authority carry out their own financial education programs. However, studies conducted worldwide have shown the limitations of financial education programs (Klapper, Lusardi, Panos, 2012; Ene & Panait, 2017; Iacovoiu, 2018). Financial knowledge is not enough to make good decisions, other aspects such as the attitude, opinions or values that each investor believes in are important. Investors' confidence in their knowledge and skills are essential for starting financial market transactions. For this reason, financial education must also be done from the perspective of behavioral finance. Financial education must not only focus on the investment process, but also on the saving process. On the one hand, financial education programs must aim to increase the savings rate and, on the other hand, they must aim to attract low-income and vulnerable people into the financial system (Klapper, Lusardi, Panos, 2012; Ene & Panait, 2017). In summary, future research can examine the effect of digitalization on financial inclusion and education focusing specifically on financial education programs, strategies, risk management and financial education for the general public.

CONCLUSIONS

This chapter sets out to investigates the genuineness or greenwashing of CSR activities of the FTNCs, and how they promote the principles of sustainable development by adopting green standards such as the Equator principles (EPs). At the end of the discourse, it was found that there is substantial evidence that credit institutions are actively involved in CSR activities and community development initiatives aimed at given back to the society. Similarly, it was found the credit institution further sustainable development through financial inclusion and adoption of the Equator Principles for determining, assessing and managing environmental and social projects. At the same time, there are incidence of greenwashing activities such as widespread misinformation on environmental compliance in sustainability reports and CSR reporting, which is designed to cover-up unethical behaviours of these credit institutions.

In view of the forgoing, it is important that the financial industry takes appropriate steps to promote transparency and accountability. The government through its regulatory authorities and professional associations should set up enforcement mechanisms to significantly reduce if not eliminate the practice of greenwashing. However, the activity of credit institutions is under the sign of maximizing profit, which is why employees have targets for the sale of products and services. Moreover, in some cases, the involvement of financial institutions in CSR programs is only a method of window dressing that aims only to improve the image among stakeholders. The practice has also demonstrated the promotion of sustainable development principles by financial institutions is only for declarative purposes, some of them being accused of greenwashing. Therefore, the chameleon behavior of financial institutions is a reality found in most countries.

The management of banks must set realistic targets for employees regarding the sale of financial products so that they have an ethical attitude towards the customers and explain to them all the risks that financial products entail. Financial institutions must also target vulnerable groups (women, people with disabilities, those without a financial history, those with low incomes, those in rural areas) and promote new instruments against poverty such as microfinance and impact investments. Therefore, financial inclusion must guide the actions of credit institutions, and their social inclusion strategies must be correlated with those of CSR and business strategy.

The international crises of the last decades have once again demonstrated the importance of the financial market and the activity of the actors that operate on this market: banks, stock exchanges, brokerage firms, insurance companies, rating agencies, investors, consulting firms. Given the multiple scandals that have affected the financial sector, the management of financial institutions must not only promote the Codes of Ethics among employees, but should exercise more rigorous management of the risks they can generate, to promote a culture of risk and have a healthy remuneration policy. The launch of new financial products should only be done after a rigorous risk assessment. Therefore, the creativity of employees must be manifested in a limited framework given the multiple social implications of using toxic financial products Financial institutions need to set realistic targets for their employees who are no longer under pressure to sell as many products as possible to customers. This practice can generate unethical behavior on the part of employees who, out of the desire to achieve the set objectives, do not explain to customers the risks posed by the traded financial products. So, the social responsibility of financial institutions must be manifested directly in the relationship with its customers, by promoting an ethical behavior of employees, and not only through CSR programs. The international financial crisis has led to a loss of consumer confidence in the financial system. Internationally, financial institutions seek to regain the trust of customers and investors by promoting principles of ethics, transparency and social responsibility. Financial inclusion programs run by banks or even national financial inclusion strategies aim at the access of certain categories of people such as women or the rural population to modern financial services. In order to forestall the phenomenon of greenwashing from eroding the good intentions of CSR programs in the operating environment, the disadvantaged stakeholders including customers of FTNCs have to be systematically empowered through financial inclusion, financial education and digitalisation integration. Thus, well informed stakeholders would be able to identify the difference between genuine CSR and greenwashing palliatives.

This study contributes to the research on sustainability and highlights the need for a framework to report sustainability initiates. Findings from this study are important to the managers interested in improving the CSR activities within their organizations and to regulatory bodies that are interested in compliance of FTNCs to sustainability reporting frameworks. The suggested areas for future research provides opportunities for more informed research on CSR activities of FTNCs.

ACKNOWLEDGMENT

The author is grateful to the KPMG Foundation at University of Regina, Saskatchewan, Canada.

REFERENCES

Amalric, F. (2005). *The Equator Principles: a step towards sustainability.* Center for Corporate Responsibility and Sustainability. Working Paper No. 01/05.

Andrei, J. V., Panait, M., & Voica, C. (2018). Challenges and approaches for the corporate social responsibility and human resource management in the financial sector. *Economics. Management and Financial Markets, 13*(3), 415–431.

Atkinson, L., & Kim, Y. (2014). I drink it anyway and I know I shouldn't": Understanding green consumers' positive evaluations of norm-violating non-green products and misleading green advertising. *Environmental Communication, 9*(1), 37–57. doi:10.1080/17524032.2014.932817

Babu, N., De Roeck, K., & Raineri, N. (2020). Hypocritical organizations: Implications for employee social responsibility. *Journal of Business Research, 114*, 376–384. doi:10.1016/j.jbusres.2019.07.034

Bagnoli, M., & Watts, S. G. (2003). Selling to socially responsible consumers: Competition and the private provision of public goods. *Journal of Economics & Management Strategy, 12*(3), 419–445. doi:10.1162/105864003322309536

Bank Act. (1991). *Bank Canada Act.* Retrieved from https://laws-lois.justice.gc.ca/PDF/B-1.01.pdf

Botosan, C. A. (2006). Disclosure and the cost of capital: What do we know? *Accounting and Business Research, 36*(sup1), 31-40.

Chen, N., Huang, H., & Lin, C. (2018). Equator principles and bank liquidity. *International Review of Economics & Finance, 55*, 185–202. doi:10.1016/j.iref.2017.07.020

Chih, H., Chih, H., & Chen, T. (2010). On the determinants of corporate social responsibility: International evidence on the financial industry. *Journal of Business Ethics, 93*(1), 115–135. doi:10.100710551-009-0186-x

Ciutacu, C., Chivu, L., & Preda, D. (2005). Company's Social Responsibility-A Challenge for Contemporary World. *Romanian Journal of Economics, 20*(1 (29)), 79–94.

COM 2001/366, GREEN PAPER Promoting a European framework for Corporate Social Responsibility, retrieved Microsoft Word - en 366-5.doc (europa.eu)

COM 2010/284, Green Paper - Corporate governance in financial institutions and remuneration policies, retrieved COM(2010)284/F1 - EN (europa.eu)

Contreras, G., Bos, J. W. B., & Kleimeier, S. (2019). Self-regulation in sustainable finance: The adoption of the equator principles. *World Development, 122*, 306–324. doi:10.1016/j.worlddev.2019.05.030

de Jong, M. D., Huluba, G., & Beldad, A. D. (2020). Different shades of greenwashing: Consumers' reactions to environmental lies, half-lies, and organizations taking credit for following legal obligations. *Journal of Business and Technical Communication, 34*(1), 38–76. doi:10.1177/1050651919874105

De Koker, L., & Jentzsch, N. (2013). Financial inclusion and financial integrity: Aligned incentives? *World Development, 44*, 267–280. doi:10.1016/j.worlddev.2012.11.002

Delmas, M. A., & Burbano, V. C. (2011). The drivers of greenwashing. *California Management Review, 54*(1), 64–87. doi:10.1525/cmr.2011.54.1.64

Dewhirst, H. (2020). *Bank of England accused of Greenwashing*. Positive Money Europe. https://positivemoney.org/2020/06/bank-of-england-governor-accused-of-greenwashing/

Dzafic, J., & Petersson, A. (2016). *Greenwashing in CSR reports-A case study of two entities.* http://www.diva-portal.org/smash/get/diva2:932386/FULLTEXT01.pdf

Ene, C., & Panait, M. (2017). The financial education-Part of corporate social responsibility for employees and customers. *Revista Romana de Economie, 44*(1), 145–154.

Financial Times. (2019). *Bankers push back against 'greenwashing' criticism.* https://www.ft.com/content/c4cefa40-f4ab-11e9-b018-3ef8794b17c6

Financial Times. (2021). *Greenwashing in finance: Europe's push to police ESG investing.* https://www.ft.com/content/74888921-368d-42e1-91cd-c3c8ce64a05e

Friedman, A. L., & Miles, S. (2001). Socially responsible investment and corporate social and environmental reporting in the UK: An exploratory study. *The British Accounting Review*, *33*(4), 523–548. doi:10.1006/bare.2001.0172

García-Sánchez, I. M., Gómez-Miranda, M. E., David, F., & Rodríguez-Ariza, L. (2019). Analyst coverage and forecast accuracy when CSR reports improve stakeholder engagement: The Global Reporting Initiative-International Finance Corporation disclosure strategy. *Corporate Social Responsibility and Environmental Management*, *26*(6), 1392–1406.

Garvin, D. A. (1983). Can industry self-regulation work? *California Management Review*, *25*(4), 37–52. doi:10.2307/41165031

Gatti, L., Seele, P., & Rademacher, L. (2019). Grey zone in–greenwash out. A review of greenwashing research and implications for the voluntary-mandatory transition of CSR. *International Journal of Corporate Social Responsibility*, *4*(1), 1–15. doi:10.118640991-019-0044-9

Gigauri, I. (2021). Corporate Social Responsibility and COVID-19 Pandemic Crisis: Evidence From Georgia. *International Journal of Sustainable Entrepreneurship and Corporate Social Responsibility*, *6*(1), 30–47. doi:10.4018/IJSECSR.2021010103

Gigauri, I., & Damenia, N. (2020). Cooperation between Social Entrepreneurs and Government to Develop Solutions to Social Problems. *Business and Economic Review*, *10*(3), 116–136. doi:10.5296/ber.v10i3.17383

Glozer, S., & Morsing, M. (2020). Helpful hypocrisy? Investigating 'double-talk' and irony in CSR marketing communications. *Journal of Business Research*, *114*, 363–375. doi:10.1016/j.jbusres.2019.08.048

Grigorescu, A., Cerchia, A. E., Oachesu, M. M., & Udroiu, F. (2017). Enhancing Internet Banking-Solutions for Customer Relationship Management. *Saudi J. Bus. Manag. Stud*, *2*, 38–43.

Hysa, E. (2011). Corruption and human development correlation in Western Balkan countries. *Euro Economica*, *30*(04), 148–157.

Hysa, E., Kruja, A., Rehman, N. U., & Laurenti, R. (2020). Circular economy innovation and environmental sustainability impact on economic growth: An integrated model for sustainable development. *Sustainability*, *12*(12), 4831. doi:10.3390u12124831

Iacovoiu, V. B. (2018). An Empirical Analysis of Some Factors Influencing Financial Literacy. *Economic Insights - Trends and Challenges*, *70*(2).

Jepson, D. (2009). Studying leadership at cross-country level: A critical analysis. *Leadership*, *5*(1), 61–80. doi:10.1177/1742715008098310

Kass, S. L., & McCarroll, J. M. (2006). The revised equator principles. *New York Law Journal*, *236*(44), 3.

Klapper, L. F., Lusardi, A., & Panos, G. A. (2012). Financial literacy and the financial crisis (No. w17930). National Bureau of Economic Research.

Koomson, I., Villano, R. A., & Hadley, D. (2020). Intensifying financial inclusion through the provision of financial literacy training: A gendered perspective. *Applied Economics*, *52*(4), 375–387. doi:10.1080/00036846.2019.1645943

Lawrence, P. (2009). Equator Principles: Or how I learned to stop worrying and love sustainability. *Impact Assessment and Project Appraisal*, *27*(1), 1, 3–6. doi:10.3152/146155109X428996

Lyon, T. P., & Montgomery, A. W. (2015). The means and end of greenwash. *Organization & Environment*, *28*(2), 223–249. doi:10.1177/1086026615575332

Macve, R., & Chen, X. (2010). The "equator principles": A success for voluntary codes? *Accounting, Auditing, &. Accountability*, *23*(7), 890–919. doi:10.1108/09513571011080171

Matei, M. (2013). *Responsabilitatea socială a corporaţiilor şi instituţiilor şi dezvoltarea durabilă a României*. Expert Publishing House.

McWilliams, A., & Siegel, D. (2001). Corporate social responsibility: A theory of the firm perspective. *Academy of Management Review*, *26*(1), 117–127.

Meyerstein, A. (2015). Are the equator principles greenwash or game changers? effectiveness, transparency and future challenges. In S. O. Idowu, & R. Schmidpeter (Eds.), Responsible investment banking: Risk management frameworks, sustainable financial innovation and softlaw standards (pp. 267-284). Springer International Publishing. doi:10.1007/978-3-319-10311-2_16

Morina, F., Hysa, E., Ergün, U., Panait, M., & Voica, M. C. (2020). The Effect of Exchange Rate Volatility on Economic Growth: Case of the CEE Countries. *Journal of Risk and Financial Management*, *13*(8), 177.

Mukherjee, A. (2012). CSR Holiday for Banks in India: Dilemma between CSR and Financial Inclusion. *Interdisciplinary Journal of Management and Behavioural Sciences*, *1*(4).

Noja, G. G., Thalassinos, E., Cristea, M., & Grecu, I. M. (2021). The Interplay between Board Characteristics, Financial Performance, and Risk Management Disclosure in the Financial Services Sector: New Empirical Evidence from Europe. *Journal of Risk and Financial Management*, *14*(2), 79.

Oni, O. (2016). *Banking on financial sector sustainability regulations*. University of Waterloo. https://uwspace.uwaterloo.ca/bitstream/handle/10012/10367/Oni_Olawuwo.pdf?isAllowed=y&sequence=1

Osellame, I. (2013). *Corporate social responsibility in Canadian banking a case study on the equator principles*. Academic Press.

Palazzo, M., Vollero, A., & Siano, A. (2020). From strategic corporate social responsibility to value creation: An analysis of corporate website communication in the banking sector. *International Journal of Bank Marketing*.

Panait, M., Radulescu, I., & Brezoi, A. (2020). Financial Markets–Under the Sign of CSR. Some Evidences Regarding Financial Education. *LUMEN Proceedings, 11*, 96-106.

Panait, M., Radulescu, I., & Voica, C. (2018). Financial Inclusion and Vulnerabilities Generated by the International Crisis. *Romanian Journal of Economics*, *47*(2), 71–81.

Park, H., & Kim, J. D. (2020). Transition towards green banking: Role of financial regulators and financial institutions. *Asian Journal of Sustainability and Social Responsibility*, *5*(1), 1–25.

Pérez, A., & del Bosque, I. R. (2009). The social role of financial companies as a determinant of consumer behaviour. *International Journal of Bank Marketing*, *27*(6), 467–485.

Pérez, A., & del Bosque, I. R. (2015). Customer values and CSR image in the banking industry. *Journal of Financial Services Marketing*, *20*(1), 46–61.

Pérez, A., López-Gutiérrez, C., García-De Los Salmones, M. D. M., & San-Martín, P. (2020). Stakeholder salience, positive CSR news and the market value of banks. Spanish Journal of Finance and Accounting. *Revista Española de Financiación y Contabilidad*, *49*(4), 483–502.

Equator Principles. (2020). *The Equator Principles July 2020 A financial industry benchmark for determining, assessing and managing environmental and social risk in projects.* Author.

Raimi, L. (2020). What are the Motivations for Undertaking Corporate Social Responsibility (CSR) Practices in the Middle-East? A Critical Literature Review. *Global Journal of Human Social Science: E-Economics, 20*(6), 58–64.

Raimi, L., Adelopo, A. O., & Yusuf, H. (2019) Corporate social responsibility and sustainable management of solid wastes and effluents in Lagos megacity Nigeria. *Social Responsibility Journal*. Doi.org/10.1108/SRJ-09-2018-0239

Raimi, L., Akhuemonkhan, I. A., & Ogunjirin, O. D. (2015). Corporate social responsibility and entrepreneurship (CSRE): Antidotes to poverty, insecurity and underdevelopment in Nigeria. *Social Responsibility Journal*, *11*(1), 56–81.

Raimi, L., & Isiaka, A. S. (2021). Regional Review: Africa: CSR in Sub-Saharan Africa: Emerging Trends, Perspectives, and Debates. In D. Crowther & S. Seifi (Eds.), *The Palgrave Handbook of Corporate Social Responsibility*. Palgrave Macmillan.

Raimi, L., Panait, M., & Hysa, E. (2021). Financial Inclusion in ASEAN Countries–A Gender Gap Perspective and Policy Prescriptions. *LUMEN Proceedings, 15*, 38-55.

Rehman, N. U., Hysa, E., & Mao, X. (2020). Does public R&D complement or crowd-out private R&D in pre and post economic crisis of 2008? *Journal of Applied Econometrics*, *23*(1), 349–371.

Robertson, B. (2011). *Corporate social responsibility programs for emissions-trading risk management in Canadian banks*. Springer Berlin Heidelberg. doi:10.1007/978-3-642-20592-7

Ruiz, A. P. (2018). Reacciones hacia la publicidad de responsabilidad social corporativa de entidades financieras. *UCJC Business & Society Review*, (60), 18-53.

Scheidler, S., Edinger-Schons, L. M., Spanjol, J., & Wieseke, J. (2019). Scrooge posing as MotherTeresa: How hypocritical social responsibility strategies hurt employees and firms. *Journal of Business Ethics*, *157*(2), 339–358.

Schepers, D. H. (2011). The equator principles: A promise in progress? *Corporate Governance (Bradford)*, *11*(1), 90–106. doi:10.1108/14720701111108871

Scholtens, B., & Dam, L. (2007). Banking on the equator. are banks that adopted the equator principles different from non-adopters? *World Development*, *35*(8), 1307–1328. doi:10.1016/j.worlddev.2006.10.013

Siano, A., Raimi, L., Palazzo, M., & Panait, M. C. (2020). Mobile Banking: An Innovative Solution for Increasing Financial Inclusion in Sub-Saharan African Countries. Evidence from Nigeria. *Sustainability*, *12*(23), 10130.

Singh, K., Misra, M., & Yadav, J. (2021). Corporate social responsibility and financial inclusion: Evaluating the moderating effect of income. *Managerial and Decision Economics*.

Stancu, A., Panait, M., Raimi, L., & Palazzo, M. (2020, December). Financial Consumers–Promoters of Sustainable Development? Evidences from Europe and Central Asia. In *Proceedings of the International Conference on Economics and Social Sciences* (pp. 841-850). Sciendo.

TerraChoice. (2009) *The six sins of greenwashing™: A study of environmental claims in North American consumer markets*. http://sinsofgreenwashing.com/index6b90.pdf

Truth in Advertising. (2020a). *Earth Day 2020: Companies accused of Greenwashing*. https://www.truthinadvertising.org/six-companies-accused-greenwashing/

Truth in Advertising. (2020b). *Overdraft fee at Sunmark Credit Union*. https://www.truthinadvertising.org/overdraft-fees-at-sunmark-credit-union/

Truth in Advertising. (2020c). *Patelco Credit Union's Overdraft fees*. https://www.truthinadvertising.org/patelco-credit-unions-overdraft-fees/

Truth in Advertising. (2020d). *Overdraft fees at Pathfinder Bank*. https://www.truthinadvertising.org/patelco-credit-unions-overdraft-fees/

UL. (2021). *Sins of Greenwashing*. https://www.ul.com/insights/sins-greenwashing

Vasile, V., Panait, M., & Apostu, S. A. (2021). Financial Inclusion Paradigm Shift in the Postpandemic Period. Digital-Divide and Gender Gap. *International Journal of Environmental Research and Public Health*, *18*(20), 10938.

Voica, M. C. (2017). Financial inclusion as a tool for sustainable development. *Romanian Journal of Economics*, *44*(1), 121–129.

Voica, M. C., & Stancu, A. (2021). Corporate Social Responsibility Reporting: Background, Evolution and Sustainability Promoter. *Sustainable Management for Managers and Engineers*, 109-155.

Volgger, M., & Huang, S. S. (2019). Scoping irresponsible behaviour in hospitality and tourism: Widening the perspective of CSR. *International Journal of Contemporary Hospitality Management*.

Vollero, A., Palazzo, M., Siano, A., & Elving, W. J. (2016). Avoiding the greenwashing trap: Between CSR communication and stakeholder engagement. *International Journal of Innovation and Sustainable Development*, *10*(2), 120–140.

Walker, K., & Wan, F. (2012). The harm of symbolic actions and green-washing: Corporate actions and communications on environmental performance and their financial implications. *Journal of Business Ethics*, *109*, 227–242.

Williams, O. (2021). Europe's Largest Bank Suspected of Greenwashing. *Forbes*. https://www.forbes.com/sites/oliverwilliams1/2021/01/10/europes-largest-bank-suspected-of-greenwashing/

Williams, T. A., & Shepherd, D. A. (2017). Mixed method social network analysis: Combining inductive concept development, content analysis, and secondary data for quantitative analysis. *Organizational Research Methods*, *20*(2), 268–298.

Wright, C., & Rwabizambuga, A. (2006). Institutional pressures, corporate reputation, and voluntary codes of conduct: An examination of the equator principles. *Business and Society Review*, *111*(1), 89–117.

ADDITIONAL READING

Durst, S., & Gerstlberger, W. (2021). Financing Responsible Small-and Medium-Sized Enterprises: An International Overview of Policies and Support Programmes. *Journal of Risk and Financial Management*, *14*(1), 10. doi:10.3390/jrfm14010010

Frączek, B. (2020). A System to Support the Transparency of Consumer Credit Offers. *Journal of Risk and Financial Management*, *13*(12), 317. doi:10.3390/jrfm13120317

Gheitani, A., Imani, S., Seyyedamiri, N., & Foroudi, P. (2019). Mediating effect of intrinsic motivation on the relationship between Islamic work ethic, job satisfaction, and organizational commitment in banking sector. *International Journal of Islamic and Middle Eastern Finance and Management*.

Kuchciak, I., & Wiktorowicz, J. (2021). Empowering Financial Education by Banks—Social Media as a Modern Channel. *Journal of Risk and Financial Management*, *14*(3), 118. doi:10.3390/jrfm14030118

Riskos, K., Dekoulou, P., & Tsourvakas, G. (2019). Contemporary Environmental CSR Strategies in the Banking Sector: The Case of Piraeus Bank. In Cases on Corporate Social Responsibility and Contemporary Issues in Organizations (pp. 90-109). IGI Global.

Valenzuela, M. C., Martínez-Martínez, S. L., & Soto, J. D. (2021). Financial Literacy and Innovation Performance in SMEs: The Mediating Effect of Risk-Taking. In *Financial Management and Risk Analysis Strategies for Business Sustainability* (pp. 58–81). IGI Global. doi:10.4018/978-1-7998-7634-2.ch004

Vollero, A., Siano, A., Palazzo, M., & Elving, W. (2011). Corporate Communication and CSR; comparing Italian and Dutch energy companies on anti-greenwashing strategies. *CSR Communication conference 2011: The 1st international CSR Communication conference.*

Windsor, D. (2013, July). Authenticity, greenwashing, and institutionalization of CSR best practices. In *Proceedings of the International Association for Business and Society* (Vol. 24, pp. 70-80). 10.5840/iabsproc2013249

Yue, X. G., Cao, Y., Duarte, N., Shao, X. F., & Manta, O. (2019). Social and Financial Inclusion through Nonbanking Institutions: A Model for Rural Romania. *Journal of Risk and Financial Management*, *12*(4), 166. doi:10.3390/jrfm12040166

KEY TERMS AND DEFINITIONS

CSR: Corporate social responsibility (CSR) is a type of business self-regulation with the aim of being socially accountable.

Financial Education: Financial education is a process where the consumers of financial products and services improve their understanding for financial market. Through the financial education programs, the consumers develop the skills and confidence in strengthening information about financial risks and occasions, make decisions on the bases of good information, are acquainted with the fact where to find help and take other effective measures for improving their wealth.

Financial Inclusion: Financial inclusion means that individuals and businesses have access to useful and affordable products and services that meet their needs—transactions, payments, savings, credit, and insurance—delivered in a responsible and sustainable way by specific institutions.

Financial Institutions: A financial institution is an intermediary between consumers and the capital or the debt markets providing banking and investment services.

Sustainable Development: Sustainable development is development that meets the needs of the present without compromising the ability of future generations to meet their own needs.

ENDNOTES

1 https://equator-principles.com/about/

2 https://cba.ca/banks-and-the-environment

3 https://laws-lois.justice.gc.ca/eng/acts/b-1.01/page-76.html

4 https://equator-principles.com/about/

5 Project related corporate loans require the client to have operational control over the majority of the loan. Also, the loan tenor must be of at least 2 years in tenor and the total aggregate loan amount and the EPFIs commitment are at least $50 million. Project-Related Finance and Project-Related Acquisition Finance requires the project to be in accordance with the Equator Principles framework, has no material change in scope of the project, and the project must be uncompleted at the time of signing the loan agreement.

APPENDIX

Table 6.

	Institution	Adoption Date	Country of Headquarters	Region of Headquarters
1	Absa Group Limited	22-Oct-09	South Africa	Africa
2	Access Bank Plc	04-Jun-09	Nigeria	Africa
3	Arab African International Bank	25-Jan-09	Egypt	Africa
4	Bank of Africa	10-May-10	Morocco	Africa
5	Ecobank Transnational Incorporated	01-Jan-12	Togo	Africa
6	Fidelity Bank Plc	01-Nov-12	Nigeria	Africa
7	FirstRand Limited	13-Jul-09	South Africa	Africa
8	Mauritius Commercial Bank Ltd.	15-May-12	Mauritius	Africa
9	Nedbank Limited	10-Nov-05	South Africa	Africa
10	Standard Bank Group	02-Feb-09	South Africa	Africa
11	Bank of Chongqing	02-Feb-21	China	Asia
12	Bank of Guizhou	30-Nov-20	China	Asia
13	Bank of Huzhou	24-Jul-19	China	Asia
14	Bank of Jiangsu	20-Jan-17	China	Asia
15	Bank Sinopac	27-Feb-20	Taiwan, R.O.C.	Asia
16	Cathay United Bank Co., Ltd	23-Mar-15	Taiwan R.O.C.	Asia
17	Chongqing Rural Commercial Bank	27-Feb-20	China	Asia
18	CTBC Bank Co., Ltd	23-Jan-19	Taiwan R.O.C.	Asia
19	DBS Bank	18-Nov-19	Singapore	Asia
20	Development Bank of Japan	01-Jul-20	Japan	Asia
21	E.SUN Commercial Bank, LTD	28-Dec-15	Taiwan R.O.C.	Asia
22	First Commercial Bank	21-Dec-20	Taiwan R.O.C.	Asia
23	IDFC FIRST Bank	03-Jun-13	India	Asia
24	Industrial Bank Co., Ltd	31-Oct-08	China	Asia
25	KB Kookmin Bank	2004-02-21	South Korea	Asia
26	Korea Development Bank	02-Jan-17	South Korea	Asia
27	Mian Yang City Commercial Bank	20-Jul-20	China	Asia
28	Mizuho Bank, Ltd.	27-Oct-03	Japan	Asia
29	MUFG Bank, Ltd	22-Dec-05	Japan	Asia
30	Nippon Life Insurance Company	01-Apr-19	Japan	Asia
31	OCBC Bank	01-Dec-20	Singapore	Asia
32	Shinhan Bank	09-Sep-20	South Korea	Asia
33	Shinsei Bank, Limited	01-Apr-20	Japan	Asia
34	Sumitomo Mitsui Banking Corporation	23-Jan-06	Japan	Asia
35	Sumitomo Mitsui Trust Bank, Limited	01-Feb-16	Japan	Asia

Continued on following page

Table 6. Continued

	Institution	Adoption Date	Country of Headquarters	Region of Headquarters
36	Taipei Fubon Commercial Bank	04-Dec-17	Taiwan R.O.C.	Asia
37	Taishin International Bank	11-Nov-19	Taiwan R.O.C.	Asia
38	The Norinchukin Bank	01-May-17	Japan	Asia
39	Yuanta Commercial Bank	08-Oct-20	Taiwan R.O.C.	Asia
40	ABN Amro	03-Aug-09	The Netherlands	Europe
41	Banco Bilbao Vizcaya Argentaria, S.A. (BBVA)	18-May-04	Spain	Europe
42	Banco Sabadell	29-Sep-11	Spain	Europe
43	Banco Santander S.A.	30-Apr-09	Spain	Europe
44	Bankia	18-Jul-18	Spain	Europe
45	Bankinter	12-Dec-16	Spain	Europe
46	Barclays plc	04-Jun-03	UK	Europe
47	BNP Paribas	24-Oct-08	France	Europe
48	CaixaBank	19-Mar-07	Spain	Europe
49	Coöperatieve Rabobank U.A.	04-Jun-03	The Netherlands	Europe
50	Credit Suisse Group	04-Jun-03	Switzerland	Europe
51	Crédit Agricole Corporate and Investment Bank	04-Jun-03	France	Europe
52	De Volksbank	25-Nov-09	The Netherlands	Europe
53	DekaBank Deutsche Girozentrale	01-Mar-11	Germany	Europe
54	Deutsche Bank AG	27-Jul-20	Germany	Europe
55	DNB	29-May-08	Norway	Europe
56	DZ Bank AG	01-Jan-13	Germany	Europe
57	Eksport Kredit Fonden	14-May-04	Denmark	Europe
58	Export Credit Norway	27-Jun-14	Norway	Europe
59	FMO (Netherlands Development Finance Company)	19-Oct-05	The Netherlands	Europe
60	Green Investment Group Limited	02-Dec-13	UK	Europe
61	HSBC Holdings plc	04-Sep-03	UK	Europe
62	ING Bank N.V.	23-Jun-03	The Netherlands	Europe
63	Instituto de Crédito Oficial (ICO)	26-Oct-16	Spain	Europe
64	Intesa Sanpaolo SpA	04-Aug-06	Italy	Europe
65	KBC Group N.V.	27-Jan-04	Belgium	Europe
66	KfW IPEX-Bank GmbH	03-Mar-08	Germany	Europe
67	La Banque Postale	15-Oct-19	France	Europe
68	LBO France	02-Mar-20	France	Europe
69	Lloyds Banking Group Plc	31-Jan-08	UK	Europe
70	Natixis	30-Dec-10	France	Europe
71	NIBC Bank N.V.	09-Nov-10	The Netherlands	Europe
72	NN Investment Partners	02-Apr-19	The Netherlands	Europe
73	Nordea Bank AB (publ)	21-Feb-07	Sweden	Europe

Continued on following page

Table 6. Continued

	Institution	Adoption Date	Country of Headquarters	Region of Headquarters
74	NWB Bank	01-Oct-20	The Netherlands	Europe
75	OP Financial Group	16-Dec-16	Finland	Europe
76	NatWest Group Plc	04-Jun-03	Scotland	Europe
77	Skandinaviska Enskilda Banken AB	03-Apr-07	Sweden	Europe
78	Société Générale	03-Sep-07	France	Europe
79	Standard Chartered PLC	08-Oct-03	UK	Europe
80	Svenska Handelsbanken AB (publ)	23-Jun-16	Sweden	Europe
81	Swedbank AB	28-Oct-19	Sweden	Europe
82	Swedish Export Credit Corporation (SEK)	21-Jun-17	Sweden	Europe
83	UK Export Finance	31-Mar-16	UK	Europe
84	UniCredit SpA	04-Jun-03	Italy	Europe
85	Banco Bradesco, S.A.	08-Sep-04	Brazil	Latin America
86	Banco de Crédito	22-Jan-13	Peru	Latin America
87	Banco de Galicia y Buenos Aires S.A.	19-Mar-07	Argentina	Latin America
88	Banco de la República Oriental del Uruguay	03-Jan-08	Uruguay	Latin America
89	Banco do Brasil	06-Jul-06	Brazil	Latin America
90	Banco Votorantim SA	30-Jun-16	Brazil	Latin America
91	Bancolombia S.A.	11-Dec-08	Colombia	Latin America
92	BTG Pactual	31-Aug-20	Brazil	Latin America
93	CAIXA Econômica Federal	18-Nov-09	Brazil	Latin America
94	CIFI (Corporacion Interamericana Para El Financiamiento de Infraestructura S.A.)	06-Apr-07	Panama	Latin America
95	Itaú Unibanco S.A.	12-Aug-04	Brazil	Latin America
96	Ahli United Bank B.S.C.	01-May-11	Kingdom of Bahrain	Middle East
97	First Abu Dhabi Bank (FAB)	20-Sep-15	United Arab Emirates	Middle East
98	Banco Mercantil del Norte S.A.	12-Mar-12	Mexico	North America
99	Bank of America Corporation	15-Apr-04	USA	North America
100	Bank of Montreal	15-Sep-05	Canada	North America
101	Bank of Nova Scotia	25-Sep-06	Canada	North America
102	Canadian Imperial Bank of Commerce (CIBC)	03-Dec-03	Canada	North America
103	CIBanco S.A.	07-Mar-12	Mexico	North America
104	Citigroup Inc.	04-Jun-03	USA	North America
105	Ex-Im Bank	31-Mar-11	USA	North America
106	Export Development Canada	25-Oct-07	Canada	North America
107	JPMorgan Chase & Co.	04-Dec-06	USA	North America
108	Manulife	11-May-05	Canada	North America
109	Royal Bank of Canada	21-Jul-03	Canada	North America

Continued on following page

Table 6. Continued

	Institution	Adoption Date	Country of Headquarters	Region of Headquarters
110	TD Bank Financial Group	12-Apr-07	Canada	North America
111	Wells Fargo Bank, N.A.	12-Jul-05	USA	North America
112	Australia and New Zealand Banking Group Limited (ANZ)	15-Dec-06	Australia	Oceania
113	Commonwealth Bank of Australia	26-May-14	Australia	Oceania
114	Export Finance Australia	03-Mar-09	Australia	Oceania
115	National Australia Bank Limited	25-Oct-07	Australia	Oceania
116	Westpac Banking Corporation	04-Jun-03	Australia	Oceania

Chapter 6
Leadership and Organizational Culture in Football: A Coaches' Perspective

Tiago Vilaça Oliveira
School of Economics and Management, Portugal

Ana Paula Ferreira
https://orcid.org/0000-0002-3417-4979
School of Economics and Management, CICS.NOVA, University of Minho, Portugal

ABSTRACT

Football coaches often play a differentiating role for the clubs, helping them to survive in a demanding mediatic, changing, and competitive environment where innovation may arise regarding leadership. This chapter seeks to unveil the perceptions of football coaches as leaders and the role of the football clubs' organizational culture in affirming this leadership. Studies on leadership constitute a broad field of organizational and management theories, highlighting the role of personality traits, as well as the organizational and social contexts surrounding the leaders' actions. As there is no significant academic literature on football coaches and leadership, it was sought to explore the coaches' perceptions as leaders, as well as the influence of the clubs' organizational culture in which they developed their activity. Based on 22 interviews with football coaches of reference clubs, this chapter highlights their difficulties, demands, and needs to deal with their professional context.

INTRODUCTION

The growing complexity of sports structures and the media coverage of the sports phenomenon (Sousa, 2018), combined with the growth in revenue generated and the increasingly demanding competitive environment, makes the world of football attractive for the development of studies in various areas.

In management, namely regarding people, sports leadership has been studied by the academic community, especially the one exerted by the coach (Moreno-Arrebola et al., 2017). Such studies make it

DOI: 10.4018/978-1-6684-2339-4.ch006

possible for organizations to replicate the practices of successful leaders in these sports contexts into their specific contexts.

Research studies carried out with football coaches reveal the variety of behaviors adopted and the different effects caused by the actions of the actors, namely their leadership styles and the link to success (e.g., Mills & Boardley, 2016; Ancelotti, Brady, & Forde, 2017; Koch, 2018; Thompson, 2019). The role of today's football coaches goes beyond the field of technical competence, especially in a media, competitive, turbulent context that exerts pressure on this type of professional to ensure that their teams have high performance (Flint et al., 2014), and where the time needed to reach excellence levels is often overlooked. Thus, it is expected that coaches spearhead a global process of evolution of their athletes and the group, influencing their development and progress favorably (Coelho, 2004; Garganta, 2004).

This breadth of tasks and functions entails a wider range of variables that influence the coach's performance in football clubs and vice versa: individual variables (technical, socio-relational, conceptual and leadership competence), relational variables (between coaches and athletes), contextual factors beyond the control of the coach, such as issues related to finances, injuries and refereeing decisions, among others (Crust & Lawrence, 2006; Vieira et al., 2017; Rego et al., 2020), and also organizational variables, such as organizational culture (Sousa, 2018).

Studying football coaches and their leadership, in their organizational contexts (clubs), but which transcend them due to their constant transfer within these same contexts seems to be an adequate field of analysis for the identification and exploration of a set of factors associated with leadership, organizational culture and relationship with performance. The conclusions of these studies can be transferred to other organizations that work in globalized, demanding, competitive and highly changeable environments, where leadership is seen as a critical function in the survival of institutions.

This chapter seeks to identify and analyze the role of football coaches' leadership through the "eye of the beholder", aiming, specifically, to:

- assess the coaches' perception as leaders;
- identify the factors that contribute to their success;
- understand the relationship between organizational culture and coaches' leadership.

LEADERSHIP AND ORGANIZATIONAL CULTURE

This initial section deals with the main concepts and approaches regarding leadership and organizational culture addressing, also, the main interactions between these constructs.

Leadership

Leadership can be seen from several perspectives. Some researchers conceptualize it as a characteristic (e.g. Lord et al., 1986; Kirkpatrick & Locke, 1991) or as a process (Vroom & Jago, 2007). Others view leadership from a relational point of view (Simonet & Tett, 2013).

As a process, leadership involves influence, groups and common goals (Silva, 2016) where someone impacts (influence) a set of individuals (group) to achieve something (goal). So, leadership is not only an individual trait or characteristic but also a transactional event that happens between the leader and his followers. As such, the leader affects, and is affected by, followers because leadership is an interactive phenomenon that aims attending some goal, and where the leader seems to be frequently considered the responsible to initiate the relationship, create communication lines, motivate, and is also responsible for maintaining the relationship (Vroom & Jago, 2007; Northouse, 2015).

The vast and diverse literature on leadership brings further complexity to the analysis (Ashford & Sitkin, 2019). Day and Antonakis (2012) identified nine key approaches to leadership: traits; behavioral; contingency; contextual; of skeptics; relational; new leadership approaches; information processing approaches; and biological/evolutionary approaches. Using another classification, leadership studies can be divided into traits, behavioral, situational, and emergent approaches.

The traits approach, which emerged at the turn of the 20th century, focuses on the idea that leaders have innate qualities that characterize them as such and, therefore, differentiate them from non-leaders. This approach has a body of research that validates it to some extent and enables the idea of a leader selection process based on personal traits, especially as psychometric tests evolve, and interest in different individual areas, such as gender or diversity, increases. However, it does not seem to be capable of establishing a definitive list of traits and does not consider the weight ascribed to the influence that the situation exerts on leadership (Day & Antonakis, 2012; Northouse, 2015).

Bernard Bass and other theorists have developed visionary and charismatic theories (Bryman et al., 1992; Bass, 1998). While it is not typically an approach based on the leader's individual features, it highly emphasizes this aspect. The concept of charisma still prompts controversy, namely due to the fact that it is sometimes associated with individuals, some other times with situations, and also with the interaction of both. It is argued that charismatic effects are more likely to occur in unfavorable and stressful contexts. Charismatic leaders shape this bond by emphasizing rewards. Throughout the whole process, leaders express high expectations towards followers and help them gain a sense of trust and self-efficacy. Charismatic leadership has its effects by highly engaging followers' self-concepts in the interest of the mission articulated by the leader (Shamir et al., 2018).

Transformational leadership, associated with the work of Burns (cited in Bryman et al., 1992) and Bass (1985), regards a type of leadership that focuses on dispositions such as emotions, values, ethics, standards, and long-term goals. Leaders need to include and adapt to the followers' needs and motivations, being acknowledged as agents of change with the implementation of new ideas. They develop an inspired and shared vision for the future, playing a critical role in organizational change through the design of new ideas. Despite being a relevant approach, its weaknesses have led to a lack of interest in it (Ghasabeh, Soosay, & Reaiche, 2015; Northouse, 2015).

The behavioral approach focuses on the leaders' behavior/actions, differentiating between task-oriented behaviors (where the leader emphasizes the definition and structuring of work regarding the goals to be attained) from relationship-oriented behaviors (where the leader is concerned with subordinates and their well-being), and in the combination of these behaviors (leadership styles) and their influence on followers. These theories, such as those developed by the universities of Ohio and Michigan, and the Blake and Mouton model, have broadened the scope of leadership research beyond its personal traits or features, being supported by a solid base of studies and a broad conceptual framework useful for understanding behaviors. However, they have not been capable of identifying a universal set of behaviors that would consistently result in effective leadership, nor have they been capable of linking the leaders' behaviors to outcomes (Day & Antonakis, 2012; Northouse, 2015).

The situational/contingency approaches value the role of the context in the adequacy of the leadership style, and where the idea of leadership and the leaders' choices are a consequence of a complex interaction of the context, information about potential leaders, and the role expected to be performed by the leader. These theories emphasize, for example, the role of the relationship between the leader and their followers and the leader-member exchanges, highlighting high-quality exchanges by the leader with members of a group considered closer to the leader (ingroup), as opposed to those of the members of the least close group of followers (outgroup), and their results in terms of trust, respect, loyalty, and formality in behaviors (Graen & Uhl-Bien, 1995; House & Aditya, 1997; Chaudry, Vidyarthi, Liden & Wayne, 2021). Fiedler (1978) puts forth the idea that the effectiveness as a leader is determined by how well your leadership style matches the situation, and there is a set of factors that must be considered for this adjustment (relationship with the followers, power position, tasks structuring). In turn, the model developed by Hersey and Blanchard (1988) focuses on the adequacy of the leader's style to the psychological and professional maturity of their followers.

While much of the research on leadership focuses on the leaders' behavior and effectiveness (Felfe & Petersen, 2007), in the 1980s, several criticisms were made to the research of this dimension. Methodological issues, conceptual problems, ambiguities in definitions and lack of coherence have been pointed out to leadership studies (e.g. Meindl, Ehrlich, & Dukerich, 1985). Studies highlight that poor company performance is often ascribed to uncontrollable external events, whereas good performance is credited to foresight and leadership quality (Meindl & Ehrlich, 1987; Puffer, 1990; Kim & Miner, 2007). The fact is that the responsibility of CEOs (leaders) on company results, even after considering contextual effects, seems to have minimal impact (House & Aditya, 1997).

The emerging approaches to leadership seem to involve an emphasis on information processing, associated with the work of Lord (2019). According to these approaches, leadership processes take place in hierarchically organized dynamic and adaptive social systems. Individuals integrate leadership-relevant information using double-processing systems that allow both automatic, local processing, and more conscious, brain-scale processing to take place concurrently. Research in the areas of cognition, information processing and emotions has added to the development of the understanding of leadership (Cerni et al., 2013; Mumford et al., 2017; Watts et al., 2019). Likewise, evolutionary and biological perspectives have been gaining attention from the academic community (Day & Antonakis, 2012).

These approaches explore the change in measurable individual variables (such as, for instance, brain waves, endocrinology and genes) to link it to leadership behaviors, decisions and social interaction processes. They can be very useful in developing process models and identifying causal effects of potentially endogenous variables on outcomes. Studies on this line have sought to establish important relationships, such as the analysis of behavioral genetics (Ilies et al., 2004), the performance of leadership roles, both in men and women (Arvey et al., 2007), and the effect of hormones, relating to with leadership, such as, for example, the effects of testosterone (Zyphur et al., 2009).

Research on leadership is expanding, profiting from new approaches and new research. Diversity, namely gender issues (Vecchio, 2002; Coleman, 2003; Abdallah & Jibai, 2020), race (Randolph-Seng et al., 2016; Miller, 2019), ethnicity (Ospina & Foldy, 2009; Chin, 2013), and sexual orientation (Barrantes & Eaton, 2018; Miles & Naumann, 2021), has also been one of the most explored lines in the area of leadership, with several studies highlighting the importance of minorities in an organizational context. Leadership models have developed around the role of the leader ethics and morals, as well as creativity, innovation and complex thinking (Knights & O'Leary, 2006; Curral et al., 2016; Uhl-Bien & Arena, 2018; Hughes et al., 2018).

It can be easily seen that leadership is a complex phenomenon that operates at different levels of analysis, involves multiple mediating and moderating factors, and takes place over substantial periods of time (Dinh et al., 2014), which makes it attractive but also difficult to study it.

Organizational Culture

Despite not having a consensual definition in the academic community, organizational culture (OC) impacts the daily life of companies (Barney, 1986; Abu-Jarad et al., 2010). Several authors address this dimension of organizations as a complex combination of shared values, beliefs, assumptions and symbols that define the way organizations behave, function and lead their business and become more or less unique (Barney,1986; Deshpande & Webster, 1989; Schein, 1996; Leithy, 2017).

The development of an OC can begin with the CEO conveying the desired culture to all employees. Each employee interprets the meaning of expressions and intentions conveyed by the CEO from the perspective of personal culture and using contextual cues. OC aggregates the collection of all the employees' interpretations (Gorton & Zentefis, 2020), and they influence and are influenced by it (Shahzad et al., 2012).

The analysis of OC goes through different dimensions that result from the diverse models and theories (Huey Yiing & Zaman Bin Ahmad, 2009), but it is envisaged as the foundation of social cohesion among employees within a company (Gorton & Zentefis, 2020).

Edgar Schein (1996) developed a model that describes OC according to three levels of visibility. The first regards artifacts and behavior patterns (e.g., organization of physical space, attitudes, ways of doing things); the second encompasses organizational values (e.g., beliefs, ethical and moral codes, ideologies); and the least visible, the level of basic assumptions, which are implicit, fundamental beliefs that guide behavior and determine how each individual perceives, thinks and feels the reality; they belong to the field of the unconscious, taken for granted, unquestionable and are the core of OC.

Deshpandé, Farley and Webster (1993) also developed a taxonomy of organizational cultures, defining four types resulting from the intersection of two axes – internal-external orientation and flexibility-stability –, and which result in the clan, adaptive, bureaucratic culture, and in the culture of market or realization. In a similar vein, Cameron and Quinn (2011) built the Competing Values Framework (CVF) model, composed of two axes, which oppose stability to flexibility and internal to external factors, resulting in four types of OC: hierarchical, market, adhocracy and clan culture.

Clan culture is related to internal focus and flexibility, being considered supportive; the organization is seen as a second family, and all members are expected to participate in the decision-making process. Adhocracy culture is related to external focus and flexibility; creativity and innovation are the main factors that add to the success of an organization. Hierarchical culture is described by internal focus and stability and is closely linked to stability in the organization's functioning, with clear policies and regulations by which the company may act. Market culture is characterized by stability and external focus; it is ruled by achievement and coordination, is task-oriented, and has clear goals and decision-making processes.

Other models exist (e.g. Groysberg et al., 2018), but regardless of the use of the OC model and of the specific relationships of each dimension with several variables, culture influences organizations, their performance in its various aspects, including financial, identity and reputation, and can determine the well-being of their members (O'Reilly, Caldwell, Chatman & Doerr, 2014; Chatman & O'Reilly, 2016). It also has positive implications for the employee's professional performance (Shahzad et al., 2012). Despite divergent results in some studies (e.g., Leithy, 2017), there seems to be a general consensus that OC (especially some types) has a decisive impact on innovation and performance (Deshpandé & Farley, 2004; Naranjo-Valencia, Jiménez-Jimenéz e Sanz-Valle, 2016; Fiordelisi et al., 2019; Tran, 2021).

Leadership and Organizational Culture

Schein (2010) sustains that OC and leadership can be seen as two sides of the same coin, insofar that leadership affects as much culture as culture influences leadership. Some researchers indicate that OC is relatively stable regardless of personnel or environmental change and that individuals have relatively little effect on culture (Alvesson et al., 2017; Schein, 2010). Here, a contingency perspective seems to be appropriate, as there are times when leadership can have a strong influence on OC and others when such efforts will not yield results (Ehrhart et al., 2014). The truth is that behaviors are highly influenced by the context and culture in which individuals are inserted. While there are aspects of leadership that seem to be universal across cultures, there is evidence that points toward the importance of culture in the way leaders emerge, are selected, developed and seen (Dickson et al., 2012).

The followers' characteristics and cultural values can influence the leadership process, affecting a leader's behavior (Matthews et al., 2021) to adapt to new realities, be capable of leading their team to face obstacles, and be prepared to turn threats into opportunities (Arruda et al., 2010). However, Groysberg and colleagues (2018) claim that the founders of organizations and their influential leaders also set in motion new cultures and imprint values and assumptions that persist for decades. Over time, the leaders of an organization can also shape the culture through conscious and unconscious actions (Schein, 1985). Especially in global organizations and in constant change, these leaders work with people from different backgrounds and cultures, regardless of whether it is in their country of origin or an expatriate context, being capable of building and maintaining a strong culture that brings together the inherent diversity (Tsai, 2011; Warrick et al., 2016).

Rego and colleagues (in Cunha et al., 2016), based on the work of Schein and Yukl (Table 1), put forth the (primary and secondary) mechanisms through which leaders can influence OC.

Table 1. Leaders' mechanisms to influence OC

<table>
<tr><td rowspan="6">Primary</td><td>Attention</td><td>Leaders convey their priorities and values by choosing the things they question, measure, comment on, praise and criticize. For example, what do they do with a subordinate who does not know what is going on with their unit? How do they deal with someone who does not share information and knowledge with colleagues?</td></tr>
<tr><td>Reactions to critical incidents and crises</td><td>Given the emotionalism that surrounds them, moments when organizational survival is threatened, when norms are obscure or challenging, when insubordination and threatening events occur carry a high potential for conveying values and assumptions. For example, in crisis situations, does the leader reduce the salaries of all members (regardless of hierarchy) or fire staff?</td></tr>
<tr><td>Role modeling</td><td>Leaders can convey values through their own actions, particularly those that demonstrate special loyalty, self-sacrifice and a spirit of service.</td></tr>
<tr><td>Allocation of rewards and status</td><td>What behaviors do the leaders reward and punish? For example, do they promote yes-men or encourage constructive criticism? Do they assign symbols of status (e.g., private parking, private toilet), or do these differences not exist?</td></tr>
<tr><td>Allocation of scarce resources</td><td>For example, how are budgets prepared? From above (the top management decides), or based on plans, projects and proposals coming from below?</td></tr>
<tr><td>Promotion, admission and dismissal criteria</td><td>What criteria are used for promotions? Age, seniority, experience, competence, (a)critical stance…? And, when there are dismissals, who are the "chosen ones": the oldest, the least committed or the most "inconvenient"?</td></tr>
<tr><td rowspan="2">Secondary</td><td>Design of the organizational structure</td><td>For example, a centralized structure reflects the belief that only the leader can determine what is best, whereas a decentralized structure denotes the belief in individual initiative and shared responsibility.</td></tr>
<tr><td>Design of systems and procedures</td><td>Budgets, planning sessions, reports, management development programs, information systems, etc. can focus the attention on certain activities and criteria while reducing role ambiguity.</td></tr>
</table>

Source: Based on Cunha et al. (2016, p .631).

There are several studies on the impact of leadership on OC (and vice versa), as well as on other organizational variables. Several researchers conclude that the CEO's leadership style influences and depends on the process of organizational adaptation and change through the establishment of a stimulating OC (Ostroff et al., 2013; Raguž & Zekan, 2017).

Al-Ali and colleagues (2017) concluded that change leadership has a positive and significant impact on three types of OC: clan, adhocracy and hierarchical. Hierarchical culture, in particular, plays an important positive role in emerging and planned change in (especially public) organizations. It was also found that this type of OC, and not that of clan or adhocracy, positively and significantly influences the management of change in the public organizations studied.

Steers and Shim (2013) suggest that, at least to some extent, the success of companies is based on different leadership styles that help create and sustain different organizational cultures. Moreover, Barreto and colleagues (2013) advocate that some cultural values can be influenced by components of the manager's leadership.

Dickson and colleagues (2012) and Li et al. (2021) confirmed the positive relationship between various leadership styles and employee engagement, and the moderating relationship of culture towards leadership and employee engagement, with servant and ethical leadership styles being desired in all contexts and not varying with cultural factors.

Chong et al. (2018) concluded that supportive leadership styles and task-focused leadership, as well as a persuasive influence strategy, are correlated with team, detail, and innovation cultures, respectively.

These were found to be significantly stronger than the other leadership styles/strategies. The study provides a tentative conclusion that different types of culture may have different levels of strength in shaping middle management and, consequently, influence subordinate outcomes.

Some authors reinforce that different types of culture are not mutually exclusive – one type of culture could be equally strong in two or more leadership styles (Nguyen & Mohamed, 2011; Xenikou & Simosi, 2006).

Despite the existing evidence, Raguž and Zekan (2017) claim that the number of studies that address the relationship between OC and specific leadership styles is still not enough.

LEADERSHIP AND ORGANIZATIONAL CULTURE IN FOOTBALL

Sports organizations and their coaches, athletes and members are, as previously mentioned, promising contexts for the study of OC. This type of organization is usually associated with specific values and a wide variety of symbols, stories, myths and rituals (Maitland et al., 2015). According to Jackson and Haigh (2008), sports has long served as an important source of collective identification and is perhaps one of the most powerful and visible symbols of national identity and nationalism. Often, the term "organizational culture" is used to describe a vision of a sports organization as a stable and unique combination of meanings (Wagstaff & Burton-Wylie, 2018).

What is certain is that there are very different realities in cultural terms in the dynamics of football, where there is a huge movement of human capital across countries and continents. Likewise, the media very intensely explore the coaches' personality traits and personal and professional success and often relate this success to the coaches' ability to react and manage teams of a transnational type (Sousa, 2018).

To analyze football coaches' leadership, it is also necessary to explore the context, that is, clubs' OC. The football context is highly influenced by the cultural reality of a country and its diplomatic relationships (Kuper, 1996), which means that it is paramount for a coach to adapt to the national and organizational cultures in which they are inserted, trying to understand the language and cultural behaviors to achieve the intended results (Ancelotti, Brady, & Forde, 2017). Carlos Carvalhal, an elite Portuguese coach, analyzes how some of the cultural traits of the Portuguese can explain his international success: "With life experience, I learned that, when there is a cultural reality, we have to adapt. We are hired to bring ideas and work to achieve certain goals, we go for them, but the way we will attain them is different from culture to culture. We have to look for other paths, often different from the ones we are used to walking. But one thing that the Portuguese are very good at is their ability to work and adapt. Perhaps for this reason, there are hundreds of Portuguese coaches succeeding in all corners of the world" (Carvalhal, 2014, p. 204).

Notwithstanding the huge media exposure of football coaches, who frequently change their place of activity and are transitory in the global and dynamic football market (Maguire, 2008), there are few specific studies either on the leadership of these professionals and their characteristics and training or on how their performance as coaches, especially in a transnational environment, relates to the surrounding culture and clubs' OC.

In this sense, this study seeks to know the perception of coaches as leaders, identify the factors that, according to their perceptions, add to the success of this leadership and identify the role of OC in the performance of the football coach's activity.

To attain these goals, 22 football coaches were interviewed (all male), holding the UEFA PRO license and/or a record that guarantees the quality of their role, exercising their activity as head coach for at least two seasons. These criteria were established to ensure that the individuals interviewed had experience in managing teams in highly demanding contexts.

The interviews were conducted in distinct ways (face-to-face, via video calls and via phone call), and, on average, each interview lasted 52.22 minutes, having been recorded with authorization and transcribed for later analysis.

The average age of the interviewees is 46.2 years, and the average number of seasons as a coach in a professional context is 12.4 seasons, both in national and international contexts. Thirteen of the 22 coaches have worked in a country other than their home country, 20 hold the highest UEFA license (UEFA PRO), and two hold the UEFA A license (level immediately prior to the PRO).

Regarding nationality, 19 are Portuguese, one is Spanish, one is Canadian, and one is English. Of the 22 interviewees, 13 had been professional football players, and 11 have an academic background (bachelor's or master's). Of those who played professionally, only two reconciled the university context, associated with an early dropout from the football career.

Given the need for trust and the difficulty of accessing these individuals, shared contacts or references were used to schedule interviews.

In the interview script, introductory questions were asked about the roles of a coach, and the role of leadership and the cultural context. Data treatment, based on the transcripts, followed the procedures of content analysis (Mendes & Miskulin, 2017), from previously established categories based on the study objectives and theoretical framework, but which changed over the course of the study, as data emerged that justified it.

Thus, and considering the goals of the study, three thematic areas were identified: coaches' perception as leaders, factors of success for the activity, and perception of the relationship between OC and leadership.

Coaches' Perception as Leaders

In this category, four dimensions emerged about what coaches think and feel about their role, in addition to the leadership dimension: decisions and discussion of ideas, relationship with players, responsibility for defeats, and adaptation to the context.

In the functions that they did not directly associate with the concept of leadership, the dimension decisions and discussion of ideas seems to be associated with the typical manager's role of decision-making, as well as with structuring. Having a high degree of autonomy and a lot of freedom to appoint their support group, something that is uncommon in the business manager (Kelly, 2008), all the interviewees admit that they always have the final word, although they understand that they must listen to other people, negotiating some aspects and including the players, but above all their assistant coaches. They also emphasize the definition of tasks and the need for delegating responsibilities, as well as the need for the group to be accountable to them.

There are things that are not negotiated. One of the things has to do with commitment, effort and dedication to their professional activity. I cannot negotiate this. I can forgive a player who is tired, who had a bad night. What is negotiable is the way we play and the way we can face the opponent. In this situation,

I like to talk to the players, not all of them, about how we can play against the opponent; basically, it is about creating a bond with the players… (E22).

[…] the leader has to clearly delegate this type of functions to each one. […] the coach obviously has the last word in his role as leader, but I think it is important to share ideas and also to delegate responsibility to our collaborators. I do not like to be involved with everything, I think there are very competent people, and I also have competent people working with me right now, and I like that they feel responsible for what they do and what they have planned, period (E5).

Another component that seems typically associated with the role of the manager, but now in a dimension of direction and leadership, is the relationship with the players. Words that designate behaviors such as straightforwardness, persuasion, proximity, respect, naturalness, responsibility and discipline stand out from the discourses. The coaches state that it is necessary to act "without masks", keeping the "genuine" nature of the relationship, as if they were "selling the idea" because:

[…] there are things, there is criticism that it is better to make in front of everyone because they act as a lesson for everyone, and sometimes there are things, there are details, particular things that in private conversations work better […] the player who enters believing that something will work because of what he trained, for what he is understanding of the game plan and for what he knows about the opponent that was also shown to him, and he believes that in this way we are going to win, I think by entering with that spirit and with that idea we are closer to achieving it than the other way around (E2).

[…] loyalty that has to be mutual. It ends up helping us have a leadership ability that helps us get closer to success (E13).

Sometimes we may be right, but we lose with how we communicate, how we demand, how we exercise our authority (E12).

The interviewees envisage the exercise of leadership as very grounded in authority, as the ability to obtain uncontested obedience from the team, in a legitimate way, although, in this case, based on the knowledge and personality of the coach as a commanding figure, but with enough charisma to be respected and obeyed. Therefore, they acknowledge that there is the need to act when something goes wrong in the team:

I think he went too far, I think he disrespected me, and I told him there in front of everyone, and it was important to tell him in front of everyone, exactly because he was one of the leaders of the locker room, he was one of the oldest players, and there you send a message, not only to him but mainly to the whole group (E4).

When speaking about responsibility for defeats, all coaches state that responsibility for the team's failures should not be taken only by themselves, maintaining that "it depends on the moment and the circumstance":

There are times when we have to take responsibility, okay. But at all times, at all moments, you are supposed to be the leader, you are the one who is in charge, but there are times when you have to give responsibility and make people feel responsible for what is happening, whether it is good or less good (E3).

They believe that responsibility should be shared, which is based on the analysis of the game "in the locker room", in a healthy discussion, rather than in a discussion to point the finger. Therefore, they claim that,

[…] in front of the television, in front of the journalists, we publicly take on responsibility because we are the ones who also have this responsibility of putting the players on the field. Therefore, that responsibility is ours but the truth is that, inside the locker room, I tell them this too on the first day to be very clear. 'I will always be your biggest defender outside, but here inside the locker room, inside these four walls, if I have to point the finger at the players, I will do so'. Thus, I think there has to be this responsibility too because otherwise, the player conforms […] (E7).

Adaptation to the context, in particular, adaptation to the clubs' culture, is seen by the coaches as something fundamental, and some emphasize the need for the development of this relationship and the intelligence of the coaches to perform an in-depth study of the places where they work so that they are capable of adapting their action and creating team spirit, especially when working in a country with a different culture:

You can have your base idea, your matrix. We are talking about the game, but then you arrive at a given club and you see the resources you have, and which do not meet your expectations or the way you want to play, you will have, in a way, to maintain your matrix, but the intelligent coach is the one who manages to get to a club and see the group they have […] how to make the best use of the means they have (E3).

In line with what researchers suggest, respondents perceive that their effectiveness depends on understanding the followers' progress, adapting their style to that level of development (Northouse, 2015). Contextual factors are seen as the variables that lead to certain leadership behaviors (or absence thereof) (Liden & Antonakis, 2009).

Concerning, specifically, leadership and the way interviewees perceive and exercise it, some highlight the need for charisma, which is in line with the traits approaches (Kirkpatrick & Locke, 1991; Lord et al., 1986) and the charismatic leadership approaches (Antonakis et al., 2016):

It is something very innate. We can improve with experience, teachings, but it is very innate. We either have leadership or we do not, it is charisma (E14).

The coaches interviewed approach leadership without, however, defining it clearly, associating it with the fact that:

[…] to lead is to create teams because when you start processes, you have different cultures, it is not just about clubs, but about cities, countries, as they also change (E11).

Or that leadership is shared:

[…] leadership of the coach, the captains, the technical staff that has a strong leadership, the sports director if there is a sports director, to me, this is vital, of the medical department, that is why I usually say that leadership has to be strong in all areas (E17).

Furthermore, coaches associate leadership with a process of influencing and convincing:

The leadership that you have, that you have to make others follow you, believe in you (E16).

And the idea of responsibility emerges:

Leadership is, in essence, taking responsibility. I am the leader because I am responsible for what is being done here. I am the one who has this idea, this project, I am the one that has drawn up this plan, I am the one that has designed things, I am the one that has made the options… (E9).

Some coaches seem to understand leadership as synonymous of management, describing the typical areas of management functions – planning, organizing, directing (leading) and controlling:

For me, leadership is about everyone realizing that you are the man who leads at the moment, without seeking to impose anything. This is the best way, you arrive and have to make people understand that there is a person there who is determined to attain the goal, and how is he determined? With natural principles to follow success, decide, organize and plan, these three aspects are pivotal. And discipline, the man of discipline must always be the coach, it can never be anyone else. When discipline is scattered, the coach is dead […] (E18).

One coach highlighted leadership as the way of relating to people and managing the group of people who are in and involved with the team:

[…] leadership is nothing more than the way how we deal daily with all the people that involve the club. The players, the staff, structure, presidents, everything. From the lawn worker to the president. Basically, it is nothing more than the way I relate daily with all the people that involve the club […] (E6).

Leadership in sports entails several dimensions (Lança, 2015). Respondents do not have academic training in management and often mix the concepts of leader and manager, focusing everything on the character of the coach. For Chiu and colleagues (2017), managers are appointed by the organization, whereas leaders are chosen by the followers. Following this reasoning, all coaches would be managers and not leaders. Throughout the study, the coaches repeatedly stuck to the character of the leader, but without offering a clear definition of it. Coaches especially associated leadership with the ability to convince others, which fits into the definition of leadership as a process advocated by several authors (Northouse, 2015; Vroom & Jago, 2007). There is also the association of this construct with responsibility, group management and communication, with achieving results and goals. Some even consider that leadership is something innate, an idea associated with the traits approach (Ahmed Khan et al., 2016). Interviewers also address the concept of charisma. The literature shows that charismatic effects are more likely to occur in unfavorable and stressful contexts, similar to the football one (Shamir et al., 2018).

Compared with the other categories that the interviewees referred to as their role/function, regarding decisions, all coaches assume that they are the "owners of the final word". However, they revealed to be thoughtful in this aspect, admitting that the exchange of ideas with the people around them, especially their technical staff, is fundamental for better decision making. The definition of tasks is taken on by the coaches and perceived as important. Sousa (2018) also advocates that the anticipation and definition of tasks allow leading with less attrition and exposure. All interviewees adopt behaviors of delegation and accountability towards their staff.

While the dialogue with the technical staff is favored, this behavior is not always adopted with the players, with coaches admitting that, despite being flexible, some aspects are not negotiable. Naturally, the coach has a higher hierarchical position than the player, having the decision of putting to play whomever he believes serves his ideas and pretensions. This fact ascribes the coach an innate authority. The coaches essentially highlighted three words to characterize their relationships with the players: respect, responsibility and discipline.

Regarding the coach's responsibility for bad performances, the interviewees reveal a given behavior depending on the situation. Sousa (2018) stresses the importance of defining a leader for the inside but also for the outside. In front of the public, they protect their players and take responsibility for the results and negative performances. However, in front of the group, they share this responsibility with their technical staff, the structure and, especially, the protagonists, their players, which indicates the idea of a sharing of responsibility.

Being up to the leader to adapt to new realities and lead the team (Arruda et al., 2010), coaches perceive adaptation to the context as necessary, seeing this attitude as a "matter of intelligence", to make the best use of the resources they have. The interviewees admit that the club's context and functioning must be analyzed to make better decisions and obtain better performance from the players.

Coaches' Perceptions of Success

In identifying the factors that contribute to the success/failure of their activity, especially as leaders, the results point that, similarly to what several studies suggest (e.g., Kim & Miner, 2007; Meindl & Ehrlich, 1987; Puffer, 1990), coaches, as leaders, tend to associate success with factors inherent in their characteristics and behaviors, but associate failure with uncontrollable external events.

Personality traits, technical competence and communication are the characteristics most often mentioned by the interviewees when talking about successful leadership in football. Coaches comment on the need to be "intelligent", "daring", and "coherent beings, transversal beings, pure beings. When you are a good person, you can converge everyone in the same direction and make everyone cooperate with you" (E6).

Despite the personality traits, competence, particularly the technical-tactical one, was mentioned as a necessary condition because

[...] you must have technical knowledge of the area. When I talk about technical knowledge, it is not just the technical part of the game, but also the tactical, physical, mental, scientific, medical, operational parts in terms of logistics, communication, marketing, having a very comprehensive knowledge to be capable of interacting with the different professionals, and motivating and adding to the dynamics some aspect that is beneficial for performance and the club (E12).

Communication is approached both as a success factor and as a factor that can interfere with it. On the one hand, it seems to operationalize traits and personality but also technical skills because:

It is easier to teach if you can communicate. The key is communication. People with different languages, cultures, and you have to communicate with everyone (E20).

On the other hand, it can hamper it:

Communication. If I was to advise a coach who is starting his career today, in the same way that he puts time and passion into studying the game, he should study another language. This will help him communicate with the player. What hinders it? It is when that communication does not work (E19).

In the interviewees' discourse, the quality of the players is added as another factor that determines the success of the coach as a leader:

[…] I usually say joking that players practically allow the coach to succeed and kill the coach very easily these days in football (E17).

Coaches also refer some relevant issues to the organizational context, namely issues associated with the structuring of clubs and some dimensions of OC that can facilitate or "hinder" the success of their work:

[…] how the club is structured. Sometimes clubs are structured around stricter, more disciplined, more supportive leadership, more available to help. Other times, clubs behave more in a more conniving, selfish way (E12).

[…] factors that hamper, for me, of course, are disorganization… I mean disorganization both at an intra and extra level. The club, staff, technical staff, structure, everything. A club that, for example, we reach certain contexts where they do not have a team regulation model, a club regulation. Some clubs have things very well closed, 'mister, here are the regulations, the schedules, the fines', when there is a given behavior, when these rules, this organization are closed, it helps a coach much more than when they are not. When they are not, the coach must have this ability to look for them himself, implement them and organize (E6).

The issue of the coach's reputation/results seems to be perceived as a potential enabler of leadership success:

I think that the coach's reputation and CV also help (E4).

For outsiders, success is the result. If you get results, you can train well or badly, or manage well or badly, if you win, you are a good coach in people's opinion (E18).

Having passion for the activity and the game is another ingredient that stands out from the menu of features that the coach as a leader needs to master:

Passion for what we do. It is important to convey passion and believe in what to do, to have an idea of the game [...] and to be capable of putting it into practice, believing in it, and if we believe, [...] it is halfway for people to believe, for players and the structure to believe [...] (E15).

The decentralization of leadership concerns, above all, interferences in the activities of the coach and the dispersion of leadership within clubs. In large organizations, such as football clubs, leadership can be weakened by interferences in the choice of players, for example. Football as a business leads to the management of players as financial assets, whose purchase and sale have profit ends that overlap the sporting objective. The coach is often set apart from these processes. This is linked to the existence of multiple leaders, shaping decentralization of the character of the leader:

[…] leadership is very dispersed, it is not focused only on the character of the coach, it is focused on several people, and the players have another way of thinking, when they realize this, they start to doubt a lot, they start to doubt a lot. And here is where problems can start […] E18.

[…] this is becoming more and more a game of buying and selling players, of marketing, and this is increasingly hindering the coach's task […] (E7).

The coach is demanded to show results and, many times, the time necessary to attain them is ignored. Respondents highlight precisely this temporal obstacle, especially regarding game schedules:

Time. We are hostages of time. The coach is always a hostage of time. We see coaches who, at the end of a week, win some five consecutive games but have a game on Sunday, Wednesday and Sunday and have little time to plan a strategy for a game, and if they have less positive results in a week, they are fired. So, time is our biggest enemy in football (E7).

Generally, clubs strengthen habits and routines over the years, as other organizations do. Coaches mention the difficulty in managing these aspects, especially when they join the club in the middle of a season, with the season underway, not having the opportunity to establish their way of doing things in the initial and pre-competitive periods. The attempt to change routines and habits rooted in clubs, even in less relevant situations, may lead to conflict and hamper the coach's leadership:

You do not know the history of all the players, especially when you get to the middle of a club, it is more difficult; when you build a team from the beginning and you start laying down the rules right from the start, you create interesting and nice group dynamics, it is easier, it is much easier (E1).

[…] they are small things, they seem like small things, but when you arrive at a place, you want to change everything. […] but when you arrive, and you want to change because these are things that are not good for the group, you end up having conflicts. Then, of course, you have conflicts once, twice and then it starts, the relationship starts to deteriorate, and it is not always easy (E2).

In short, the factors that contribute to being a successful leader in this context are associated with personality traits, technical-tactical skills and the communication of the leader. All respondents emphasize characteristics, knowledge and communication. In line with what several studies suggest (Kim &

Miner, 2007; Meindl & Ehrlich, 1987; Puffer, 1990), coaches, as leaders, tend to associate their success with factors inherent in their traits and behaviors. Some additional aspects were also mentioned, such as the quality of the players, their reputation and the results presented, the passion for the game and the organizational conditions.

However, when coaches were asked about the factors that make it difficult to achieve success, the emphasis is on contextual factors, such as joining a team in the middle of the championship, the habits and routines established in the clubs, the players, the club's organization and the need to show immediate results. Crust and Lawrence (2006) also support this idea, highlighting the impact that variables outside the coaches' direct control have on their degree of success.

The Role of Organizational Culture in Coaches' Leadership

The coaches interviewed agree that the coach's performance is steered by contextual elements that involve their action and explain their decisions. From their perspective, the club's OC is something that shapes their action and to which they have to adapt. The context is a significant part and can influence the coach's work and collective success:

[…] you have to use the club's culture in the locker room. Your leadership has a lot to do with the club. Club culture can influence leadership (E19).

[…] when we arrive at a club, we have to think that this group is ours. Adapt to the club to give more (E21).

Frontiera (2010) demonstrated that leaders have a keen awareness of the cultural elements of their organization. Matthews and colleagues (2021) reinforce the idea that cultural features and values influence the leadership process. Warrick (2017), similarly to the interviewees, advocates that a leader's effectiveness depends on their adaptation to the culture.

Therefore, coaches become worried by the constraints caused by the industrialization of football, which involves guidelines, goals and processes that are quite detached from the clubs' contexts of action, from the values, and that put the players individually considered at the forefront of their strategy, rather than the group. They consider that organized clubs can strengthen, but also weaken, the coaches' leadership and performance:

[…] because there are clubs that are very organized, very rigorous, very methodical in their processes that strengthen the coaches, they help a lot. And there are others that weaken performance (E12).

The coach lives in a dynamic and perverse cultural climate (Lança, 2015). In this aspect, coaches are quite critical and claim that the clubs' goals should be more focused on collective projects:

They are very important, but most of them have lost them […], we see more and more investors coming to the clubs […]. There may be exceptions, but investors focus a lot on the immediate, on the financial results, and fail to realize that the most important thing is not that, the basis of everything is not that, that the basis of everything is the values upheld and that is the basis for success. That is why we see a lot of people investing in clubs, but then they do not succeed, of course, there are always be some that do, but maybe they are the ones who do things better, and I think the values are being lost […] for those who are on the top, when the player is more important than the whole process itself, than the club itself, than the development of the structure itself, there is no success (E1).

Although the influence of OC on the coach's leadership is understood, the truth is that most coaches interviewed believe that the culture can be changed through their action. Coaches who do not believe in the impact of their leadership are in line with Alvesson and colleagues (2017), who argue that individuals have relatively little effect on culture. However, some authors advocate that leadership styles have a strong effect on OC because employees tend to act in a way that mirrors their leaders (Ostroff et al., 2013; Raguž & Zekan, 2017).

And, in this sample, there are those who assume that "[...] the coach agglutinates the act of putting passion, dynamics, emotion [...]" (E14), highlighting the club's culture, avoiding "losing values", but some sustain that "[...] you will not be capable of changing [...] you may even be capable of changing for a while, but then you end up not being capable of changing and I think that is mostly it" (E1).

Thus, even believing in the role as influencers, these coaches assume that this can only be accomplished with integration into the club and with the support of the club's structure, emphasizing the integration of players, seen as elements of the collective project. In other words, to take place, the process of change involves, according to the coaches' standpoint, both coaches, players and club managers, in a clear commitment:

They can, but I think that what I told you before is what seems to be central to me. [...] That is, if you arrive at a club that makes things in a certain way, and you, who have a completely different way of thinking and doing things, start doing it the way the club always did, you are not doing anything there. As someone said, to think like me, I am here already (E9).

This idea that there are times when leadership can have a strong influence on OC and other times when such efforts will not yield results, depending on organizational contingencies, is advocated by several authors (e.g. Ehrhart et al., 2014). Hence, before the change, these coaches emphasize the need for adaptation to avoid shocks and impositions that are misunderstood by the players themselves, which also entails a certain maturity and ability to relativize. To lead the club:

[...] is adapting, we have to adapt; as coaches, we have to understand that the players are the ones who play, we cannot play their game, we cannot. We have to help them play their game better, but we cannot play their game, they are the ones who play, and if we clash our idea and their game, there is no chance of a good performance; this harmony is also up to us sometimes. It depends on us colliding and taking our intransigence to the extreme regarding details that I think we later, with the growth, as coaches, end up ... like now, I have admitted it (E2).

Coaches mention, therefore, acculturation, the need to understand from within the culture and how the club operates, while, at the same time, imprinting their entity as coaches. Each "place" is a cultural challenge that these coaches deem necessary to understand and incorporate for them to be capable of performing their work and having high performance. Dimitrova (2019) stresses that this change depends on several contextual factors, such as the features of the internal organizational environment, top management, other members of the organization, trends in the macro and microenvironment and the business culture that shapes the organization.

CONCLUSIONS AND FUTURE RESEARCH DIRECTIONS

This study highlights that, regarding leadership and their success, these coaches seem to see themselves as managers, believing that their leadership is sustained in personal characteristics (e.g. ability to convince others, charisma, communication skills, …), formal authority, competence and adaptation to the team and clubs' context.

The results of this study also seem to indicate that, from the coaches' standpoint, culture and leadership are inextricably connected, influencing each other. Regarding the impact of their leadership, the coaches who participated in this study are divided, although a slight majority believe that it can influence clubs' OC. This influence can be exerted by primary mechanisms as attention and reactions to critical incidents, expressed in features like the need of exerting authority and the issue of responsibility in face of failure. Respondents who believe that it is possible to change emphasize that it is a challenging process, which requires time and results to sustain it. They also argue that it depends on the degree of rigidity of the culture and how it is rooted. They maintain that the outcome plays a critical role as a potential catalyst for organizational change.

However, the coach's effectiveness and sports results do not rely exclusively on this relationship. Gomes and colleagues (2008) also highlight that the coach's effectiveness stems from "[…] a combination between who the coach is as a person (e.g., personal and professional goals, availability to help others, etc.), the type of athletes they prepare (e.g., gender, age, personal preferences, etc.), and the context in which they perform their duties (e.g., collective or individual modality, competitive level, etc.)" (Gomes et al., 2008, p. 489).

The interviewed coaches mentioned the constraints caused by the industrialization of football, and of the several members, organizations, that influence the daily life in clubs. It would be interesting to tap the influence these "players" have in the development of the clubs' OC and coaches' leadership.

REFERENCES

Abdallah, J., & Jibai, S. (2020). Women in leadership: Gender personality traits and skills. *Business Excellence and Management*, *10*(1), 5–15. doi:10.24818/beman/2020.10.1-01

Abu-Jarad, I. Y., Yusof, N., & Nikbin, D. (2010). A review paper on organizational culture and organizational performance. *International Journal of Business and Social Science*, *1*(3), 26–46.

Ahmed Khan, Z., Nawaz, A., & Khan, I. (2016). Leadership theories and styles: A literature review. *Journal of Resources Development and Management*, *16*, 1–7.

Al-Ali, A. A., Singh, S. K., Al-Nahyan, M., & Sohal, A. S. (2017). Change management through leadership: The mediating role of organizational culture. *The International Journal of Organizational Analysis*, *25*(4), 723–739. doi:10.1108/IJOA-01-2017-1117

Alvesson, M., Kärreman, D., & Ybema, S. (2017). Studying culture in organizations: Not taking for granted the taken-for-granted. In A. Wilkinson, S. Armstrong, & M. Lounsbury (Eds.), *The Oxford Handbook of Management* (pp. 103–126). Oxford University Press.

Ancelotti, C., Brady, C., & Forde, M. (2017). *Quiet leadership: Winning hearts, minds and matches.* Penquin Books.

Antonakis, J., Bastardoz, N., Jacquart, P., & Shamir, B. (2016). Charisma: An ill-defined and ill-measured gift. *Annual Review of Organizational Psychology and Organizational Behavior, 3*(1), 293–319. doi:10.1146/annurev-orgpsych-041015-062305

Arruda, Â. F., Chrisóstomo, E., & Rios, S. S. (2010). A importância da liderança nas organizações. *Revista Razão Contábil & Finanças, 1*(1).

Arvey, R., Zhang, Z., Avolio, B., & Krueger, R. (2007). Developmental and genetic determinants of leadership role occupancy among females. *The Journal of Applied Psychology, 92*(3), 693–706. doi:10.1037/0021-9010.92.3.693 PMID:17484551

Ashford, S. J., & Sitkin, S. B. (2019). From problems to progress: A dialogue on prevailing issues in leadership research. *The Leadership Quarterly, 30*(4), 454–460. doi:10.1016/j.leaqua.2019.01.003

Barney, J. B. (1986). Organizational culture: Can it be a source of sustained sompetitive sdvantage? *Academy of Management Review, 11*(3), 656–665. doi:10.5465/amr.1986.4306261

Barrantes, R. J., & Eaton, A. A. (2018). Sexual orientation and leadership suitability: How being a gay man affects perceptions of fit in gender-stereotyped positions. *Sex Roles, 79*(9), 549–564. doi:10.100711199-018-0894-8

Bass, B. M. (1985). *Leadership and performance beyond expectations.* Free Press.

Bass, B. M. (1998). *Transformational leadership: industrial, military, and educational impact.* Lawrence Erlbaum Associates Publishers.

Bryman, A., Gillingwater, D., & McGuinness, I. (1992). Leadership and organisational transformation. *Management Research News, 15*(5/6), 26–27. doi:10.1108/eb028221

Cameron, K. S., & Quinn, R. E. (2011). *Diagnosing and changing organizational culture: Based on the competing values framework.* John Wiley & Sons.

Carvalhal, C. (2014). *Entre Linhas* [Between Lines]. Prime Books.

Cerni, T., Curtis, G., & Colmar, S. (2013). Information-processing and leadership: An introductory review and the cognitive leadership model (CLM). *Proceedings of the International Conference on Management, Leadership and Governance.*

Chatman, J. A., & O'Reilly, C. A. (2016). Paradigm lost: Reinvigorating the study of organizational culture. *Research in Organizational Behavior, 36*, 199–224. doi:10.1016/j.riob.2016.11.004

Chaudhry, A., Vidyarthi, P. R., Liden, R. C., & Wayne, S. J. (2021). Two to tango? Implications of alignment and misalignment in leader and follower perceptions of LMX. *Journal of Business and Psychology, 36*(3), 383–399. doi:10.100710869-020-09690-8

Chin, J. L. (2013). Diversity leadership: Influence of ethnicity, gender, and minority status. *Open Journal of Leadership, 2*(01), 1–10. doi:10.4236/ojl.2013.21001

Chiu, C.-Y., Balkundi, P., & Weinberg, F. J. (2017). When managers become leaders: The role of manager network centralities, social power, and followers' perception of leadership. *The Leadership Quarterly, 28*(2), 334–348.

Chong, M., Shang, Y., Richards, M., & Zhu, X. (2018). Two sides of the same coin? Leadership and organizational culture. *Leadership and Organization Development Journal, 39*(8), 975–994. doi:10.1108/LODJ-05-2017-0122

Coelho, O. (2004). *Pedagogia do desporto: Contributos para uma compreensão do desporto juvenil* [Pedagogy of sport: Contributions to an understanding of youth sport]. Livros Horizonte.

Coleman, M. (2003). Gender and the orthodoxies of leadership. *School Leadership & Management, 23*(3), 325–339. doi:10.1080/1363243032000112810

Crust, L., & Lawrence, I. (2006). A review of leadership in sport: Implications for football management. *Athletic Insight, 8*(4), 28–48.

Cunha, M. P., Rego, A., Cunha, R. C., Cabral-Cardoso, C., & Neves, P. (2016). *Manual de Comportamento Organizacional e Gestão* [Handbook of Organizational Behavior and Management]. RH Editora.

Curral, L., Gomes, C., Marques-Quinteiro, P., & Lind, P. (2016). *Caos e Complexidade: Novos Conceitos para a Gestão das Organizações* [Chaos and Complexity: New Concepts for Managing Organizations]. RH Editora.

Day, D., & Antonakis, J. (2012). Leadership: Past, present, and future. In J. Antonakis & D. V. Day (Eds.), *The nature of leadership* (pp. 3–25). Sage Publications, Inc. doi:10.1093/oxfordhb/9780199928309.013.0022

Deshpandé, R., & Farley, J. U. (2004). Organizational culture, market orientation, innovativeness, and firm performance: An international research odyssey. *International Journal of Research in Marketing, 21*(1), 3–22. doi:10.1016/j.ijresmar.2003.04.002

Deshpandé, R., Farley, J. U., & Webster, F. E. Jr. (1993). Corporate culture, customer orientation, and innovativeness in Japanese firms: A quadrad analysis. *Journal of Marketing, 57*(1), 23–37. doi:10.1177/002224299305700102

Deshpandé, R., & Webster, F. E. Jr. (1989). Organizational culture and marketing: Defining the research agenda. *Journal of Marketing, 53*(1), 3–15. doi:10.1177/002224298905300102

Dickson, M. W., Castaño, N., Magomaeva, A., & Den Hartog, D. N. (2012). Conceptualizing leadership across cultures. *Journal of World Business, 47*(4), 483–492. doi:10.1016/j.jwb.2012.01.002

Dimitrova, Y. (2019). Corporate Culture Change Management. *Economic Alternatives, 2*, 296–312.

Dinh, J. E., Lord, R. G., Gardner, W. L., Meuser, J. D., Liden, R. C., & Hu, J. (2014). Leadership theory and research in the new millennium: Current theoretical trends and changing perspectives. *The Leadership Quarterly, 25*(1), 36–62. doi:10.1016/j.leaqua.2013.11.005

Ehrhart, M., Schneider, B., & Macey, W. (2014). *Organizational climate and culture: An introduction to theory, research, and practice*. Routledge.

Felfe, J., & Petersen, L.-E. (2007). Romance of leadership and management decision making. *European Journal of Work and Organizational Psychology*, *16*(1), 1–24. doi:10.1080/13594320600873076

Fiedler, F. E. (1978). The Contingency Model and the Dynamics of the Leadership Process. In L. Berkowitz (Ed.), Advances in Experimental Social Psychology (Vol. 11, pp. 59–112). Academic Press. doi:10.1016/S0065-2601(08)60005-2

Fiordelisi, F., Renneboog, L., Ricci, O., & Lopes, S. S. (2019). Creative corporate culture and innovation. *Journal of International Financial Markets, Institutions and Money*, *63*, 101–137. doi:10.1016/j.intfin.2019.101137

Flint, S. W., Plumley, D. J., & Wilson, R. J. (2014). You don't know what you're doing! The impact of managerial change on club performance in the English Premier League. *Managing Leisure*, *19*(6), 390–399. doi:10.1080/13606719.2014.910000

Frontiera, J. (2010). Leadership and organizational culture transformation in professional sport. *Journal of Leadership & Organizational Studies*, *17*(1), 71–86. doi:10.1177/1548051809345253

Garganta, J. (2004). Atrás do palco. Nas oficinas do Futebol [Behind the stage. At the Football Workshops]. In Futebol de muitas cores e sabores. Reflexões em torno do desporto mais popular do mundo [Football of many colors and flavors. Reflections around the most popular sport in the world] (pp. 227–234). Campo das Letras.

Ghasabeh, M. S., Reaiche, C., & Soosay, C. (2015). The emerging role of transformational leadership. *Journal of Developing Areas*, *49*(6), 459–467. doi:10.1353/jda.2015.0090

Gorton, G. B., & Zentefis, A. K. (2020). *Corporate culture as a theory of the firm.* National Bureau of Economic Research Working Paper Series, No. 27353. doi:10.3386/w27353

Graen, G. B., & Uhl-Bien, M. (1995). Relationship-based approach to leadership: Development of leader-member exchange (LMX) theory of leadership over 25 years: Applying a multi-level multi-domain perspective. *The Leadership Quarterly*, *6*(2), 219–247. doi:10.1016/1048-9843(95)90036-5

Groysberg, B., Lee, J., Price, J., & Cheng, J. (2018). The leader's guide to corporate culture. *Harvard Business Review*, *96*(1), 44–52.

Hersey, P., & Blanchard, K. H. (1988). *Management and Organizational Behavior.* Prentice-Hall.

House, R. J., & Aditya, R. N. (1997). The social scientific study of leadership: Quo Vadis? *Journal of Management*, *23*(3), 409–473. doi:10.1177/014920639702300306

Huey Yiing, L., & Zaman Bin Ahmad, K. (2009). The moderating effects of organizational culture on the relationships between leadership behaviour and organizational commitment and between organizational commitment and job satisfaction and performance. *Leadership and Organization Development Journal*, *30*(1), 53–86. doi:10.1108/01437730910927106

Hughes, D. J., Lee, A., Tian, A. W., Newman, A., & Legood, A. (2018). Leadership, creativity, and innovation: A critical review and practical recommendations. *The Leadership Quarterly*, *29*(5), 549–569. doi:10.1016/j.leaqua.2018.03.001

Ilies, R., Gerhardt, M. W., & Le, H. (2004). Individual differences in leadership emergence: Integrating meta-analytic findings and behavioral genetics estimates. *International Journal of Selection and Assessment 12*(3), 207–219. doi:10.1111/j.0965-075X.2004.00275.x

Kelly, S. (2008). Understanding the role of the football manager in britain and ireland: A weberian approach. *European Sport Management Quarterly*, *8*(4), 399–419. doi:10.1080/16184740802461652

Kim, J.-Y., & Miner, A. S. (2007). Vicarious learning from the failures and near-failures of others: Evidence from the U.S. commercial banking industry. *Academy of Management Journal*, *50*(3), 687–714. doi:10.5465/amj.2007.25529755

Kirkpatrick, S., & Locke, E. (1991). Leadership: Do Traits Matter? *The Executive*, *5*, 48–60. doi:10.2307/4165007

Knights, D., & O'Leary, M. (2006). Leadership, ethics and responsibility to the other. *Journal of Business Ethics*, *67*(2), 125–137. doi:10.100710551-006-9008-6

Koch, C. (2018). Lead like ... Gareth Southgate. *Director*, *71*(3), 28. http://search.ebscohost.com/login.aspx?direct=true&db=bth&AN=128242031&site=eds-live

Kuper, S. (1996). *Football Against the Enemy*. Orion Books.

Lança, R. (2015). *Coach to Coach*. Prime Books.

Leithy, W. (2017). Organizational culture and organizational performance. *International Journal of Economics & Management Sciences*, *6*(4), 1–66. doi:10.4172/2162-6359.1000442

Li, P., Sun, J.-M., Taris, T. W., Xing, L., & Peeters, M. C. W. (2021). Country differences in the relationship between leadership and employee engagement: A meta-analysis. *The Leadership Quarterly*, *32*(1), 101458. doi:10.1016/j.leaqua.2020.101458

Liden, R. C., & Antonakis, J. (2009). Considering context in psychological leadership research. *Human Relations*, *62*(11), 1587–1605. doi:10.1177/0018726709346374

Lord, R. G. (2019). Leadership and information processing: A dynamic system, dual-processing perspective. In Leader Thinking Skills (pp. 46–70). Routledge.

Lord, R. G., de Vader, C. L., & Alliger, G. M. (1986). A meta-analysis of the relation between personality traits and leadership perceptions: An application of validity generalization procedures. *The Journal of Applied Psychology*, *71*(3), 402–410. doi:10.1037/0021-9010.71.3.402

Maguire, J. (2008). 'Real politic' or 'ethically based': Sport, globalization, migration and nation-state policies. *Sport in Society*, *11*(4), 443–458. doi:10.1080/17430430802019375

Maitland, A., Hills, L. A., & Rhind, D. J. (2015). Organisational culture in sport – A systematic review. *Sport Management Review*, *18*(4), 501–516. doi:10.1016/j.smr.2014.11.004

Matthews, S. H., Kelemen, T. K., & Bolino, M. C. (2021). How follower traits and cultural values influence the effects of leadership. *The Leadership Quarterly*, *32*(1), 101497. doi:10.1016/j.leaqua.2021.101497

Meindl, J. R., & Ehrlich, S. B. (1987). The romance of leadership and the evaluation of organizational performance. *Academy of Management Journal, 30*(1), 1, 91–109. doi:10.2307/255897

Meindl, J. R., Ehrlich, S. B., & Dukerich, J. M. (1985). The Romance of Leadership. *Administrative Science Quarterly*, *30*(1), 78–102. doi:10.2307/2392813

Mendes, R. M., & Miskulin, R. G. S. (2017). A análise de conteúdo como uma metodologia [Content analysis as a methodology]. *Cadernos de Pesquisa, 47(*165), 1044-1066. https://dialnet.unirioja.es/servlet/oaiart?codigo=6132756

Miles, J. A., & Naumann, S. E. (2021). Gender, sexual orientation, academic self-concept and leadership. *Higher Education, Skills and Work-Based Learning, 11*(5), 983-1001. doi:10.1108/HESWBL-05-2020-0101

Miller, P. W. (2019). 'Tackling' race inequality in school leadership: Positive actions in BAME teacher progression – evidence from three English schools. *Educational Management Administration & Leadership*, *48*(6), 986–1006. doi:10.1177/1741143219873098

Mills, J. P., & Boardley, I. D. (2016). Expert Premier League soccer managers' use of transformational leadership behaviours and attitude towards sport integrity: An intrinsic case study. *International Journal of Sports Science & Coaching*, *11*(3), 382–394. doi:10.1177/1747954116645011

Moreno-Arrebola, R., Martinez Martinez, A., Zurita Ortega, F., & San Roman Mata, S. (2017). La influencia del liderazgo del entrenador en los deportistas. *European Journal of Child Development. Education and Psychopathology*, *5*(2), 73. doi:10.30552/ejpad.v5i2.49

Mumford, M. D., Todd, E. M., Higgs, C., & McIntosh, T. (2017). Cognitive skills and leadership performance: The nine critical skills. *The Leadership Quarterly*, *28*(1), 24–39. doi:10.1016/j.leaqua.2016.10.012

Naranjo-Valencia, J. C., Jiménez-Jiménez, D., & Sanz-Valle, R. (2016). Studying the links between organizational culture, innovation, and performance in Spanish companies. *Revista Latinoamericana de Psicología*, *48*(1), 30–41. doi:10.1016/j.rlp.2015.09.009

Nguyen, H., & Mohamed, S. (2011). Leadership behaviors, organizational culture and knowledge management practices: An empirical investigation. *Journal of Management Development*, *30*(2), 206–221. doi:10.1108/02621711111105786

Northouse, P. G. (2015). *Leadership: Theory and Practice*. SAGE Publications.

O'Reilly, C. A. III, Caldwell, D. F., Chatman, J. A., & Doerr, B. (2014). the promise and problems of organizational culture: CEO personality, culture, and firm performance. *Group & Organization Management*, *39*(6), 595–625. doi:10.1177/1059601114550713

Ospina, S., & Foldy, E. (2009). A critical review of race and ethnicity in the leadership literature: Surfacing context, power and the collective dimensions of leadership. *The Leadership Quarterly*, *20*(6), 876–896. doi:10.1016/j.leaqua.2009.09.005

Ostroff, C., Kinicki, A., & Muhammad, R. (2013). Organizational Culture and Climate. In Handbook of Psychology *(Vol 12*, pp. 643–676). John Wiley & Sons, Inc.

Puffer, S. M. (1990). Attributions of charismatic leadership: The impact of decision style, outcome, and observer characteristics. *The Leadership Quarterly*, *1*(3), 177–192. doi:10.1016/1048-9843(90)90019-E

Raguž, I., & Zekan, S. (2017). Organizational culture and leadership style: key factors in the organizational adaptation process. *Ekonomski Vjesnik/Econviews - Review of Contemporary Business, Entrepreneurship and Economic Issues, 30*(1). Retrieved from https://hrcak.srce.hr/ojs/index.php/ekonomski-vjesnik/article/view/4603

Randolph-Seng, B., Cogliser, C. C., Randolph, A. F., Scandura, T. A., Miller, C. D., & Smith-Genthôs, R. (2016). Diversity in leadership: Race in leader-member exchanges. *Leadership and Organization Development Journal*, *37*(6), 750–773. doi:10.1108/LODJ-10-2014-0201

Rego, A., Mota, T., Cunha, M., & Seabra, A. (2020). *O treinador como "dominador e amigo"? Uma abordagem ao amor duro na liderança* [The coach as "dominator and friend"? An approach to tough love in leadership]. Católica Porto Business School. Available at https://www.catolicabs.porto.ucp.pt/Files/PortalReady/v000/Docs/PDF/amor_duro.pdf

Schein, E. H. (1996). Culture: The missing concept in organization studies. *Administrative Science Quarterly*, *2*(41), 229–240. doi:10.2307/2393715

Schein, E. H. (2010). *Organizational Culture and Leadership*. Jossey-Bass.

Shahzad, F., Luqman, A., Khan, A., & Shabbir, L. (2012). Impact of organizational culture on organizational performance: An overview. *Interdisciplinary Journal of Contemporary Research in Business*, *3*(9), 975–985.

Shamir, B., House, R. J., & Arthur, M. B. (2018). The motivational effects of charismatic leadership: A self-concept based theory. In I. Katz, G. Eilam-Shamir, R. Kark, & Y. Berson (Eds.), *Leadership Now: Reflections on the Legacy of Boas Shamir* (pp. 9–29). Emerald Publishing Limited. doi:10.1108/S1479-357120180000009009

Silva, A. (2016). What is leadership? *Journal of Business Studies Quarterly*, *8*(1), 1–5. PMID:29355200

Simonet, D., & Tett, R. (2013). Five perspectives on the leadership–management relationship: A competency-based evaluation and integration. *Journal of Leadership & Organizational Studies*, *20*(2), 199–213. doi:10.1177/1548051812467205

Sousa, G. (2018). *Decidir como um treinador - A ciência e a prática no futebol e na vida* [Deciding like a coach - The science and practice in football and life]. Federação Portuguesa de Futebol.

Steers, R. M., & Shim, W. S. (2013). Strong leaders, strong cultures: Global management lessons from Toyota and Hyundai. *Organizational Dynamics*, *42*(3), 217–227. doi:10.1016/j.orgdyn.2013.06.007

Thompson, H. (2019). These times. *New Statesman (London, England)*, *5491*(148), 25. http://search.ebscohost.com/login.aspx?direct=true&db=bth&AN=138948208&site=eds-live

Tran, Q. H. N. (2021). Organisational culture, leadership behaviour and job satisfaction in the Vietnam context. *The International Journal of Organizational Analysis*, *29*(1), 136–154. doi:10.1108/IJOA-10-2019-1919

Tsai, Y. (2011). Relationship between organizational culture, leadership behavior and job satisfaction. *BMC Health Services Research, 11*(1), 98. doi:10.1186/1472-6963-11-98 PMID:21569537

Uhl-Bien, M., & Arena, M. (2018). Leadership for organizational adaptability: A theoretical synthesis and integrative framework. *The Leadership Quarterly, 29*(1), 89–104. doi:10.1016/j.leaqua.2017.12.009

Vecchio, R. P. (2002). Leadership and gender advantage. *The Leadership Quarterly, 13*(6), 643–671. doi:10.1016/S1048-9843(02)00156-X

Vieira, L., Pizzo, G., Contreira, A., Lazier-Leão, T., Moreira, C., Gaion, P., & Nascimento, J. R. Junior. (2017). Associação entre motivação e coesão de grupo no futebol profissional: O relacionamento treinador-atleta é um fator determinante? [Association between motivation and group cohesion in professional football: the coach-athlete relationship is a determining factor?]. *Revista de Psicología del Deporte, 27*, 51–57.

Vroom, V., & Jago, A. (2007). The role of the situation in leadership. *The American Psychologist, 62*(1), 17–24. doi:10.1037/0003-066X.62.1.17 PMID:17209676

Wagstaff, C., & Burton-Wylie, S. (2018). Organisational culture in sport: A conceptual, definitional and methodological review. *Sport and Exercise Psychology Review, 1*(14), 32–52.

Warrick, D. D. (2017). What leaders need to know about organizational culture. *Business Horizons, 60*(3), 395–404. doi:10.1016/j.bushor.2017.01.011

Warrick, D. D., Milliman, J. F., & Ferguson, J. M. (2016). Building high performance cultures. *Organizational Dynamics, 45*(1), 64–70. doi:10.1016/j.orgdyn.2015.12.008

Watts, L. L., Steele, L. M., & Mumford, M. D. (2019). Making sense of pragmatic and charismatic leadership stories: Effects on vision formation. *The Leadership Quarterly, 30*(2), 243–259. doi:10.1016/j.leaqua.2018.09.003

Xenikou, A., & Simosi, M. (2006). Organizational culture and transformational leadership as predictors of business unit performance. *Journal of Managerial Psychology, 21*(6), 566-579. doi:10.1108/02683940610684409

Zyphur, M. J., Narayanan, J., Koh, G., & Koh, D. (2009). Testosterone–status mismatch lowers collective efficacy in groups: Evidence from a slope-as-predictor multilevel structural equation model. *Organizational Behavior and Human Decision Processes, 110*(2), 70–79. doi:10.1016/j.obhdp.2009.05.004

Chapter 7
Market and Entrepreneurial Orientation Strategies in the Wood Furniture Industry in Mexico

Paulina Aburto-De La Llave
Syngenta, Mexico

Connie Atristain-Suárez
https://orcid.org/0000-0002-1883-6194
Universidad Panamericana, Mexico

Andrea Uribe-Palau
Universidad Panamericana, Mexico

ABSTRACT

The wood furniture industry plays a predominant role in the Mexican economy. This research takes as its axis to analyze the competitiveness factors of wooden furniture manufacturing SMEs based on market orientation (MO) and business orientation (EO) strategies that influence consumers' perception and purchase preference in Mexico. Are the wooden furniture manufacturing SMEs aligned with competitive strategies? Do companies in this sector implement MO-EO-oriented strategies? The study showed that Mexican consumers generally recognize the wood furniture industry in Mexico, assign it an authenticity value, and associate it with a positive image. When customers perceive that the company is committed to creating superior value for the customer through the MO and EO, they show loyalty and purchase preference behaviours.

DOI: 10.4018/978-1-6684-2339-4.ch007

INTRODUCTION

The Wood Furniture Industry in Mexico currently has several structural deficiencies that have led it, to some extent, to a critical situation. The sector is characterized as uncompetitive and inefficient; however, the domestic market has aggressive international competition, mainly from Asia, Chile, and the United States. According to Chapela (2012), the main comparative disadvantages are the following: (1) more modern technology; (2) more modern transport and lower transportation costs; (3) substantial advances in the standardization of products on the market and (4) services supply associated with the product. These deficiencies make the competitive landscape more complicated for the industrialization and marketing of wood furniture in Mexico.

The furniture manufacturing industry is located in the secondary sector, within the following activity classification: Manufacture and repair of furniture, mainly wood, manufacture of parts and furniture parts (INEGI-DENUE, 2018). The manufacturing sector is divided into 21 industries, one of which corresponds to the manufacture of furniture. According to the National Statistical Directory of Economic Units (2018), its importance corresponds to 6.51% of the total industry, resulting in a national total of 34,526 economic units. This subsector is divided into four categories: (1) manufacture of furniture except for office and bookshelf; (2) manufacture of office furniture and bookshelf; (3) manufacture of window coverings and rods, and mattresses.

The local Mexican offer is characterized by a separation between the world of design and production at the industrial level. The vast majority of companies engaged in the manufacture of furniture in Mexico are located within the classification of small and medium-sized enterprises (SMEs) and direct their products primarily to the medium and low purchasing power segments. In contrast, companies that produce designer furniture in large volumes turn out to be a minority (Legua, 2015). It is considered an industry characterized by having many companies where family workshops and small manufacturers-owners prevail, intensive labor, little automation, and small-scale production. (Lozano and Trinidad, 2016).

Mexico's Economy Ministry (2017) notes that from 2013 to 2017, the total amount of direct investment to the Furniture Industry in Mexico channeled 140 million, United States is the principal investor with 51.7%, followed by the United Kingdom with 16.9%, Sweden with 16.1% and Canada with 8.8%. Furniture exports represent 7.4% of the total manufactured goods exported in Mexico (INEGI, 2018). According to the Bank of Mexico (2017), exports from the Furniture Industry have a degree of commercial importance for Mexico, of 0.4% of total trade.

The Furniture Industry in Mexico follows the guideline of value chains led by buyers. That is why the purchasing power of society plays a decisive role in purchasing behavior and the personalization of goods and services (Sahui, 2008). Regarding the consumption of wooden furniture in Mexico, with data from the National Survey of Household Expenses (INEGI, 2013), Mexican families allocated 4.5% of their annual expenses to the purchase of furniture and accessories, carpets, and other materials, raising the percentage for 2016 to 5.6%.

The marketing strategies most used by the Wood Furniture Industry in the Mexican national market are department stores. According to the National Association of Supermarket and Department Stores (ANTAD by its acronym in Spanish, 2019), the total sales of these stores were $326 000 MP. From the study carried out by Legua (2015) under the supervision of the Economic and Commercial Office of the Embassy of Spain in Mexico City, the authors can highlight based on the data provided in the study that the leading distribution channels in our country for this type of product, three stand out mainly:

1) Department stores: 16% of total furniture sales in Mexico.
2) Shops specialized in the design furniture: Located in the best areas of Mexico City, Guadalajara, and Monterrey and semi-urban holiday centers.
3) Own stores and showrooms of international brands: They mainly market collections of international and local firms.

Another way to market wooden furniture is through information technology and electronic commerce (Jones, Motta, and Aldrette, 2016); however, some of the factors that have influenced the loss of competitiveness in the Mexican Wooden Furniture Industry are the distribution channels and the lack of use of technology (Valenzo and Ortiz, 2007). For this industry, implementing a competitive strategy of TI and e-commerce is a critical factor in activating participation in domestic markets.

The manufacturing sector plays a predominant role in Mexico's economy, particularly the Wood Furniture Industry, which is of significant importance within the Mexican industrial fabric due to many small and medium-sized companies. This research takes as an axis of analysis the competitiveness factors of SME manufacturers of wooden furniture based on the Market Orientation (MO) and Entrepreneurial Orientation (EO) strategies that influence the perception and preference of purchase of consumers in Mexico. In addition, a model is proposed for the wooden furniture industry that provides strategic direction to achieve competitive advantage and growth of the SME business within the Wooden Furniture Industry in Mexico.

All small and medium-sized businesses have the potential to be competitive, regardless of their size or how long they have been in business. Knowing the strategies that improve competitiveness allows achieving better results; likewise, the implementation of the strategy is of the utmost importance. However, it lies in its link with business results, however, are the wooden furniture manufacturing SMEs aligned with competitive strategies? Do companies in this sector implement MO-EO-oriented strategies? The general objective of this research is to identify the competitiveness factors of wooden furniture manufacturing SMEs, based on the MO and EO strategies that influence the perception and purchase preference of consumers in Mexico.

The specific objectives include:

1) Analyze the effect of the implementation of MO and EO strategies on consumer perception of the industry of wooden furniture in Mexico.
2) Analyze the effect of the consumer's perception of MO and EO actions in purchase preference.

BACKGROUND

Perception-Related to Market Orientation (MO) and Entrepreneurial Orientation (EO) Strategies

Since the 1950s, it has been talked about customer-oriented philosophy. This concept has evolved to give rise to MO. It has had a more significant presence in the last 25 years, focusing on customers, customer value, customer relations, and competition. Subsequently, in the 1980s, the concept of EO was introduced for companies dedicated to proactive innovation in the market and that take risks to outperform their competitors. Some studies say that by simultaneously implementing both orientations in a company, a powerful brand is generated, and sales and profits are maximized.

The concept of Market Orientation (MO) highlights three perspectives: behavioral, cultural, and capacity. The first perspective of MO is attributed to the behavioral perspective of Kohli and Jaworski (1990). The organization's ability to generate and disseminate market intelligence, as well as its responsiveness. For these authors, market intelligence includes customers' current and future needs and external market factors, such as competition, regulation, technology, and other environmental forces. Likewise, they highlight that it is the responsibility of all areas of the organization to generate and collect market information and disseminate it effectively in the organization.

The second perspective relates to more fundamental characteristics of the organization in terms of values and norms. From this perspective, Narver and Slater (1990) defined Market Orientation as "the organizational culture that more effectively and efficiently creates the necessary behaviors, for the superior value creation for buyers and, therefore, continuous superior business performance." The MO has three dimensions: Customer orientation (it is sufficient understanding of the target buyers to be able to create superior value for them continually); Competitive orientation (is the understanding of the strengths and weaknesses in the short term and the capabilities and long-term strategies of the primary current and potential competitors); and Inter-functional coordination (is the coordinated use of a company's resources to create superior value for target customers).

The third perspective focuses on resource capabilities (Kyriakopoulos and Moorman 2004). The organization's capacity level is the linkage degree that the company has with its external environment for the customers' needs collection and the information dissemination of the market and customer obtained within the organization to reach the market promptly. Thus, companies can compete by anticipating market needs before competitors and creating and maintaining good relationships with customers, channel members, and suppliers.

Market Orientation (MO) has been classified as an elementary strategy and has frequently been related to high business performance. Especially when factors such as globalization, rapid technological advances, economic crises, and many competitors that each business faces are involved (Hajipour, Rahimi, and Hooshmand, 2013; Grinstein, 2008). Market Orientation (MO) is established to reflect the organization's priorities to satisfy customers' needs for existing and potential and create value (Day, 1994). In addition, the implementation of the MO strategies reflects a committed organization to identifying and exploiting new market opportunities (Lumpkin and Dess, 1996).

Miller (1983;2011) introduced the concept of Entrepreneurial Orientation (EO). He identified three dimensions for Entrepreneurial Orientation, which are innovation (is the activity where it is sought to introduce new goods, services, and technology); proactivity (implies taking initiatives in the development of creative actions and the search for new market opportunities); and risk-taking (reflects a company's willingness to allocate available resources to opportunities that could be in conjunction with the possibility of costly failure). He points out that an entrepreneurial company is dedicated to innovation in the product market, takes risks, and is the first to propose proactive innovations to outperform competitors. An EO firm usually maintains productive learning by finding and exploring value-added opportunities (Chaston and Scott, 2012). The entrepreneurial attitude of companies is an essential condition to be able to compete in a globalized and changing economic environment (Van Doorn et al., 2013). The EO promotes creativity and innovations within companies (Engelen et al., 2015).

EO and MO are two strategic orientations that firms may utilize to develop successful new products and remain competitive within the marketplace (Morgan, Anokhin, Kretinin, and Frishammar, 2015;

Miles and Arnold, 1991). The importance of Market Orientation and Entrepreneurial Orientation has been emphasized in the field of Organizational Theory. In large part of the research on MO and EO, it is considered that organizations need to adopt both guidelines to create synergies simultaneously and contribute to the success of the company (Baker and Sinkula, 2009; Schindehutte, Morris and Kocak, 2008; Miles and Darroch, 2008; George and Zahra, 2002). Therefore, firms that develop their strategic framework based on MO and EO orientations will be better positioned to achieve their long-term objectives, client satisfaction and easily survive in the competitive market. These orientations consequently culminate in improved organization performance (Vega-Vázquez et al., 2016).

Companies have to plan their marketing strategies to provide more value for the customer in a way that enables them to retain the customers and increase their loyalty. (Kim et al. 2004). The following factors must be addressed in the strategy to take care of the consumer relationship (1) Personalization – it is used as a tactic and try to enhance the customers' satisfaction in order to make more profit; (2) Satisfaction – is significantly taken into account as assurance for customer retention.; (3) Communication – is considered an important characteristic that is also believed to be a core competency.; and (4) Customer Loyalty is an essential factor that has to be developed if want to maintain the company and develop its profitability. (Bagherzad, Chavosh,and Hosseinikhah, 2011).

Knowing what the image and reputation that consumers have about the company and the factors that positively or negatively influence the consumer's perception of both the image and the reputation is fundamental. Since companies are not only entities productive, they also exercise constant communication with their audiences (Walesska and Alvarado, 2009). For example, one of the critical factors influencing the consumers' decision-making when buying is Corporate Social Responsibility. It gives consumers a better perception of the company, better reputation, and image (Bigné and Curras, 2008; Marín, Espinal, López, Maldonado, Mira, and Pertusa, 2011); hence,

Market Orientation (MO)

Hypothesis One: Consumer-oriented actions of SMEs positively impact the industry image consumers' perception.

Hypothesis Two: Competitively oriented actions of SMEs positively impact the industry image consumers' perception.

Hypothesis Three: The inter-functional coordination actions of SMEs positively impact the industry image consumers' perception.

Entrepreneurial Orientation (EO):

Hypothesis Four: The innovative actions of SMEs positively impact the industry image consumers' perception.

Hypothesis Five: The Proactive actions of SMEs positively impact the industry image consumers' perception.

Hypothesis Six: The Risk-taking actions of SMEs positively impact the industry image consumers' perception.

Consumer's Perception of MO-EO-Related to Purchasing Preferences: In Situ and Online

The theory of consumer perception attempts to explain consumer behavior by analyzing the motivations for the purchase or absence (Jacoby and Olson, 1985; Bishop, 1984; Doyle, 1984; Sawyer and Dickson, 1987; Schechter, 1984). It is focused on three main areas: price perception, perception of quality, and perception of a benefit that affects its quality of life (Krishnan, 1996; Zeithaml, 1988; Walters and McKenzie, 1988). The perception can be positive or negative, depending on whether the brand, product, or service experience turned out to be satisfactory or unsatisfactory (Richins, 1983). In addition to the above, a factor that significantly influences consumer perception is word-of-mouth communication (Maxham and Netemeyer, 2003).

Purchase intention indicates the possibility that consumers will plan to purchase a particular product in the future, which is the basis of the exhibited purchasing behavior (Martins et al., 2019). Previous studies indicate that the increase in a client's perception of the product reflects the increase in the possibility of purchase. When the consumers have a positive perception, this forms a positive brand commitment, which propels consumers to take an actual purchase action (Wu et al., 2011). Cognitive effects, within which perception is located, can be exploited by subjects in the purchasing process (Zielke, 2008; Bigné and Andreu, 2004). Companies can also leverage them to achieve their advertising goals, such as publicizing a brand or providing essential information to initiate a purchasing process (Lavidge and Steiner, 1961).

The variables in the literature to evaluate cognitive effects, which are decisive when making decisions by a consumer, consist of repetition of the appearance of the mark, recognition of the location, notoriety of the mark, and record of the site (Balasubramanian, Karrh and Patwardhan, 2006). Technological advances and the Internet have driven e-commerce to significantly impact organizational operations (DeLone and McLean, 2004). In addition, its growing development has provided significant advantages for both organizations and consumers (Devaraj, Fan, and Kohli, 2002). E-commerce has generated a convenience-based alternative to offering products and services, i.e., their online purchase and sale.

There are two types of online shoppers, those "goal-oriented" (GO), specific product consumer, and "experimental consumers" (EC), consumers looking for experiences (Wolfinbarger and Gilly, 2001). GO, also called utility buyers, are task-oriented, efficient, rational, and premeditated in their purchases. His interests focus on electronic retail purchasing with four attributes: convenience and accessibility, selection, available product information, and lack of sociability. On the other hand, the EC's behavior, mainly focusing on their hobbies, enjoy online buying new additions for their collections, and the greater their joy of finding new products, the greater the satisfaction of buy, even greater than that of a GO.

When an individual must decide what to buy and not buy, one of the most influential variables in this decision is the price, which is influenced not only by the price seen in the product but also by the perception of the retailer's price image (Hamilton and Chernev, 2013). Knowing in real-time the price of a product gives the consumer some bargaining power in front of the retailer, which can improve the consumer experience, since, if reaching an agreement with the trader would avoid having to go to another store even if it has a lower price (Molinillo and Viano-Pastor, 2015).

The consumer has more information about the characteristics of the products and prices thanks to the "Smartphone," which allows evaluating more alternatives. However, the purchase decision is more complex since the consumer has a more significant number of decisions (Verkasalo, López-Nicolás, Molina-Castillo, and Bouman, 2010). The merchant must facilitate the evaluation of alternatives by providing complete information, interacting with the consumer, using regular buyers and recommendation systems, or participating in geolocation platforms, among other strategies to remain competitive (Molinillo and Viano-Pastor, 2015).

Today, consumers show more participatory and active behavior in purchase decisions. Their degree of connection and linkage through social networks and the Internet allows them to be permanently informed about a remarkable number of alternatives. In many cases, they know more about the marketed products or services than it is thought. More informed and prepared, this new consumer demands a more significant value generation in products and services (Sánchez, Iniesta and Holbrook, 2009). Thus,

Market Orientation (MO)

Hypothesis Seven: The greater the consumer's perception of market-oriented actions, the greater online purchase preference.

Hypothesis Eight: The greater the consumer's perception of market-oriented actions, the greater in-situ purchase preference.

Entrepreneurial Orientation (EO):

Hypothesis Nine: The greater the consumer's perception of entrepreneurial-oriented actions, the greater online purchase preference.

Hypothesis Ten: The greater the consumer's perception of entrepreneurial-oriented action, the greater in-situ purchase preference

CONCEPTUAL FRAMEWORK

This study takes as independent variables the Market Orientation (MO) and the Entrepreneurial Orientation (EO) of the SMEs manufacturing of wooden furniture. It seeks to identify the purchasing preferences according to the consumer perception of these two Constructs. In other words, this analyses consumers' perceptions (mediator variable) of the implementation of MO-EO strategies by SMEs manufacturing wooden furniture, and that influences the behavior's purchase (dependent variable). Figure 1 presents the conceptual framework of consumers' purchase preference based on their perception of SMEs' MO and EO strategies manufacturing wooden furniture.

Figure 1. Conceptual framework of consumers' purchase preference based on their perception of MO-EO strategies of SMEs manufacturing wooden furniture
Source: Authors' elaboration

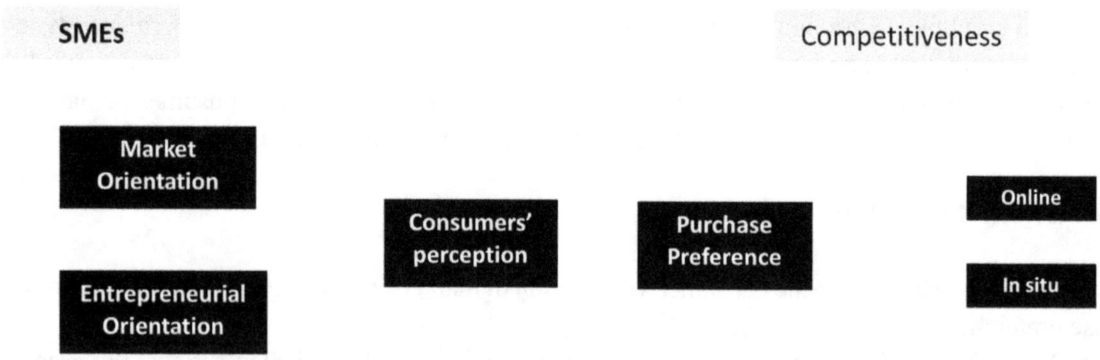

RESEARCH DESIGN

The research design is defined as non-experimental, transactional, and the approach is correlational/causal. The research is non-experimental, as it is done without the intentional manipulation of variables. Transactional, by its temporal dimension when the data is collected. It is Correlational/Causal since it is intended to describe the relationships between two or more variables in a given time, as well as their causality (Hernández et al., 2018). This research aims to provide information on the purchase preference of consumers according to their perception of market orientation and entrepreneurial orientation carried out by the SMEs of the Wood Furniture Industry in Mexico.

The sample has been chosen by convenience or intentional sampling since the individuals of the population have been directly and intentionally selected. The segment of the population that was considered for the sample consisted of men and women, residents of Mexico City, over 18 years of age, with a medium socioeconomic level and who had bought wooden furniture at least once in their lives.

An instrument was developed with a Likert scale of five-point, where 1 is "Strongly Disagree," 2 is "Disagree," 3 is "Neutral," 4 is "Agree," and 5 is "Strongly Agree." The instrument was tested on a sample of 25 wooden furniture consumers. The instrument was personally administered to know the consumers' reactions, review the questions' wording, and make the corrections. Later, the instrument was sent to 494 consumers of wooden furniture through different electronic means: an online platform, "Survey Monkey," email, and social network Facebook. Of the sample, only 189 consumers completed the survey, i.e., there was a response rate of 38.26 percent (see Table 1). The response rate obtained from the sample in question is above the average of online surveys. The higher number of online surveys achieve much lower response rates than paper-based ones, averting 33% online versus 56% on paper (Nulty, 2008). The sample was made up of only those consumers of wooden furniture who answered the instrument in its entirety.

Table 1. Summary of the electronic means used, data, and response rate.

Middle	Number of Questionnaires Sent	Returned	Outside the Sample Profile	Answered	Completed the Survey	Response Rate (%)
Email	316	69	38	198	11	3.48
Platform Web Link	156	0		0	156	100
Social networks "Facebook."	22	0		0	22	100
Import	494	69	38	198	189	38.26

Source: Authors' elaboration

Variables' Operationalization

The variables selected for the study were widely referenced and classified in the following constructs: Market Orientation, Entrepreneurial Orientation, Consumer Perception, and Purchase Preference. The independent variables of the research are the Market Orientation and Entrepreneurial Orientation of the manufacturing SMEs of wooden furniture and are measured through three dimensions each:

- **Market Orientation:**
  - Customer Orientation
  - Competitive Orientation
  - International Coordination
- **Entrepreneurial Orientation:**
  - Innovation
  - Proactivity
  - Assumption to Risk

The dependent variable is the purchase preference of wooden furniture and is measured through two dimensions:

- **Purchase Preference:**
  - Online
  - In Situ

The following Tables 2 and 3 present a breakdown of the reagents that were used to measure each of the Conceptual Model constructs presented in the literature review.

Table 2. Reagents breakdown to measure each of the Market Orientation and Entrepreneurial Orientation constructs

INDEPENDENT VARIABLE					
Market Orientation			**Entrepreneurial Orientation**		
Client (OMCL)	**Competition (OMCO)**	**Interfunctional Coordination (OMIC)**	**Innovation (EOI)**	**Proactivity (EOP)**	**Risk Oriented (EORO)**
17. Furniture made in Mexico is child-proof (safe and resistant).	14. The wooden furniture made in Mexico is elegant.	2. The wooden furniture price made in Mexico is the most important thing.		5. The purchasing place (store) of the wooden furniture made in Mexico is the most important.	10. The wooden furniture price made in Mexico is comparable to imported (foreign) furniture.
18. The wooden furniture made in Mexico is durable.	15. The wooden furniture made in Mexico is fine.	3. The wooden furniture quality made in Mexico is the most important thing.	21. The wooden furniture made in Mexico is innovative.	6. Promotions (offers, discounts, interest-free months) offered by the place of wooden furniture purchase made in Mexico is the most important thing	
24. The wood furniture made in Mexico covers all Mexican consumers' needs.	16. The wooden furniture made in Mexico is comfortable.	4. The wooden furniture design made in Mexico is the most important thing.	22. The wooden furniture made in Mexico is just as innovative as imported (foreign) furniture.	20. The wooden furniture supply made in Mexico is adequate (sufficient supply for demand).	
	23. The wooden furniture made in Mexico is multifunctional (multipurpose).			26. Advertising in mass media (TV, radio, magazine, newspaper, billboard, etc.) of wooden furniture made in Mexico influences my purchase decision.	
	25. The wooden furniture made in Mexico covers all foreign consumers' needs.			27. Online advertising (Facebook, Twitter, YouTube, Instagram, etc.) of wooden furniture made in Mexico influences my purchasing decision.	

Source: Authors' elaboration

Table 3. Reagents breakdown to measure each of the Consumer Perception and Purchase Preference constructs

DEPENDENT VARIABLE				
Consumer Perception			**Purchase Preference**	
Market Orientation (CPMO)	**Entrepreneurial Orientation (CPEO)**	**MO-EO balance (CPMOEO)**	**Online (PPOL)**	**In Situ (PPIS)**
11. The wood furniture made in Mexico is made according to the tastes and preferences of Mexican consumers.	7. It is evident when wooden furniture is made in Mexico.	7. It is evident when wooden furniture is made in Mexico.	13. I prefer buying wooden furniture made in Mexico.	13. I prefer buying wooden furniture made in Mexico.
12. The wood furniture made in Mexico is made according to the tastes and preferences of foreign consumers.	8. The wooden furniture price made in Mexico is affordable (attainable).	8. The wooden furniture price made in Mexico is affordable (attainable).	30. I prefer to showcase (look without buying) wooden furniture made in Mexico online (via a website).	28. It is important to test the wooden furniture made in Mexico before buying it.
19. The wooden furniture made in Mexico is of high quality.	9. The wooden furniture price made in Mexico goes according to its value.	9. The wooden furniture price made in Mexico goes according to its value.	32. I prefer buying wooden furniture made in Mexico online (via a website).	29. I prefer to showcase (look without buying) wooden furniture made in Mexico in situ (go to the furniture store).
		11. The wood furniture made in Mexico is made according to the tastes and preferences of Mexican consumers.		31. I prefer buying wooden furniture made in Mexico in situ (in a furniture store).
		12. The wood furniture made in Mexico is made according to the tastes and preferences of foreign consumers.		
		19. The wooden furniture made in Mexico is of high quality.		

Source: Authors' elaboration

According to their perception of the implementation of Market and Entrepreneurial Orientation's strategies of SMEs manufacture wooden furniture, the participants responded to the survey questions. It about knows if companies have formally established procedures or programs within the company that guide them to serve their market or carry out a job or activity in a programmed way that allows them to innovate, take risks, and be more competitive, among others. The following control variables were included, which determine the sample typology, and are Sex (men and women, residents of Mexico City); Age (over 18 years old); Socioeconomic level (with a medium socioeconomic level) and, Consumers (at least once in their life they have bought wooden furniture).

For this research, Pearson's Chi-square nonparametric test was used to measure independence between variables (Ritchey, 2008; Siegel and Castellan, 2012). The goal is to check if there is a dependency between the qualitative variables that define rows and columns, i.e., if for every $i = 1, ..., k$ and $j = 1, .., m$, it is verified that the probability of the result corresponding to the combination $A_i \cap B_j$ is the product of the corresponding marginal probabilities. $P(A_i)$ is the probability of the outcome i for the row variable and $P(B_j)$ the one of the result j for the column variable.

$$P(A_i \cap B_j) = P(A_i) \cdot P(B_j)$$

A more simplified notation:

$$P(Ai \cap Bj) = pij$$

$$P(Ai) = pi\cdot$$

$$P(Bj) = pj$$

The values of pi and pj were estimated from the values observed in the contingency table, by ni/N y nj/N respectively.

The Null Hypothesis of Independence: for any combination of results of the row and column variables (i, j).

$$H_0: p_{ij} = (p_i)\cdot(p_j) \text{ for every: } i = 1, \ldots, k\ j = 1, .., m$$

The Alternative Hypothesis, which implies dependence, can be formulated by saying that one of the equalities in the null hypothesis is false.

The observed values are nij. The expected values under the null hypothesis of independence are calculated as follows:

$$e_{ij} = (N)(p_{ij}) = (N)(p_i)(p_j) = (N)(n_i/N)(n_j/N) = (n_i)(n_j)/N$$

The contrast statistic is calculated as follows:

$$\chi^2_{exp} = \sum_{i=1}^{k}\sum_{j=1}^{m}\frac{\left(n_{ij} - e_{ij}\right)^2}{e_{ij}}$$

The asymptotic distribution under the null hypothesis is a χ^2 with $(k - 1) \cdot (m - 1)$ degrees of freedom. The degrees of freedom can be understood intuitively, understanding that the number of parameters that are estimated are $(k - 1)$ y $(m - 1)$, since the probability of the last class of each characteristic is set once the rest have been estimated. Therefore, applying the formula for the degrees of freedom, we obtain:

Degrees of freedom = number of classes – number of estimated parameters – 1

Degrees of freedom = (k) (m) – (k – 1) – (m – 1) – 1 = (k – 1) · (m – 1)

The decision criterion is the same as in the general case:

We reject the null hypothesis if:

$$\chi^2_{exp} \geq \chi^2_{\pm}$$

Where the last term is the critical value associated with a distribution χ^2, with $(k-1)\cdot(m-1)$ degrees of freedom, such that it leaves to its right a probability equal to α.

The validity condition is that the expected frequencies e_{ij} are greater than 0.05, which is the level of significance. The level of significance (maximum error limit) is associated with the verification of a hypothesis. That is, it is the maximum amount of error that we are willing to accept, in order to accept the alternative hypothesis as valid. Complementarily to the significance level, the confidence level refers to the confidence that we must reach to generalize our result or our conclusion, regardless of the hypothesis that we have raised. A high probability will give us the peace of mind that what we have found or concluded is close to reality and not due to chance. The confidence level is conventionally expressed as a percentage; thus, a confidence level of 95 per cent corresponds to a significance level of 5 per cent (Ritchey, 2008; Kerlinger, 2002).

The confidence level is the probability that the parameter to estimate is in the confidence interval. The confidence level *(p)* is designated by $1-\alpha$, and is usually taken as a percentage (Ritchey, 2008; Kerlinger. 2002). The most common confidence levels are 90, 95 and 99 per cent. The significance level is designated by means of α. The critical value *(k)* as $z_{\alpha/2}$

$$P(Z>z_{\alpha/2}) = \alpha/2$$

$$P[-z_{\alpha/2} < z < z_{\alpha/2}] = 1 - \alpha$$

Table 4. Significance level

$1-\alpha$	$\alpha/2$	$z_{\alpha/2}$
0.90	0.05	1.645
0.95	0.025	1.96
0.99	0.005	2.575

In a distribution $N(\mu, \sigma)$ the characteristic interval corresponding to a probability $p = 1 - \alpha$ is:

$$(\mu - z_{\alpha/2}\,\sigma,\ \mu + z_{\alpha/2}\,\sigma)$$

The contingency table, also known as a crosstab or as a two-way table, has the objective of representing, in summary, the relationship between different categorical variables. The table allows measuring the interaction between two variables to understand the results of an investigation better. The report it offers also shows Pearson's Chi-square Statistics, which represents the degree of correlation between the variables that use the chi-square, the p-value, and the degree of freedom. The research was done with a confidence level of 95 per cent which corresponds to a significance level of 5 per cent. The SPSS Statistics 24 program was used for the analysis of independence tests with Pearson's Chi-square.

RESULTS

With the contingency tables used in the research, the relationship between two categorical variables was evaluated, mainly their independence. This independence test focuses on analyzing the similarity of the distribution of one of the variables, whatever the level of the other that is examined. This results in a contingency table in which the frequencies of the rows and columns are approximately proportional. With Pearson's Chi-square test, the differences observed between each of the analyzed groups of variables attributable to chance were contrasted. Pearson's Chi-square test contrasted the observed differences between each of the analyzed groups of randomly attributable variables. With the statistical analysis of Pearson's Chi-square, the significance level was determined in each of the variable relationships. The results obtained are presented in Tables 5, 6, and 7.

From the results presented in Table 5 regarding Market Orientation and Consumer Perception, it can be seen that H_1 and H_2 are accepted, with a significance level of $p<0.05$. As for H_1, customer-oriented actions have eight categorical relationships, and H_2, in the framework of competition-oriented actions, has 15 categorical relationships, all with a significance level of less than 0.05. In other words, the greater customer-oriented actions perceived by consumers, the greater the impact on consumer perception regarding the Wood Furniture Industry; likewise, greater actions with a competitive orientation perceived by consumers have a greater positive impact on consumer perception regarding the Wood Furniture Industry.

On the other hand, H_3 which measures interfunctional coordination actions has nine categorical relationships, of which only three have a significance level of less than 0.05, so H_3 is not accepted, that is, the interfunctional coordination actions carried out by SMEs manufacturing wooden furniture do not present a correlation with the perception of consumers about the image of the industry.

Table 5. Consumer Perception of Market Orientation of SMEs manufacturing wooden furniture

		Consumer Perception of Market Orientation (CPMO)								
		CPMO	**Significance**	**Hypothesis**	**CPMO**	**Significance**	**Hypothesis**	**CPMO**	**Significance**	**Hypothesis**
		(11,17)	Sig.							
		(11,18)	Sig.							
		(11,24)	Sig.							
		(12,17)	Sig.							
	MOCU (Customer-oriented actions)	(12,18)	No Sig.	H_1						
		(12,24)	Sig.							
		(19,17)	Sig.							
		(19,18)	Sig.							
		(19,24)	Sig.							
					(11,14)	Sig.				
					(11,15)	Sig.				
					(11,16)	Sig.				
					(11,23)	Sig.				
					(11,25)	Sig.				
					(12,14)	Sig.				
					(12,15)	Sig.				
Market Orientation (MO)	**MOCO (Competitively oriented actions)**				(12,16)	Sig.	H_2			
					(12,23)	Sig.				
					(12,25)	Sig.				
					(19,14)	Sig.				
					(19,15)	Sig.				
					(19,16)	Sig.				
					(19,23)	Sig.				
					(19,25)	Sig.				
								(11,02)	No Sig.	
								(11,03)	No Sig.	
								(11,04)	Sig.	
								(12,02)	Sig.	
	MOIC (Interfunctional coordination actions)							(12,03)	No Sig.	H_3
								(12,04)	No Sig.	
								(19,02)	No Sig.	
								(19,03)	No Sig.	
								(19,04)	Sig.	

Source: Authors' elaboration

Table 6 shows the results achieved in Pearson's Chi-square test for hypotheses 4, 5, and 6 regarding Entrepreneurial Orientation and Consumer Perception. The three hypotheses are accepted since the degree of correlation between the variables presents a level of significance of $p < 0.05$ in most of them (80 per cent). In H_4, the innovative actions have 6 categorical relationships, in which 4 of them have an acceptable level of significance, meaning the innovative actions of SMEs has a positive correlation with the industry image consumers' perception. In H_5, the proactive actions have 15 categorical relationships, of which 12 have an acceptable level of significance, meaning the proactive actions of SMEs has a positive correlation with the industry image consumers' perception. In H_6, the risk-taking actions have 3 categorical relationships which all have an acceptable level of significance, meaning the risk-taking actions of SMEs has a positive correlation with the industry image consumers' perception. Therefore, it can be concluded that the innovative, proactive, and risk-taking actions of SMEs have has a positive correlation with consumer perception regarding the image of the wooden furniture industry. Likewise, it is essential to highlight that the indicators of the Entrepreneurial Orientation dimension are strategic determinants for Small and Medium Enterprises (SMEs) that manufacture wooden furniture.

Table 6. Consumer Perception of Entrepreneurial Orientation of SMEs manufacturing wooden furniture

		Consumer Perception of Entrepreneurial Orientation (CPEO)								
		CPEO	**Significance**	**Hypothesis**	**CPEO**	**Significance**	**Hypothesis**	**CPEO**	**Significance**	**Hypothesis**
Entrepreneurial Orientation (EO)	**EOIA (Innovative actions)**	(07,21)	No Sig.	H_4						
		(07,22)	Sig.							
		(08,21)	Sig.							
		(08,22)	Sig.							
		(09,21)	Sig.							
		(09,22)	No Sig.							
	EOPA (Proactive actions)				(07,05)	Sig.	H_5			
					(07,06)	Sig.				
					(07,20)	Sig.				
					(07,26)	Sig.				
					(07,27)	Sig.				
					(08,05)	Sig.				
					(08,06)	No Sig.				
					(08,20)	Sig.				
					(08,26)	Sig.				
					(08,27)	Sig.				
					(09,05)	No Sig.				
					(09,06)	No Sig.				
					(09,20)	Sig.				
					(09,26)	Sig.				
					(09,27)	Sig.				
	EORA (Risk-taking actions)							(07,10)	Sig.	H_6
								(08,10)	Sig.	
								(09,10)	Sig.	

Source: Authors' elaboration

For the dimension, Consumers' Perception of MO and EO strategies carried out by the manufacturing SMEs of wooden furniture vs. Online or In Situ Purchase Preferences are accepted H_7-H_{10} since the degree of correlation between the variables presents a level of significance of $p < 0.05$ in most of them. The consumers' perception of the implementation of MO-EO strategies has a positive correlation with purchase preference both online and In situ; however, there is a greater preference for buying Online than In Situ; 72% vs. 63% (see Table 7).

Table 7. Purchase Preference concerning Consumers' Perception of Market Orientation strategies and Entrepreneurial Orientation of wood furniture manufacturing SMEs

		Purchase Preference (PP)					
		PPOL (Online)	**Significance**	**Hypothesis**	**PPIS (In Situ)**	**Significance**	**Hypothesis**
	CPMO	(13,11)	Sig.		(13,11)	Sig.	
		(13,12)	No Sig.		(13,12)	No Sig.	
		(13,19)	No Sig.		(13,19)	Sig.	
		(30,11)	No Sig.		(28,11)	Sig.	
		(30,12)	Sig.	H_7	(28,12)	Sig.	
		(30,19)	Sig.		(28,19)	Sig.	
		(32,11)	Sig.		(29,11)	No Sig.	H_8
		(32,12)	Sig.		(29,12)	No Sig.	
		(32,19)	Sig.		(29,19)	Sig.	
					(31,11)	Sig.	
					(31,12)	No Sig.	
					(31,19)	Sig.	
Consumer Perception of MO and EO	**CPEO**	(13,07)	No Sig.		(13,07)	No Sig.	
		(13,08)	Sig.		(13,08)	Sig.	
		(13,09)	Sig.		(13,09)	Sig.	
		(30,07)	Sig.		(28,07)	No Sig.	
		(30,08)	Sig.	H_9	(28,08)	Sig.	
		(30,09)	Sig.		(28,09)	Sig.	
		(32,07)	No Sig.		(29,07)	Sig.	H_{10}
		(32,08)	Sig.		(29,08)	Sig.	
		(32,09)	Sig.		(29,09)	Sig.	
					(31,07)	No Sig.	
					(31,08)	No Sig.	
					(31,09)	No Sig.	

Source: Authors' elaboration

The results obtained for the dimension of the Online or In Situ Purchase Preference with respect to the Consumer Perception of the Market Orientation (MO) strategies carried out by the manufacturing SMEs of wooden furniture can be observed that H_7 and H_8 are accepted, due to the higher number of categorical relationships (67%) with a significance value below 5%. This means that the preference for online or in Situ purchase has a positive correlation by consumers' perception of the Market Orientation (OM) actions implemented by wood furniture manufacturing SMEs (See table 7).

On the other hand, from the dimension of the Preference to purchase online or in Situ with respect to the Consumer Perception of Entrepreneurial Orientation (EO), we can see in Table 7 that hypotheses H_9 and H_{10} are accepted, considering the largest number of categorical relationships with a significance value of Pearson's Chi-square below 5%. This explains that the preference for online or in Situ purchase has a positive correlation by consumers' perception of the Entrepreneurial Guidance (OE) actions implemented by wood furniture manufacturing SMEs.

LIMITATIONS AND FUTURE RESEARCH

The principal limitation of the study is the size of the sample, despite the study carrying out more than one hundred and eighty consumers. The study results are characteristic of the particularities of the participants; therefore, they cannot be generalized. It is suggested that future research focuses on understanding the reasons why variables associated with the unaccepted hypothesis have little significance in Mexican consumer decision-making. In addition, future research could be based on a cross-industry comparison, that is, the influence of the MO and EO strategies of companies across industries and on consumer behaviour in Mexico. It would be interesting to see possible differences between consumer preferences of products and services from different sectors. Future research can also focus on measuring these two strategies (MO-EO) and how they influence the wooden furniture manufacturing industry competitiveness. As well as its economic contribution both regionally and internationally. As a last idea, it would be interesting to analyze how these two strategies (MO-EO) within SMEs wooden furniture have undergone modifications after the COVID-19 pandemic and how the digital transformation has manifested itself in this industry.

CONCLUSION

The overall objective of this study was to identify the competitiveness factors of SME manufacturers of wooden furniture, based on the MO and EO strategies that influence the perception and preference of purchase of consumers in Mexico. The study showed that Mexican consumers generally recognize the Wood Furniture Industry in Mexico, assign it an authenticity value, and associate it with a positive image. In general, small and medium-sized wood furniture manufacturing companies have a positive corporate image and good reputation. When customers perceive that the company is committed to creating superior value for the customer through the MO and EO, they tend to show loyalty and purchase preference behaviors. In addition, the Mexican consumer has a degree of preference for acquiring Mexican furniture rather than acquiring it from foreign companies.

While the preference of the shopping channel does not depend directly on the Market Orientation or Entrepreneurial Orientation, the results indicate that the consumer is attracted to online sales. The study showed that most Mexican consumers are interested in online shopping of wooden furniture (72%). Nevertheless, many consumers also showed a willingness to buy wooden furniture In Situ (63%). Therefore, SMEs manufacturing wooden furniture need to consider rushing to e-commerce by the consumers' interest in the wooden furniture in online purchases.

The way of marketing wooden furniture through information technology and electronic commerce benefits companies to the extent that it allows them to reduce transaction costs and increase the speed and efficiency of organizational processes and operations. Also, the Internet and e-commerce are tools that enable better communication and interaction with stakeholders, provides information about products, and offer the possibility of online shopping (Jones, Motta, and Aldrette, 2016).

Therefore, a large market of potential customers for the Wood Furniture Industry in Mexico and the industrial enterprises have great potential to achieve sustained growth. Thus, if these companies manage to position themselves online, they could notice a significant improvement in consumer satisfaction and attracting potential customers. SMEs could therefore exploit e-commerce as a competitive strategy and become significant industry players. Moreover, SMEs belonging to the Wood Furniture Industry in Mexico, immersed in a globalized market, would open the market through the channel of preference for most consumers around the world.

REFERENCES

ANTAD. (2019). *INDICANTAD, 2018. Datos a septiembre de 2019.* https://antad.net/indicadores/indicantad/

Bagherzad, A., Chavosh, A., & Hosseinikhah, S. (2011). The influence of relationship marketing tactics on customer loyalty in B2C relationships–the role of communication and personalization. *European Journal of Economics, Finance and Administrative Science*, (31), 49–56.

Baker, W. E., & Sinkula, J. M. (2009). Complementary effects of market orientation and entrepreneurial orientation on profitability in small businesses. *Journal of Small Business Management*, *47*(4), 443–464. doi:10.1111/j.1540-627X.2009.00278.x

Balasubramanian, S. K., Karrh, J. A., & Patwardhan, H. (2006). Audience response to product placements: An integrative framework and future research agenda. *Journal of Advertising*, *35*(3), 115–141. doi:10.2753/JOA0091-3367350308

Bank of Mexico. (2017). *Balanza Comercial de Mercancías de México.* SNIEG. Información de Interés Nacional. Información disponible en: https://www.banxico.org.mx/SieInternet/consultarDirectorioInternetAction.do?accion=consultarCuadro&idCuadro=CE37&locale=es

Bigné, E. & Curras, R. (2008). ¿Influye la imagen de Responsabilidad Social en la Intención de compra? El papel de la identificación del consumidor con la empresa. *Universia Business Review*, *19*, 10-23.

Bigné, J. E., & Andreu, L. (2004). Modelo cognitivo-afectivo de la satisfacción en servicios de ocio y turismo. *Cuadernos de Economía y Dirección de Empresa*, *21*, 89–120.

Bishop, W. R. J. (1984). Competitive Intelligence. *Progressive Grocer*, (March), 19–20.

Chapela G. (2012). Problemas y Oportunidades en el mercado para las empresas Sociales Forestales en México. Agencia de los Estados Unidos para el Desarrollo Internacional (USAID). *Proyecto Competitividad de las Empresas Sociales Forestales.*

Chaston, I., & Scott, G. J. (2012). Entrepreneurship and open innovation in an emerging economy. *Management Decision*, *50*(7), 1161–1177. doi:10.1108/00251741211246941

Day, G. S. (1994). The capabilities of market-driven organizations. *Journal of Marketing*, *58*(4), 37–52. doi:10.1177/002224299405800404

DeLone, W. H., & McLean, E. R. (2004). Measuring e-Commerce Success: Applying the DeLone & McLean Information Systems Success Model. *International Journal of Electronic Commerce*, *9*(1), 31–47. doi:10.1080/10864415.2004.11044317

Devaraj, S., Fan, M., & Kohli, R. (2002). Antecedents of B2C Channel Satisfaction and Preference: Validating e-Commerce Metrics. *Information Systems Research*, *13*(3), 316–333. doi:10.1287/isre.13.3.316.77

Doyle, M. (1984). New Ways of Measuring Value. *Progressive Grocer-Value, Executive Report*, 15-19.

Engelen, A., Gupta, V., Strenger, L., & Brettel, M. (2015). Entrepreneurial orientation, firm performance, and the moderating role of transformational leadership behaviors. *Journal of Management*, *41*(4), 1069–1097. doi:10.1177/0149206312455244

George, G., & Zahra, S. A. (2002). *Being entrepreneurial and being market-driven – Exploring the interaction effects of entrepreneurial and market orientation on firm performance. In Frontiers of Entrepreneurship Research.* Babson College.

Grinstein, A. (2008). The relationships between market orientation and alternative strategic orientations: A meta-analysis. *European Journal of Marketing*, *42*(1/2), 115–134. doi:10.1108/03090560810840934

Hajipour, B., Rahimi, F., & Hooshmand, M. (2013). Market orientation: Review and new research agenda. *Journal of Management Research*, *5*(1), 92–111.

Hamilton, R., & Chernev, A. (2013). Low prices are just the beginning: Price image in retail management. *Journal of Marketing*, *77*(6), 1–20. doi:10.1509/jm.08.0204

Hernández, R., Fernández, C., & Baptista, P. (2018). Metodología de la investigación. McGraw-Hill.

INEGI. (2013). Instituto Nacional de Estadística y Geografía. *Encuesta Nacional de Gastos de los Hogares*. https://www.inegi.org.mx/est/contenidos/proyectos/encuestas/hogares/regulares/engasto/2012/doc/engasto_resultados.pdf

INEGI. (2018). Instituto Nacional de Estadística y Geografía. *Directorio Estadístico Nacional de Unidades Económicas*. http://www.beta.inegi.org.mx/app/mapa/denue/

INEGI DENUE. (2018). *Directorio Estadístico Nacional de Unidades Económicas 2018, Información para la actualización e incorporación de unidades económicas al DENUE; datos a abril de 2018*. http://www3.inegi.org.mx/rnm/index.php/catalog/477

Jacoby, J., & Olson, J. C. (1985). *Perceived Quality: How Consumers View Stores and Merchandise*. Lexington Books.

Jones, C., Motta, J., & Alderete, M. V. (2016). Gestión estratégica de tecnologías de información y comunicación y adopción del comercio electrónico en Mipymes de Córdoba, Argentina. *Estudios Gerenciales*, *32*(138), 4–13. doi:10.1016/j.estger.2015.12.003

Kim, M. K., Park, M. C., & Jeong, D. H. (2004). The effects of customer satisfaction and switching barrier on customer loyalty in Korean mobile telecommunication services. *Telecommunications Policy*, *28*(2), 145–159. doi:10.1016/j.telpol.2003.12.003

Kohli, A. K., & Jaworski, B. J. (1990). Market orientation: The construct, research propositions, and managerial implications. *Journal of Marketing*, *54*(2), 1–18. doi:10.1177/002224299005400201

Krishnan, H. (1996). Characteristics of memory associations: A consumer-based brand equity perspective. *International Journal of Research in Marketing*, *13*(4), 389–405. doi:10.1016/S0167-8116(96)00021-3

Kyriakopoulos, K., & Moorman, C. (2004). Tradeoffs in marketing exploitation and explorations Strategies: The overlooked role of marketing orientation. *International Journal of Research in Marketing*, *21*(3), 219–240. doi:10.1016/j.ijresmar.2004.01.001

Lavidge, R. J., & Steiner, G. A. (1961). A model for predictive measurements of advertising effectiveness. *Journal of Marketing*, *25*(6), 59–62. doi:10.1177/002224296102500611

Legua J. (2015). *El mercado del Hábitat y el Diseño en México.* Oficina Económica y Comercial de la Embajada de España en México, ICEX España importaciones e inversiones.

Lozano K. & Trinidad J. (2016). Estrategias locales para recuperar la competitividad en el mercado internacional: el caso del sector mueblero de Jalisco, México a partir del 2008. *Panorama de la investigación en las ciencias económico administrativas*, 194.

Lumpkin, G. T., & Dess, G. G. (1996). Clarifying the entrepreneurial orientation construct and linking it to performance. *Academy of Management Review*, *21*(1), 135–172. doi:10.2307/258632

Marín, L., Espinal, M., López, S., Maldonado, R., Mira, L., & Pertusa, M. (2011). La percepción de los consumidores sobre la RSC en la región de Murcia, los efectos de la crisis económica. *Servicios de Estudios colegios de economistas Región de Murcia*, *9*, 5857.

Martins, J., Costa, C., Oliveira, T., Gon∩alves, R., & Branco, F. (2019). How Smartphone Advertising Influences Consumers' Purchase Intention. *Journal of Business Research*, *94*, 378–387. doi:10.1016/j.jbusres.2017.12.047

Maxham, J. G. III, & Netemeyer, R. G. (2003). Firms reap what they sow: The effects of shared values and perceived organizational justice on customers' evaluations of complaint handling. *Journal of Marketing*, *67*(1), 46–62. doi:10.1509/jmkg.67.1.46.18591

Mexico's Economy Ministry. (2017). Inversión Extranjera Directa en México y el mundo. *Carpeta de Información Estadística.* http://www.economia.gob.mx/files/comunidad_negocios/ied/analisis_publicaciones/Otros por ciento20estudios/carpeta_informacion_estadistica_1115.pdf

Miles, M., & Darroch, J. (2008). A commentary on current research at the marketing and entrepreneurship interface. *Journal of Small Business Management*, *46*(1), 46–49. doi:10.1111/j.1540-627X.2007.00230.x

Miles, M. P., & Arnold, D. R. (1991). The relationship between marketing orientation and entrepreneurial orientation. *Entrepreneurship Theory and Practice*, *15*(4), 49–65. doi:10.1177/104225879101500407

Miller, D. (1983). Revisited a reflection on EO research and some suggestions for the future. *Entrepreneurship Theory and Practice*, *35*(5), 873–894. doi:10.1111/j.1540-6520.2011.00457.x

Molinillo, S. & Viano-Pastor, A. (2015). El papel del smartphone en la experiencia de compra offline. *Dos Algarves: A Multidisciplinary e-Journal, 26*(2), 68-86.

Morgan, T., Anokhin, S., Kretinin, A., & Frishammar, J. (2015). The dark side of the entrepreneurial orientation and market orientation interplay: A new product development perspective. *International Small Business Journal*, *33*(7), 731–751. doi:10.1177/0266242614521054

Narver, J. C., & Slater, S. F. (1990). The effect of a market orientation on business profitability. *Journal of Marketing*, *54*(4), 20–34. doi:10.1177/002224299005400403

Nulty, D. (2008). The adequacy of response rates to online and paper surveys: What can be done? *Assessment & Evaluation in Higher Education*, *33*(3), 301–314. doi:10.1080/02602930701293231

Richins, M. L. (1983). Negative word-of-mouth by dissatisðed consumers: A pilot study. *Journal of Marketing*, *47*(4), 68–78. doi:10.1177/002224298304700107

Ritchey, F. J. (2008). Estadística para las Ciencias Sociales. Editorial McGraw Hill.

Sahui, J. A. (2008). *Factores que influyen en la conducta del consumidor. Una aproximación desde las ciencias sociales*. Universidad Modelo.

Sánchez, R., Iniesta, M. A., & Holbrook, M. B. (2009). The Conceptualisation and Measurement of Consumer Value in Services. *International Journal of Market Research*, *51*(1), 93–113. doi:10.2501/S1470785308200328

Sawyer & Dickson, P. R. (1987). Psychological Perspectives on Consumer Response to Sales Promotion. In Research on Sales Promotion: Collected Papers. Cambridge, MA: Marketing Science Institute.

Schechter, L. (1984). A Normative Conception of Value, Progressive Grocer. *Executive Report*, 12-14.

Schindehutte, M., Morris, M. H., & Kocak, A. (2008). Understanding market-driving behavior: The role of entrepreneurship. *Journal of Small Business Management*, *46*(1), 4–26. doi:10.1111/j.1540-627X.2007.00228.x

Siegel, S., & Castellan, N. J. (2012). *Estadística No Paramétrica: Aplicada a las ciencias de la conducta*. Editorial Trillas.

Valenzo, M., & Ortiz, C. (2007, July-December). Competitividad de la industria exportadora de mueble en la región centro Occidente de México. *Inceptum*, *2*(3), 155–186.

Van Doorn, S., Jansen, J. J., Van den Bosch, F. A., & Volberda, H. W. (2013). Entrepreneurial orientation and firm performance: Drawing attention to the senior team. *Journal of Product Innovation Management*, *30*(5), 821–836. doi:10.1111/jpim.12032

Vega-Vázquez, M., Cossío-Silva, F.-J., & Revilla-Camacho, M.-Á. (2016). Entrepreneurial orientation–hotel performance: Has market orientation anything to say? *Journal of Business Research*, *69*(11), 1–6. doi:10.1016/j.jbusres.2016.04.085

Verkasalo, H., López-Nicolás, C., Molina-Castillo, F. J., & Bouman, H. (2010). Analysis of Users and Non-Users of Smartphone Applications. *Telematics and Informatics*, *27*(3), 242–255. doi:10.1016/j.tele.2009.11.001

Walesska, M., & Alvarado, A. (2009). Imagen y reputación corporativa. Estudio empírico de operadoras de telefonía en España. *Teoría y Praxis*, (6), 9–29.

Walters, R. G., & McKenzie, S. B. (1988). A structural equations analysis of the impact of price promotions on store performance. *JMR, Journal of Marketing Research*, *25*(1), 51–63. doi:10.1177/002224378802500105

Wolfinbarger, M., & Gilly, M. (2001). Shopping Online for Freedom, Control, and Fun. *California Management Review*, *43*(2), 34–55. doi:10.2307/41166074

Wu, P. C. S., Yeh, G. Y., & Hsiao, C. (2011). The Effect of Store Image and Service Quality on Brand Image and Purchase Intention for Private Label Brands. *Australasian Marketing Journal*, *19*(1), 30–39. doi:10.1016/j.ausmj.2010.11.001

Zeithaml, V. A. (1988). Consumer perceptions of price, quality, and value: A means-end model and synthesis of evidence. *Journal of Marketing*, *52*(3), 2–22. doi:10.1177/002224298805200302

Zielke, S. (2008). Exploring asymmetric effects in the formation of retail price satisfaction. *Journal of Retailing and Consumer Services*, *15*(5), 335–347. doi:10.1016/j.jretconser.2007.08.005

ADDITIONAL READING

Álvarez, L. I., Santos, M. L., & Vázquez, R. (2000). Análisis cultural y operativo de la orientación al mercado. Efectos moderadores en la relación OM-Resultados. *Revista Española de Investigación de Marketing-Esic*, *4*(1), 7–41.

Ashraf, A., Thongpapanl, N., & Auh, S. (2014, September). The Application of the Technology Acceptance Model under Different Cultural Contexts: The Case of Online Shopping Adoption. *Journal of International Marketing*, *22*(3), 68–93. doi:10.1509/jim.14.0065

Levin, A. M., Levin, I. P., & Heath, C. E. (2003). Product Category Dependent Consumer Preferences for Online and Offline Shopping Features and their Influence on Multi-channel Retail Alliances. *Journal of Electronic Commerce Research*, *4*(3), 85–93.

Lynch, P. D., Kent, R. J., & Srinivasan, S. (2001). The Global Internet Shopper: Effects of Trust, Affect, and Site Quality in Shopping Tasks for Low- and High-touch Products in Twelve Countries. *Journal of Advertising Research*, *41*(3), 15–23. doi:10.2501/JAR-41-3-15-23

Martins, I., Uribe, F., & Mesa, D. (2012). Contribución de la orientación emprendedora a la rentabilidad de las PyMEs: Un análisis contingente considerando la función del entorno. *Ecos de Economía*, *16*(35), 45–71. doi:10.17230/ecos.2012.35.3

Perea, T., Benedict, M., Dellaert, G. C., & de Ruyter, K. (2004, February). What drives consumers to shoponline? A literature review. *International Journal of Service Industry Management*, *15*(1), 102–121. doi:10.1108/09564230410523358

KEY TERMS AND DEFINITIONS

Competitiveness: An organizations ability to deliver value and quality through their products or services.

Innovation: It is the activity where it is sought to introduce new goods, services, and technology.

Manufacture: It is an automatic process consisting of putting several parts together.

Online: Referring to something connected to the Internet.

Perception: A process made by the consumer in which they discover, analyze and interpret information to create meaning or knowledge.

Purchase Preference: It is a clear inclination from the consumer regarding a product, service, or brand.

Technology: It is the use of science to solve problems or create new tools.

Value: It is an equivalent return in goods, products, services or experiences.

Chapter 8
The Influence of Culture on Innovation and CSR Practices

Carla Marisa Magalhães
Universidade Lusófona do Porto, Portugal & TRIE, Portugal

ABSTRACT

CSR practices derive directly from strategic actions of companies but, more indirectly, from other factors that influence organizational behavior, such as culture. But talking about culture is something too comprehensive since we have national and organizational culture. At the level of CSR practices, will the framework of values of the leadership and the employees of an organization or the norms and values of the country where it operates be more influential? What is the role played by an innovative culture in the implementation of CSR? How can companies promote corporate social innovation? To answer these questions, this chapter will present some studies and discussions to contribute to the reflection of this issue not only to sensitize organizations to the importance of assuming CSR behaviors in an innovative way but also to analyze the role that culture has at this level.

INTRODUCTION

Corporate Social Responsibility (CSR), despite not being a recent topic, is increasingly appealing for organizations to position themselves in the market, and for many, it is no longer an option, but an obligation. Organizations that look at their CSR practices as a corporate strategy and frame them in their relationship with the environment and society in general, with great emphasis on their internal audience, certainly have a more continuous and sustainable way of facing this situation and assume its social, environmental, and economic responsibilities, which are the target of increasing pressure exerted by its stakeholders.

But what makes some companies really different from others in this and other matters? Why do some companies look at CSR as a way of being and an organizational philosophy and others look at it as something accessory or merely opportunistic and instrumental?

We believe that what makes the difference is culture, namely organizational culture, as it is what truly separates and brings companies and people together and brings these companies and people closer to their stakeholders.

DOI: 10.4018/978-1-6684-2339-4.ch008

In the same way, while there are organizational cultures that promote CSR practices, others do so in a more innovative way, that is, not limited to introducing more classic and traditional practices into their strategy, but also doing it in a more innovative model, which allows us to anticipate that these practices will be more sustainable and aligned with the dynamics of the market, which is increasing and is based on often unpredictable contexts, and in relation to which a fast and consistent adaptation is necessary.

In other words, within the current paradigm, being socially responsible should not be an option and, in addition, companies should start looking at these practices as a way for them to promote their innovation and vice versa, that is, as a way of becoming responsible for their activity, in an innovative way, which encourages creative practices, based on a dynamic organizational culture that bets on innovation and is constantly evolving with this.

In this way, this chapter intends to promote a reflection on these issues and on how they can be aligned, to influence more sustainable organizational behaviors. Indeed, there are already several studies that reflect on the influence of culture on CSR practices, and the intention is to extend this reflection to the field of innovation, both at the level of organizational culture and at the level of those practices.

CORPORATE SOCIAL RESPONSIBILITY

Talking about CSR is far from being consensual since the topic is the subject of debate, opinions are divided, giving rise to different currents of thought. So, basically – despite the variety of existing classifications – we have on the one hand a more instrumental thinking and on the other a more substantive thinking. In line with the first current, the base is conservative, defending the idea that a company's social responsibility is nothing more than obtaining profit, in the sense of satisfying shareholders and generating efficiency in the search for resources that are often scarce. For the apologists of this current, the entrepreneur who generates profit and pays wages is already socially responsible. From this perspective, transforming a company's social responsibility into something that does not generate profit is negative for the company itself, as it creates a cost that does not benefit shareholders and, therefore, reduces the effectiveness and primary objective of that company (Barry, 2002; Friedman, 1962, 1970; Levitt, 1958; Marrewijk, 2003; McWilliam & Siegel, 2001; Zwetsloot, 2003).

From this point of view, the following question arises: Should a company only be accountable to its shareholders? This issue has behind the idea that corporate social responsibility should go beyond generating profit and shareholder satisfaction, seeking to take into account all its stakeholders (Carroll, 1991; Donaldson & Preston, 1995; Evan & Freeman, 1993; Freeman, 1984; Freeman & Reed, 1983; Gamble & Kelly, 2001; Kaler, 2003; Phillips et al, 2003; Stoney & Winstanley, 2001; Wilson, 2003; Wood, 1991) and preferably, in a substantive way, that is, based on ethical values and not economic interests (Kok et al, 2001; Mintzberg, Simons & Kunal, 2002; Wood, 1991).

In this way, CSR is assumed as a response of companies to the impact they have on society, due to the use of resources that are public and common to all, a situation that, together with the ability that companies have to satisfy certain society's needs, which cannot be fully satisfied by the government, generates a certain social expectation regarding the conduct of organizations (although this expectation varies according to the size and impact caused by the companies), which end up assuming relationships of power in society, depending on the socially responsible actions they take (Hemphill, 2004; Maignan & Ferrell, 2000, 2001a, 2001b; Matten, Crane & Chapple, 2003; Peterson, 2004; Waddock & Smith, 2000; Warhurst, 2001; Wood & Lodgson, 2002).

Furthermore, the instrumental point of view has received several criticisms, especially from academics, who argue that social actions with economic ends hardly survive in the long term, as they do not have the sustainability to perpetuate themselves. According to the defenders of this idea, CSR is only really internalized if it is based on the formation of a culture that supports it and, for that, the entire production chain of the company must be involved in this process. Therefore, CSR does not simply derive from the decision of business leaders who decide to automatically transform a company into a socially responsible institution. A long process of raising awareness and creating a socially responsible culture is necessary, which must be incorporated by all stakeholders, being essential, for this, the collaboration of society in general: citizens/customers/employees and all organizations from the public power, private power, and the so-called Third Sector, in an integrative manner (Carroll & Schwartz, 2003; Phillips et al, 2003; Stoney & Winstanley, 2001; Vos, 2003).

We cannot, still, neglect the fact that the behaviors that guide CSR depend not only on the type of motivation of the company but also on its cultural context. Therefore, the adoption (or not) of socially responsible practices by companies is, internally and externally, related to this context (Balmer, Fukukawa & Gray, 2007; Franke & Nadler, 2008; Genest, 2005; Hemingway & Maclagan, 2004; MacDonald, 2000; Moon & Franke, 2000; Robertson & Crittenden, 2003; Sanyal, 2005; Scholtens & Dam, 2007; Smith & Hume, 2005; Thomas & Mueller, 2000; Thome & Saunders, 2002). However, there is no consensus on which type of culture most influences management behaviors – it will be the national (Hofstede, 1980, 1997, 2001) or organizational one (Genest, 2005; Schein, 1985)?

NATIONAL AND ORGANIZATIONAL CULTURE

The existing literature on culture unfolds in several theories that seek to justify its importance within the administrative sphere, sometimes from a national point of view, sometimes from an organizational point of view.

The national culture was identified by Black (2005) as being the most decisive element in the variations of international relations at the industry level, being also a force of diversity in the labor market. National culture is thus seen as a basis for institutional stability and for diversity in intra and inter-organizational relationships.

Trompenaars and Hampden-Turner (1998) look at culture as a product of the interaction between people and, at the same time, as a determining component of that interaction. The authors thus define culture as a common system of meanings that shows us what we should pay attention to, how to act, and what we should value.

Hofstede (1997), in turn, states we have patterns of thought, feelings, and potential action, which are the result of continuous learning that begins in childhood, but human behavior is only partially determined by our mental software due to the human being's ability to deviate from these patterns by reacting in new, creative, destructive, or unexpected ways. For the author, this mental software starts in the family and continues in the life of the neighborhood, school, youth groups, the workplace, and the community. And it is precisely this mental software that is called culture.

Still within the national culture, but in a more intermediate view with organizational culture, we have a group of authors who are interested in contextual factors (of national scope) with an impact on organizational culture. One of these authors is precisely Hofstede (2001), who defines culture as a collective mental program that distinguishes and separates members into groups and categories from each other, based on their different characteristics. This mental program makes the individuals of a group behave in accordance with others within the same group.

Hofstede (1980) also demonstrates the importance of adapting management practices to specific contexts, that is, to culture. In the late 1960s and early 1970s, the author, together with his group of researchers, carried out a study on the existing cultural differences in 52 countries, seeking to understand how these cultural differences in different countries could influence the work in companies and administrative management, and trying to identify competitive differences between nations that could be present in a country's cultural values (Hofstede et al, 1991). Thus, based on the IBM company and its subsidiaries across 52 countries, Hofstede surveyed thousands of administrators and employees. In this way, it encompassed people who worked for the same company, who occupied similar positions and positions, and who, theoretically, should have the same pattern of behavior. However, some differences were found in the behavior and attitudes of these people, which, according to Hofstede, stems from differences in the national culture.

According to the author, organizational cultures must reflect the national culture in which they are inserted. The author states that societies differ from one another in terms of, mainly, four cultural dimensions: distance from power; individualism / collectivism; masculinity / femininity; and uncertainty avoidance / tolerance (Hofstede, 1980). He later identified a fifth dimension, Confucian dynamism (or long-term/short-term orientation), that could explain the success of eastern nations such as China and Japan (Hofstede et al, 1991).

The dimensions of culture identified in Hofstede's studies (1980, 1997) also supported the work of House et al (1999, 2002, 2004), within the scope of the Globe Project - Global Leadership and Organizational Behavior Effectiveness - which was characterized as a research program aimed at exploring cultural values and practices in a variety of countries and identifying their impact on organizational practices and leadership characteristics. The study was conducted in 62 nations and had the collaboration of resident investigators from the different participating countries. It involved 17,000 questionnaires answered by middle managers from approximately 825 companies in the food, financial, and telecommunications areas. The general objective of this project was to develop a theory to describe, understand and predict the impact of specific cultural variables on leadership and on organizational processes and the effectiveness of those processes. National cultures were examined across eight dimensions: achievement or performance orientation, affiliation or humanistic orientation, future orientation, assertiveness, power distance or hierarchical distance, control or avoidance of uncertainty, gender equality (masculinity dimension or Hofstede's femininity), and collectivism or individualism (in the institutional and group scope).

Trompenaars and Hampden-Turner (1998) were other authors who focused on contextual (national) factors with an impact on organizational culture. They developed their research focusing on the relationship between people at the organizational level, in different cultures. The authors established seven cultural dimensions, five of which concern the relationship between people: universalism / particularism, individualism / communitarianism, neutrality / expressiveness, specificity / diffusion, realization / predetermination. The other two dimensions concern past / present / future orientation and internal / external control.

Another approach at this level was made by Gesteland (1999), who proposed the following dimensions: emphasis on business / relationships, formal / informal cultures, monochronic / polychronic cultures and expressive / reserved cultures.

Charles Hampden-Turner (1992), is another author who is situated in theories that analyze national factors in the context of organizational culture. Thus, according to this author, there are four factors that have a great influence on the organizational culture of different countries: tensions between managers and other employees, hierarchy in communication, synthetic or analytical thinking style, and individualism or cooperativism.

In addition to these authors, other authors have currently defended the prevalence of national culture in organizational practices (Ansah & Louw, 2019; Cavusgil et al., 2014; Davies & Olmedo-Cifuentes, 2016; Hua & Wei, 2016; Klimas, 2016; López-Duartea, González-Loureirob, Vidal-Suáreza, & González-Díaza, 2016; McLeary & Cruise, 2015; Nazariana, Atkinsonb & Foroudi, 2017; Torelli et al., 2015; Verbekea, Kanob & Yuanc, 2016).

Schein (1985) argues that the concept of culture is largely related to the behavior of the leader, who can manipulate culture to create a more effective organization. Hence, an organizational culture starts with leaders, who impose their values, artifacts, and assumptions. Therefore, culture defines leadership, and leaders can create and change cultures. Although the leader is not the only one who influences the culture of a group, he is the one who provides the survival of the group whenever there is difficulty in adapting to the current culture.

In the case of the more idealistic and metaphorical visions of the concept, the sense of social construction of organizational identity is reinforced, that is, instead of having a culture, the organization is a culture, resulting from the interaction between the members of the organization, in a continuous process of construction and reconstruction of organizational reality. This analytical perspective of culture, whether it is based on a more idealistic or a more metaphorical view, allows us to apprehend its dynamic character, providing its understanding from the meanings shared by individuals in the search for the interpretation of reality and the definition of strategies for organizational action (Allaire & Firsirottu, 1984).

Still, according to the most metaphorical view, the dynamic nature of culture results in challenges for the management of organizations, especially when it is necessary to deal with fast-changing social contexts. Thus, the understanding of culture in the form of organizational consensus is a challenge in favor of approaches that incorporate some sense of ambiguity to the concept, ambiguity that is part of the several groups existing in the organization, along with the divergences. Therefore, the consensus and coherence in the symbolic manifestations would be limited to the members of such groups and not to the whole organization (Frost et al, 1991; Martin, 1992).

But, in addition to the authors mentioned so far, with great relevance for Schein, as he is one of the greatest representatives of the vision that defends the great influence of organizational culture, other authors have also focused on this issue, namely through the impact that leadership has on organizations (Clark, Hartline & Jones, 2009; Gerhart & Fang, 2005; Gill & Mathur, 2007; Hemingway & Maclagan, 2004; Lolowang, Troena, Djazuli & Aisjah, 2019; López-Duartea, González-Loureirob, Vidal-Suáreza & González-Díaza, 2016; Pothukuchi et al, 2002; Riketta, 2002). In fact, there are several studies that emphasize the central influence that leadership has on organizational performance (Mulyania, Wahyono, & Yanto, 2021; Salman, 2011; Sharkie, 2009). This is particularly relevant when we are in the presence of transformational leadership (Walumbwa & Lawler, 2003). Furthermore, leaders must not only be able to influence and manage the organizations they lead but also to do so in multinational contexts (Javidan & Carl, 2004; Zander & Romani, 2004).

Thus, both in organizational and national terms, culture has been the object of study by several researchers in the field. Its recognized importance has served as a basis for solving many organizational problems, insofar as the culture of a company is, undeniably, one of the decisive factors for the success or failure of its activity.

But, how about the influence of culture on CSR practices?

CULTURE AND CORPORATE SOCIAL RESPONSIBILITY

There are several authors who focus on the study of the relationship between culture and CSR (Alexandridis, Hoepner, Huang & Oikonomou, 2022; Chen, Lim, Nam & Phillips, 2020; Espasandín-Bustelo, Ganaza-Vargas & Diaz-Carrion, 2021; Jaakson, Vadi & Tamm, 2009; Soschinski, Giordani, KLann, & Brizolla, 2021).

A study conducted by Magalhães (2011) showed that organizational culture tends to be more influential in CSR practices than national culture, although it is not the only element that influences these practices. The author studied the influence of culture (national and organizational) on the CRS practices of two Portuguese companies and two Brazilian companies, in the metalworking sector, and observed that organizational culture tends to be more relevant, although with some inconclusive results. This greater influence was explained by the importance of leadership values and actions and the influence of the stakeholders, namely the employees that, sometimes, can influence the organizational culture more than the leadership itself. Thus, on the one hand, we have a situation in which the strength of the leader within an organization can make a difference (Gerhart & Fang, 2005; Hemingway & Maclagan, 2004; Pothukuchi et al, 2002) and, on the other hand, we have a situation in which there is a recognition that the strength of the workforce should be the most considered in the scope of organizational culture, for the benefit of the company itself (Ackfeldt & Wong, 2006; Ashmos & Duchon, 2000; Cameron, Bright & Caza, 2004; Christopher et al, 2004; Eisenberger et al, 2001; Gavin & Mason, 2004; Haller & Hadler, 2006; Herrbach & Mignonac, 2004; Kernbach & Schutte, 2005; Kets de Vries, 2001; Kets de Vries & Florent-Treacy, 2002; Milliman, Czaplewski & Ferguson, 2003; Sergeant & Frenkel, 2000; Wright & Cropanzano, 2004). This study also concluded that the management practices of companies, namely in the scope of CSR, are not always more influenced by the organizational culture, insofar as, when companies are faced with contextual factors (such as socio-economic or environmental a merger/acquisition process), these factors may exert more influence on management practices (namely social) and/or even suggest a greater influence on the part of the national culture. To conclude about the problem raised here, the following questions were asked: What makes companies get involved with Social Responsibility actions? And what is the purpose of this involvement? This study showed that the motivations of companies are mainly instrumental (Barry, 2002; Friedman, 1962, 1970; Levitt, 1958; Marrewijk, 2003; McWilliam & Siegel, 2001; Zwetsloot, 2003), as companies recognize that by practicing socially responsible actions, they improve their image (Kim, Yin & Lee, 2020), ensure future conditions for their survival and development (Carroll, 1980) and have more positive relationships with stakeholders (internal and external), which contributes to increased profits (Carroll, 1998, 1999). Thus, sometimes, CSR is even inevitable, because without it companies can hardly impose themselves in the market. Therefore, the objective of involvement with social actions is, above all, instrumental. This study also tried to answer the following questions: What is the reason why companies engage in certain CSR activities and not in others? Does the adoption of some practices result from deliberate action, individual initiative, the influence of internal and/or external stakeholders, or fortuitous factors? In fact, it was concluded that most of the time companies follow specific criteria to choose the social actions with which they are involved, the most common criteria being the needs detected and the influence of stakeholders. Indeed, it was seen that, on the one hand, companies are concerned with meeting the needs of society, as they themselves need a society that is developed in technological, social, and economic terms. On the other hand, situations may arise in which companies are forced to intervene socially, due to the pressures they suffer from stakeholders, who end up playing a fundamental role here, insofar as companies are aware that they need to have a good relationship with stakeholders, as their survival depends on their will (Stoney & Winstanley, 2001; Carroll & Schwartz, 2003; Phillips et al, 2003; Vos, 2003). Hence the fact that they defend an integrative concept of CSR, combined with an instrumental motivation. But to what extent does culture, namely organizational culture, help to explain these different behaviors? What cultural factors (national in scope) shape the way companies in the same country look at CSR? There are traits in organizational cultures that make them prone to CSR practices. In the case of Portuguese organizational culture, for example, it was highlighted the lack of social and environmental awareness as a characteristic that negatively interferes with socially responsible behaviors. Regarding national culture, society, in general, seems to be more open to the issue of CSR than companies themselves. This is based on certain national cultural traits, such as solidarity and religiosity, which make people worry about "the other" (Magalhães, 2011).

But if the culture does have a strong effect on CSR practices, namely organizational culture, how can we manage these practices in a more innovative and creative way? In other words, how can a culture based on innovation stimulate and/or shape CSR practices?

INNOVATIVE CULTURES

Dobni (2008) highlights that the culture of innovation refers to a multidimensional context that includes the intention to be innovative, the infrastructure that supports innovation, operational-level behavior needed to influence the market and value orientation, and the environment to implement the innovation.

Apekey, McSorley, Tilling and Siriwardena (2011), on the other hand, identify seven dimensions in innovation culture: risk, resources, information, goals, tools, rewards, and relationships.

According to Janiunaite and Petraite (2010), in turn, innovation culture is related to knowledge management and is a determinant of sustainable innovation, as well as the effective performance of an organization.

In relation to aspects of organizational culture that favor innovation, McLean (2005) highlights the following: organizational encouragement; supervisor encouragement; group encouragement work; freedom / autonomy and resources.

Regarding resources, Le Bas and Lauzikas (2010) note that an architecture that supports the innovation strategy requires financial resources, human resources, and a culture of innovation. The culture of innovation encompasses organizational architecture and its main culprits, tools management in practice, selection and implementation of new ideas, the role of the leader, and creativity of employees, among other things.

Communication is an aspect present in many studies on innovation facilitators. Johannessen and Olsen (2011), for example, consider essential communication skills. For this, it must involve value creation and be oriented for the development, transfer, and integration of knowledge. The culture of innovation needs coordination and interaction mechanisms that encompass exchanges of information and contribute to the development of knowledge — which is the basis for innovation. The capabilities of communication constitute a system that preserves the creation of value and combines economic/technical, communication management, and social and cultural communication.

Martins and Martins (2002) adopted a systemic model to identify the interaction between organizational subsystems (objectives, structure, management, technology, and psychosociological) that comprises different levels between individuals and groups and other organizations and the external environment for identifying the determinants that influence innovation from a cultural perspective and to compare the findings with the model based on the studied literature. The model proposed by Martins and Martins (2002) is comprehensive and has elements pointed out in different studies whose focus on promoting innovation. However, it is worth noting that, when explaining the factors that make up the model, there is a superposition of aspects, such as the evidenced in the factors: behavior that encourages innovation, work environment, and support from the manager. According to these authors, the determinants of organizational culture that influence innovation are strategy, structure, support behavior, and communication.

Brettel and Cleven (2011) adopt the resource-based view and present an adoption model of innovation in which the culture of innovation is seen as a facet of the corporate culture. They seek analyze the impact of opening the firm to external knowledge on the culture of innovation, which is considered as a latent factor reflected in the following four dimensions: orientation towards new technologies, learning orientation, willingness to take risks, and market orientation future.

Uzkurt, Kumar and Ensari (2013), about a construct similar to the culture of innovation, called it readiness for innovation. The most feature important aspects of the propensity to innovate refer to culture, structure, and organizational climate, being the organizational culture and the attitudes of the managers the central aspects for the innovation.

The several aspects that are consistent with a culture of innovation are also addressed by Bruno-Faria and Fonseca (2014). Among the aspects that favor the construction of a culture of innovation or culture variables associated with innovation, the following stand out: leadership (Bravo-Ibarra & Herrera, 2009; Çakar & Erturk, 2010; Jaskyte & Dressler, 2005; Lin & McDonough, 2011; Twati & Gammack, 2006; Zairi & Al-Mashari, 2005); information sharing (Das, 2003; Dombrowski et al., 2007; Martins & Terblanche, 2003; Ismail &; Abdmajid, 2007); and risk management (Apekey McSorley, Tilling & Siriwardena, 2011; Brettel & Cleven, 2011; Das, 2003; Valencia, Valle, & Jimenez, 2010).

At this point, it is also important to talk about the concept of corporate social innovation, which is the concept that links innovation with CSR (Szutowski & Ratajczak, 2016).

CORPORATE SOCIAL INNOVATION

Corporate social innovation has been the subject of several studies (Saka-Helmhout, Chappin & Rodrigues, 2021).

Mustapha, Jali and Muhamad (2021) point out that there is no consensual definition of corporate social innovation. In fact, even literature itself is not solid at this level (Baker & Mehmood, 2015; Nicholls et al., 2016; Van der Have & Rubalcaba, 2016). This may be because there are more studies in social innovation than in corporate social innovation itself (Martinez, O'Sullivan, Smith & Esposito, 2017; Tabares, 2020).

Corporate social innovation, despite being a relatively recent concept in the literature, has in its genesis strategic partnerships and joint creation of value (Altuna et al., 2015; Babu et al., 2020; De Silva, Al-Tabbaa & Khan, 2019; Doherty, Haugh & Lyon, 2014; Herrera, 2015; Lashitew, Bals & Tulder, 2020; Lusch, Vargo & Gustafsson, 2016; Vargo & Lusch, 2017). However, for these partnerships to be effective, organizations must promote a healthy institutional environment, without instability, excessive bureaucracy, corruption, and difficulty in accessing resources, markets, or products (Doh et al., 2017; Rivera-Santos, et al., 2012). At this level, it is important to mention that access to local resources by organizations is always a sensitive issue, as a society must recognize legitimacy in organizational activities (Verleye et al., 2018). This legitimacy can be promoted not only by society but also by other organizations (Deephouse & Suchman, 2008).

According to Saul (2011), corporate social innovation involves four key elements: intentional business strategy; leverage of core business; creating new value (economic and social); and positive social change. Supported by these elements, corporate social Innovation changes the way the company sees social change, turning simple philanthropy into a business opportunity. Saul (2011) believes that social innovation discovers ways to create business opportunities profit from uncontrollable social issues: education, health care, global development, hunger, and so on.

Jali, Abss and Ariffin, (2017) state that CSR is a traditional philanthropy and an old paradigm that is no longer enough to deal with the present economy. Thus, corporate social innovation is a new emerging paradigm that can provide an innovation that will lead to real changes in improving the well-being of people's lives and economic and technological growth.

Dionisio and Vargas (2020) mention that new paradigms could reinvent institutions as they open new possibilities for solving social problems by giving businesses a new way to innovate, through corporate social innovation actions.

A study carried out by Mirvis, Herrera, Googins and Albaredad (2016) shows that much of the knowledge exchanged in corporate social innovation is tacit knowledge that companies develop from shared interactions and experiences.

According to Mirvis, Herrera, Googins and Albareda (2016), corporate social innovation concerns several organizational valences, such as the strength of management, marketing, technology, and economic, social, and environmental sustainability.

Other authors associate corporate social innovation with a culture of success within an organization (Esen, & Maden-Eyiusta, 2019).

On the other hand, corporate social innovation implies a series of organizational skills, such as managing well its resources, namely capital, but also its partnerships (Lashitew et al., 2020; Porter & Kramer, 2011; Quelin et al., 2019).

Corporate social innovation can also be seen as a kind of social entrepreneurship (Hockerts, 2007).

Szutowski and Ratajczak (2016) suggested that the impact of innovation on the CSR practices depended on the type and degree of innovation and the way CSR affected innovation depended on CSR features such as type of reaction, degree of development, and field of activity.

But what is the role played by culture in this context?

CULTURE, CORPORATE SOCIAL RESPONSIBILITY AND CORPORATE SOCIAL INNOVATION

One way indicated in the literature to implement corporate social innovation involves CSR practices, combined with a strong organizational culture and leadership capable of influencing the company to achieve these goals (Brin & Nehme, 2019; Demil & Lecocq, 2010; Foss & Saebi, 2017).

In fact, culture can have a moderating effect on corporate social innovation, as some authors attest (Bock et al., 2012; Chen, 2022).

Companies, to adopt corporate social innovation strategies, must assume the leading role of leadership, in the reformulation of the corporate culture, creating and capturing values that meet these strategies.

These values can include the dissemination of CSR practices, as a way of promoting sustainability and competitive advantages (Camilleri, 2017; Košičiarová, Kádeková & Štarchoň, 2021).

Leaders need to face the several constraints that companies face, namely the lack of resources, and CSR can and should be a strategy to face these challenges, improve relations with stakeholders and increase competitive advantage. For this, it is important that new CSR practices are identified and adopted, which redefine the company's relationship with society and with the context in which it operates (Hendry, 2003; Porter & Kramer, 2006).

Indeed, there are several studies that show that leaders can disseminate CSR practices throughout their organizations, namely through role modeling (Russell et al., 2016), influencing employees to accept and adopt these practices (Glavas, 2016; Gond et al., 2017; Rupp & Mallory, 2015). In this way, companies can disseminate a culture of CSR, improving their image, which will certainly allow them to attract and retain people who add value, improve their market share, and enhance their relationships with their stakeholders (Jones et al., 2017). In this way, the ground is open for corporate social innovation.

However, leaders must understand what type of human capital they work with, as it may take longer in some companies than in others to adapt to new CSR strategies, particularly regarding corporate social innovation (Schein, 2010). This also derives from the leadership style, since the personality of the leader is very important to influence the corporate culture (Mayer et al., 2009) and, consequently, the CSR and innovation practices at that level. The ethics and morals of leaders should also be considered in this case (Mayer et al., 2009; Mulki et al., 2009; Schminke et al., 2005), because we can see leadership styles that are more conducive to the practices mentioned above.

Still, on this topic, it is important to mention that the link between culture and innovation is reported by several authors, namely regarding creativity (Eisenbeiss, Knippenberg & Boerner, 2008; Martins & Terblanche, 2003), technological innovation (Tuan and Venkatesh, 2010) and at the product level (Patterson et al., 2005).

Therefore, leaders must continually seek innovation, not only as a natural means of progression but also to anticipate or circumvent possible threats (Martins & Terblanche, 2003). However, sustaining an innovative culture implies strong leadership that promotes a corporate culture based on a logic of continuous strategic innovation (Chesbrough, 2010). This is the real key to corporate social innovation.

CONCLUSION

In conclusion, we can see that companies can no longer rely solely on their more traditional CSR practices, often based on the emerging needs of their stakeholders, namely those with whom they have a more direct relationship. It is necessary to innovate these practices, in the sense, not only to react to adversities that may arise but also to be able to anticipate these same adversities and prepare in advance for such scenarios, which are increasingly unpredictable.

We are currently experiencing great challenges, promoted by crises of different natures, boosted by factors such as the pandemic or a war that has no borders, and, for the same reason, companies must be prepared to face the emerging and urgent needs that arise, adopting practices of CSR that are increasingly innovative and capable of responding quickly and, as far as possible, sustainably, to these same needs.

By introducing CSR practices and an innovative culture, leaders can more easily promote corporate social innovation practices to create a strong corporate culture and enhance the image and positioning of the organizations they represent. With so much instability and competition, companies need to reinvent themselves and add value to their stakeholders and learn from them, through strategic partnerships that can be beneficial to both parties.

It is also important to note that, since CSR or corporate social innovation is not something regulated, its promotion is based on the beliefs of the leaders who manage the companies and the contexts in which they are inserted. There are several leadership styles that enhance these practices, such as coaching leadership (Berg & Karlsen, 2016, 2020), democratic leadership (Ray & Ray, 2012; Sharma & Singh, 2013), ethical leadership (Brown & Treviño, 2006; Kim & Thapa, 2018), laissez-faire leadership (Wong & Giessner, 2016; Pereira et al., 2021), relational leadership (Akram et al, 2016; Sánchez et al., 2020), transactional leadership (Aga, 2016; Changar & Atan, 2021), transformational leadership (Alonso-Almeida et al., 2017; Manzoor et al., 2019), servant leadership (Eva, 2019; Lythreatis et al., 2021), among others.

Indeed, as long as companies have true institutional leaders, capable of listening to others, namely their employees (Matos & Machado, 2014), and with the ability to continually adapt to the environment, leadership (or organizational culture) is the real driving force of an organization and, consequently, of its innovative practices, namely in terms of CSR (Waldman & Siegel, 2008). Thus, it is through a culture that the company can disseminate its CSR practices (Johnstone, 2018) and its innovative capacity (Fietz & Günther, 2021).

In this way, the more committed a company is to boosting CSR strategies, the more driving it will be of a culture that promotes corporate social innovation. Thus, the relationship between corporate culture and CSR is a key factor for the dissemination of innovative practices, from a social point of view (Chen, 2022).

REFERENCES

Ackfeldt, A., & Wong, V. (2006). The antecedents of prosocial service behaviors: An empirical investigation. *Service Industries Journal, 26*(7), 727–745. doi:10.1080/02642060600898237

Aga, D. (2016). Transactional Leadership and Project Success: The Moderating Role of Goal Clarity. *Procedia Computer Science, 100*, 517–525. doi:10.1016/j.procs.2016.09.190

Akram, T., Lei, S., Hussain, S. T., Haider, M. J., & Akram, M. W. (2016). Does relational leadership generate organizational social capital? A case of exploring the effect of relational leadership on organizational social capital in China. *Future Business Journal, 2*(2), 116–126. doi:10.1016/j.fbj.2016.06.001

Alexandridis, G., Hoepner, A., Huang, Z., & Oikonomou, I. (2022). Corporate social responsibility culture and international M&As. *The British Accounting Review, 54*(1), 101035. doi:10.1016/j.bar.2021.101035

Alonso-Almeida, M., Perramon, J., & Bagur-Femenias, L. (2017). Leadership styles and corporate social responsibility management: Analysis from a gender perspective. *Business Ethics (Oxford, England), 26*(2), 147–161. doi:10.1111/beer.12139

Altuna, N., Contri, A. M., Dell'Era, C., Frattini, F., & Maccarrone, P. (2015). Managing social innovation in for-profit organizations: The case of Intesa Sanpaolo. *European Journal of Innovation Management, 18*(2), 258–280. doi:10.1108/EJIM-06-2014-0058

Ansah, M., & Louw, L. (2019). The influence of national culture on organizational culture of multinational companies. *Cogent Social Sciences, 5*(1), 1–15.

Apekey, T. A., McSorley, G., Tilling, M., & Siriwardena, A. N. (2011). Room for improvement? Leadership, innovation culture and uptake of quality improvement methods in general practice. *Journal of Evaluation in Clinical Practice*, *17*(2), 311–318. doi:10.1111/j.1365-2753.2010.01447.x PMID:20438607

Ashmos, D. P., & Duchon, D. (2000). Spirituality at work: A conceptualization and measure. *Journal of Management Inquiry*, *9*(2), 134–145. doi:10.1177/105649260092008

Babu, M. M., Dey, B. L., Rahman, M., Roy, S. K., Alwi, S. F. S., & Kamal, M. M. (2020). Value co-creation through social innovation: A study of sustainable strategic alliance in telecommunication and financial services sectors in Bangladesh. *Industrial Marketing Management*, *89*, 13–27. doi:10.1016/j.indmarman.2020.06.003

Baker, S., & Mehmood, A. (2015). Social innovation and the governance of sustainable places. *Local Environment*, *20*(3), 321–334. doi:10.1080/13549839.2013.842964

Balmer, J. M., Fukukawa, T., & Gray, E. R. (2007). The Nature and Management of Ethical Corporate Identity: A Commentary on Corporate Identity, Corporate Social Responsibility and Ethics. *Journal of Business Ethics*, *76*(1), 7–15. doi:10.100710551-006-9278-z

Barry, N. (2002). The stakeholder concept of corporate control is illogical and impractical. *The Independent Review*, *6*(4), 541-554.

Berg, M., & Karlsen, J. (2020). Coaching leadership style: A learning process. *International Journal of Knowledge and Learning*, *13*(4), 356. doi:10.1504/IJKL.2020.10033158

Berg, M. E., & Karlsen, J. T. (2016). A study of coaching leadership style practice in projects. *Management Research Review*, *39*(9), 1122–1142. doi:10.1108/MRR-07-2015-0157

Black, B. (2005). Comparative industrial relations theory: The role of national culture. *International Journal of Human Resource Management*, *16*(7), 1137–1158. doi:10.1080/09585190500143980

Bock, A. J., Opsahl, T., George, G., & Gann, D. M. (2012). The effects of culture and structure on strategic flexibility during business model innovation. *Journal of Management Studies*, *49*(2), 279–305. doi:10.1111/j.1467-6486.2011.01030.x

Bravo-Ibarra, E. R., & Herrera, L. (2009). Capacidad de innovación y configuración de recursos organizativos. *Intagible Capital*, *5*(3), 301–320.

Brettel, M., & Cleven, N. J. (2011). Innovation culture, collaboration with external partners and NPD performance. *Creativity and Innovation Management*, *20*(4), 253–272. doi:10.1111/j.1467-8691.2011.00617.x

Brin, P., & Nehme, M. N. (2019). Corporate social responsibility: Analysis of theories and models. *Eureka: Social and Humanities*, *5*, 22–30. doi:10.21303/2504-5571.2019.001007

Brown, M., & Treviño, L. (2006). Ethical leadership: A review and future directions. *The Leadership Quarterly*, *17*(6), 595–616. doi:10.1016/j.leaqua.2006.10.004

Bruno-Faria, M., & Fonseca, M. (2014). Culture of Innovation: Concepts and Theoretical Models. *RAC. Revista de Administração Contemporânea*, *18*(4), 372–396. doi:10.1590/1982-7849rac20141025

Çakar, N. D., & Ertürk, A. (2010). Comparing innovation capability of small and medium-sized enterprises: Examining the effects of organizational culture and empowerment. *Journal of Small Business Management*, *48*(3), 325–359. doi:10.1111/j.1540-627X.2010.00297.x

Cameron, K. S., Bright, D., & Caza, A. (2004). Exploring the relationships between organizational virtuousness and performance. *The American Behavioral Scientist*, *47*(6), 766–790. doi:10.1177/0002764203260209

Camilleri, M. A. (2017). Corporate sustainability and responsibility: Creating value for business, society and the environment. *Asian Journal of Sustainability and Social Responsibility*, *2*(1), 59–74. doi:10.118641180-017-0016-5

Carroll, A. B. (1980). Social Responsibility as an Objective of Business: Evolving Toward a Model of Corporate Social Performance In Business Policy and Strategic Management (pp. 62-70). McGraw-Hill.

Carroll, A. B. (1991). The Pyramid of Corporate Social Responsibility: Toward the Moral Management of Corporate Stakeholders. *Business Horizons*, *34*(4), 39–48. doi:10.1016/0007-6813(91)90005-G

Carroll, A. B. (1998). The four faces of corporate citizenship. *Business & Society*, *100*(1), 1–7.

Carroll, A. B. (1999). Corporate social responsibility: Evolution of a definitional construct. *Business & Society*, *38*(3), 268–295. doi:10.1177/000765039903800303

Carroll, A. B., & Schwartz, M. (2003). Corporate Social Responsibility: A Three-Domain Approach. *Business Ethics Quarterly*, *13*(4), 503–530. doi:10.5840/beq200313435

Cavusgil, S. T., Knight, G., & Riesenberger, J. R. (2014). *International Business – The New Realities* (3rd ed.). Pearson Education.

Changar, M., & Atan, T. (2021). The role of transformational and transactional leadership approaches on environmental and ethical aspects of CSR. *Sustainability*, *13*(3), 1411. doi:10.3390u13031411

Chen, C. (2022). *The mediating effect of corporate culture on the relationship between business model innovation and corporate social responsibility: A perspective from small- and medium-sized enterprises*. Asia Pacific Management Review. doi:10.1016/j.apmrv.2022.01.001

Chen, J., Lim, S., Nam, H., & Phillips, J. (2020). Local culture as a corporate social responsibility multiplier: Confucian values' mediation between firm policies and employees' attitude. *Asia-Pacific Journal of Business Administration*, *12*(3-4), 387–407. doi:10.1108/APJBA-04-2019-0088

Chesbrough, H. (2010). Business model innovation: Opportunities and barriers. *Long Range Planning*, *43*(2), 354–363. doi:10.1016/j.lrp.2009.07.010

Christopher, A. N., Victoria Kuo, S., Abraham, K. M., Noel, L. W., & Linz, H. E. (2004). Materialism and affective well-being: The role of social support. *Personality and Individual Differences*, *37*(3), 463–470. doi:10.1016/j.paid.2003.09.015

Clark, R. A., Hartline, M. D., & Jones, K. C. (2009). The effects of leadership style on hotel employees' commitment to service quality. *Cornell Hospitality Quarterly*, *50*(2), 209–231. doi:10.1177/1938965508315371

Das, G. S. (2003). Preparedness for innovation: An Indian perspective. *Global Business Review*, *4*(1), 27–39. doi:10.1177/097215090300400103

Davies, G., & Olmedo-Cifuentes, I. (2016). Corporate misconduct and the loss of trust. *European Journal of Marketing, 50*(7/8), 1426–1447. doi:10.1108/EJM-11-2014-0729

De Silva, M., Al-Tabbaa, O., & Khan, Z. (2019). Business model innovation by international social purpose organizations: The role of dynamic capabilities. *Journal of Business Research, 125*, 733–749. doi:10.1016/j.jbusres.2019.12.030

Deephouse, D. L., & Suchman, M. (2008). Legitimacy in organizational institutionalism. In R. Greenwood, C. Oliver, K. Sahlin, & R. Suddaby (Eds.), *The Sage Handbook of Organizational Institutionalism* (pp. 49–77). London Sage Publications. doi:10.4135/9781849200387.n2

Demil, B., & Lecocq, X. (2010). Business model evolution: In search of dynamic consistency. *Long Range Planning, 43*(2/3), 227–246. doi:10.1016/j.lrp.2010.02.004

Dionisio, M., & Vargas, E. (2020). Corporate social innovation: A systematic literature review. *International Business Review, 29*(2), 101641. doi:10.1016/j.ibusrev.2019.101641

Dobni, C. B. (2008). Measuring innovation culture in organizations: The development of a generalized innovation culture construct using exploratory factor analysis. *European Journal of Innovation Management, 11*(4), 539–559. doi:10.1108/14601060810911156

Doh, J., Rodrigues, S., Saka-Helmhout, A., & Makhija, M. (2017). International business responses to institutional voids. *Journal of International Business Studies, 48*(3), 293–307. doi:10.105741267-017-0074-z

Doherty, B., Haugh, H., & Lyon, F. (2014). Social enterprises as hybrid organizations: A review and research agenda. *International Journal of Management Reviews, 16*(4), 417–436. doi:10.1111/ijmr.12028

Dombrowski, C., Kim, J. Y., Desouza, K. C., Brabanza, A., Papagari, S., Baloh, P., & Jha, S. (2007). Elements of innovative cultures. *Knowledge and Process Management, 14*(3), 190–202. doi:10.1002/kpm.279

Donaldson, T., & Preston, L. (1995). The Stakeholder theory of the corporation: Concepts, evidence and implications. *Academy of Management Review, 20*(1), 65–91. doi:10.2307/258887

Eisenbeiss, S. A., Knippenberg, D., & Boerner, S. (2008). Transformational leadership and team innovation: Integrating team climate principles. *The Journal of Applied Psychology, 93*(6), 1438–1446. doi:10.1037/a0012716 PMID:19025260

Eisenberger, R., Armeli, S., Rexwinkel, B., Lynch, P. D., & Rhoades, L. (2001). Reciprocation of perceived organizational support. *The Journal of Applied Psychology, 86*(1), 42–51. doi:10.1037/0021-9010.86.1.42 PMID:11302232

Esen, A., & Maden-Eyiusta, C. (2019). Delineating the concept of corporate social innovation: Toward a multidimensional model. *International Journal of Entrepreneurship and Innovation Management, 23*(1), 23–45. doi:10.1504/IJEIM.2019.096497

Espasandín-Bustelo, F., Ganaza-Vargas, J., & Diaz-Carrion, R. (2021). Employee happiness and corporate social responsibility: The role of organizational culture. *Employee Relations, 43*(3), 609–629. doi:10.1108/ER-07-2020-0343

Eva, N., Robin, M., Sendjaya, S., van Dierendonck, D., & Liden, R. C. (2019). Servant Leadership: A systematic review and call for future research. *The Leadership Quarterly, 30*(1), 111–132. doi:10.1016/j.leaqua.2018.07.004

Evan, W. M., & Freeman, R. E. (1993). A stakeholder theory of the modern corporation: Kantian capitalism. In Ethical theory and business (pp. 75-84). Prentice-Hall.

Fietz, B., & Günther, E. (2021). Changing organizational culture to establish sustainability. *Control Management Review, 65*(3), 32–40. doi:10.100712176-021-0379-4

Foss, N., & Saebi, T. (2017). Business models and business model innovation: Between wicked and paradigmatic problems. *Long Range Planning, 51*(1), 9–21. doi:10.1016/j.lrp.2017.07.006

Franke, G. R., & Nadler, S. S. (2008). Culture, economic development and national ethical attitudes. *Journal of Business Research, 61*(3), 254–264. doi:10.1016/j.jbusres.2007.06.005

Freeman, E. (1984). *Strategic Management: A Stakeholder Approach.* Pitman Publishing.

Freeman, E., & Reed, D. L. (1983). Stockholders and stakeholders: A new perspective on corporate governance. *California Management Review, 25*(3), 88–106. doi:10.2307/41165018

Friedman, M. (1962). *Capitalism and Freedom.* The University Chicago Press.

Friedman, M. (1970). The Social Responsibility of Business is to Increase its Profits. *New York Times Magazine, 13*, 122-126.

Frost, P. (1991). *Reframing Organizational Culture.* Sage Publications.

Gamble, A., & Kelly, G. (2001). Shareholder Value and the Stakeholder Debate in the UK. *Corporate Governance, 9*(2), 110–117. doi:10.1111/1467-8683.00235

Gavin, J. H., & Mason, R. O. (2004). The virtuous organization: The value of happiness in the workplace. *Organizational Dynamics, 33*(4), 379–392. doi:10.1016/j.orgdyn.2004.09.005

Genest, C. M. (2005). Cultures, organizations and philanthropy. *Corporate Communications, 10*(4), 315–327. doi:10.1108/13563280510630115

Gerhart, B., & Fang, M. (2005). National culture and human resource management: Assumptions and evidence. *International Journal of Human Resource Management, 16*(6), 971–986. doi:10.1080/09585190500120772

Gesteland, R. R. (1999). *Cross-Cultural Business Behavior.* Copenhagen Business School Press.

Gill, A. S., & Mathur, N. (2007). Improving employee dedication and pro-social behavior. *International Journal of Contemporary Hospitality Management, 19*(4), 328–334. doi:10.1108/09596110710747661

Glavas, A. (2016). Corporate social responsibility and employee engagement: Enabling employees to employ more of their whole selves at work-. *Frontiers in Psychology, 7*, 796. doi:10.3389/fpsyg.2016.00796 PMID:27303352

Gond, J.-P., Akremi, A. E., Swaen, V., & Babu, N. (2017). The psychological microfoundations of corporate social responsibility: A person-centric systematic review. *Journal of Organizational Behavior, 38*(2), 225–246. doi:10.1002/job.2170

Haller, M., & Hadler, M. (2006). How social relations and structures can produce happiness and unhappiness: An international comparative analysis. *Social Indicators Research, 75*(2), 169–216. doi:10.100711205-004-6297-y

Hampden-Turner, C. (1992). *Creating corporate culture: from discord to harmony*. Addison-Wesley Publishing Company Inc.

Hemingway, C. A., & Maclagan, P. W. (2004). Managers' personal values as drivers of corporate social responsibility. *Journal of Business Ethics, 50*(1), 33–44. doi:10.1023/B:BUSI.0000020964.80208.c9

Hemphill, T. (2004). Monitoring global corporate citizenship: Industry self-regulation at a crossroads. *Journal of Corporate Citizenship, 14*(14), 81–95. doi:10.9774/GLEAF.4700.2004.su.00009

Hendry, J. (2003). Environmental NGOs and business: A grounded theory of assessment, targeting, and influencing. *Business & Society, 42*(2), 267–276. doi:10.1177/0007650303042002005

Herrbach, O., & Mignonac, K. (2004). How organizational image affects employee attitudes. *Human Resource Management Journal, 14*(4), 76–88. doi:10.1111/j.1748-8583.2004.tb00134.x

Herrera, M. E. B. (2015). Creating competitive advantage by institutionalizing corporate social innovation. *Journal of Business Research, 68*(7), 1468–1474. doi:10.1016/j.jbusres.2015.01.036

Hockerts, K. (2007). Social Entrepreneurship. In W. Visser, D. Matten, M. Pohl, & N. Tolhurst (Eds.), *The A to Z of Corporate Social Responsibility: A Complete Guide to Concepts, Codes and Organisations*. John Wiley.

Hofestede, G. (1980). Culture's consequences: International differences in work-related values. *Sage (Atlanta, Ga.)*.

Hofestede, G. (1994). Management scientists are human. *Management Science, 40*(1), 4–13. doi:10.1287/mnsc.40.1.4

Hofestede, G. (1997). *Cultures and Organizations: Software of the minds*. MacGraw-Hill.

Hofestede, G. (2001). *Culture's consequences: Comparing values, behaviors, institutions and organizations across nations* (2nd ed.). Sage.

Hofstede, G. (1991). Cultural Routes of Economic Performance: A Research Note. *Strategic Management Journal, 12*(S1), 165–173. doi:10.1002mj.4250120912

Hofstede, G., & Bond, M. H. (1988). The Confucius Connection: From cultural roots to economic growth. *Organizational Dynamics, 16*(4), 5–21. doi:10.1016/0090-2616(88)90009-5

House, R. J. (1996). Path-goal theory of leadership: Lessons, legacy, and a reformulated theory. *The Leadership Quarterly, 7*(3), 323–352. doi:10.1016/S1048-9843(96)90024-7

House, R. J. (1999). Cultural influences on Leadership and Organizations: Project Globe. *Advances in Global Leadership, 1*, 171–233.

House, R. J. (2004). Culture, leadership and organizations: The Globe study of 62 societies. *Sage (Atlanta, Ga.).*

House, R. J., Javidan, M., Hanges, P., & Dorfman, P. (2002). Understandig cultures and implicit leadership theories across the globe: An introduction to project Globe. *Journal of World Business, 37*(1), 3–10. doi:10.1016/S1090-9516(01)00069-4

Hua, W., & Wei, P. (2016). National culture, population age, and other country factors in volume–Price volatility relationship. *Global Finance Journal, 32*, 1044–1283.

Ismail, W. K. W., & Abdmajid, R. (2007). Framework of the culture of innovation: A revisit. *Journal Kemanusiaan, 9*, 38–49.

Jaakson, K., Vadi, M., & Tamm, K. (2009). Organizational culture and CSR: An exploratory study of Estonian service organizations. *Social Responsibility Journal, 5*(1), 6–18. doi:10.1108/17471110910939962

Jali, M., Abss, Z., & Ariffin, A. (2017). Corporate Social Responsibility and Corporate Social Innovation: A Conceptual Understanding. *SHS Web of Conferences, 34*, 1-9.

Janiunaite, B., & Petraite, M. (2010). The relationship between organizational innovative culture and knowledge sharing in organization: The case of technological innovation implementation in a telecommunication organization. *Socialiniai Mokslai, 3*(69), 14–23.

Jaskyte, K., & Dressler, W. W. (2004). Studying culture as an integral aggregate variable: Organizational culture and innovation in a group of nonprofit organizations. *Field Methods, 16*(3), 265–284. doi:10.1177/1525822X03262281

Javidan, M., & Carl, D. E. (2004). East meets west: A cross-cultural comparison of charismatic leadership among Canadian and Iranian executives. *Journal of Management Studies, 41*(4), 665–691. doi:10.1111/j.1467-6486.2004.00449.x

Johannessen, J.-A., & Olsen, B. (2011). Projects as communicating systems: Creating a culture of innovation and performance. *International Journal of Information Management, 31*(1), 30–37. doi:10.1016/j.ijinfomgt.2010.04.006

Johnstone, L. (2018). Theorising and modelling social control in environmental management accounting research. *Social and Environmental Accountability Journal, 38*(1), 30–48. doi:10.1080/0969160X.2017.1422778

Jones, D. A., Willness, C. R., & Glavas, A. (2017). When corporate social responsibility (CSR) meets organizational psychology: New frontiers in micro-CSR research, and fulfilling a quid pro quo through multilevel insights. *Frontiers in Psychology, 8*, 520. doi:10.3389/fpsyg.2017.00520 PMID:28439247

Kaler, J. (2003). Differentiating stakeholder theories. *Journal of Business Ethics, 46*(1), 71–83. doi:10.1023/A:1024794710899

Kernbach, S., & Schutte, N. S. (2005). The impact of service provider emotional intelligence on customer satisfaction. *Journal of Services Marketing*, *19*(7), 438–444. doi:10.1108/08876040510625945

Kets de Vries, M. R. (2001). Creating authentizotic organizations: Well-functioning individuals in vibrant companies. *Human Relations*, *54*(1), 101–111. doi:10.1177/0018726701541013

Kets de Vries, M. R., & Florent-Treacy, E. (2002). Creating high commitment organizations. *Organizational Dynamics*, *30*(4), 295–309. doi:10.1016/S0090-2616(02)00067-0

Kim, M., Yin, X., & Lee, G. (2020). The effect of CSR on corporate image, customer citizenship behaviors, and customers' long-term relationship orientation. *International Journal of Hospitality Management*, *88*, 1–8. doi:10.1016/j.ijhm.2020.102520

Kim, M. S., & Thapa, B. (2018). Relationship of ethical leadership, corporate social responsibility and organizational performance. *Sustainability*, *10*(2), 447. doi:10.3390u10020447

Klimas, P. (2016). Organizational culture and coopetition: An exploratory study of the features, models and role in the Polish Aviation Industry. *Industrial Marketing Management*, *53*, 91–102. doi:10.1016/j.indmarman.2015.11.012

Kok, P., Van der Weile, T., McKenna, R., & Brown, A. (2001). A corporate social responsibility audit within a quality management framework. *Journal of Business Ethics*, *31*(4), 285–297. doi:10.1023/A:1010767001610

Košičiarová, I., Kádeková, Z., & Štarchoň, P. (2021). Leadership and motivation as important aspects of the international company's corporate culture. *Sustainability*, *13*(7), 3916. doi:10.3390u13073916

Lashitew, A. A., Bals, L., & Tulder, R. V. (2020). Inclusive business at the base of the pyramid: The role of embeddedness for enabling social innovations. *Journal of Business Ethics*, *162*(2), 421–448. doi:10.100710551-018-3995-y

Le Bas, C., & Lauzikas, M. (2010). Determinants of innovation culture and major impacts on the innovation strategy: The case of the information technology sector in Lithuania. *Social Sciences Studies*, *4*(8), 125–139.

Lee, Y., & Kramer, A. (2016). The role of purposeful diversity and inclusion strategy (PDIS) and cultural tightness/looseness in the relationship between national culture and organizational culture. *Human Resource Management Review*, *26*(3), 198–208. doi:10.1016/j.hrmr.2016.01.001

Levitt, T. (1958). The dangers of social responsibility. *Harvard Business Review*, *36*(5), 41–50.

Lin, H., & McDonough, E. F. III. (2011). Investigating the role of leadership and organizational culture in fostering innovation ambidexterity. *IEEE Transactions on Engineering Management*, *58*(3), 497–509. doi:10.1109/TEM.2010.2092781

Lolowang, N., Troena, E., Djazuli, A., & Aisjah, S. (2019). The effect of leadership and organizational culture on employee performance that is educated by motivation (study on the implementation empowerment programs in Jayapura city). *Problems and Perspectives in Management*, *17*(1), 268–277. doi:10.21511/ppm.17(1).2019.23

López-Duartea, C., González-Loureirob, M., Vidal-Suáreza, M., & González-Díaza, B. (2016). International strategic alliances and national culture: Mapping the Field and developing a research agenda. *Journal of World Business*, *51*(4), 511–524. doi:10.1016/j.jwb.2016.05.001

Lusch, R. F., Vargo, S. L., & Gustafsson, A. (2016). Fostering a trans-disciplinary perspectives of service ecosystems. *Journal of Business Research*, *69*(8), 2957–2963. doi:10.1016/j.jbusres.2016.02.028

Lythreatis, S., Mostafa, A. M. S., Pereira, V., Wang, X., & Giudice, M. D. (2021). Servant leadership, CSR perceptions, moral meaningfulness and organizational identification- evidence from the Middle East. *International Business Review*, *30*(5), 101772. doi:10.1016/j.ibusrev.2020.101772

Macdonald, G. (2000). Cross-cultural Methodological Issues in Ethical Research. *Journal of Business Ethics*, *27*(1/2), 89–104. doi:10.1023/A:1006406505398

Magalhães, C. (2011). Responsabilidade social e regulada no sector metalomecânico brasileiro e português: um estudo comparativo e cultural. In Responsabilidade Social: Uma visão ibero-americana (pp. 81-109). Almedina.

Maignan, I., & Ferrell, O. C. (2000). Measuring corporate citizenship in two countries: The case of the United States and France. *Journal of Business Ethics*, *23*(3), 283–297. doi:10.1023/A:1006262325211

Maignan, I., & Ferrell, O. C. (2001a). Antecedents and benefits of corporate citizenship: An investigation of French businesses. *Journal of Business Research*, *51*(1), 37–51. doi:10.1016/S0148-2963(99)00042-9

Maignan, I., & Ferrell, O. C. (2001b). Corporate citizenship as a marketing instrument: Concepts, evidence and research directions. *European Journal of Marketing*, *35*(3-4), 457–484. doi:10.1108/03090560110382110

Manzoor, F., Wei, L., Nurunnabi, M., Subhan, Q. A., Shah, S. I. A., & Fallatah, S. (2019). The impact of transformational leadership on job performance and CSR as mediator in SMEs. *Sustainability*, *11*(2), 436. doi:10.3390u11020436

Martin, J. (1992). *Cultures in Organizations - three perspectives*. Oxford University Press.

Martinez, F., O'Sullivan, P., Smith, M., & Esposito, M. (2017). Perspectives on the role of business in social innovation. *Journal of Management Development*, *36*(5), 681–695. doi:10.1108/JMD-10-2016-0212

Martins, E., & Martins, N. (2002). An organizational culture model to promote creativity and innovation. *SA Journal of Industrial Psychology*, *28*(4), 58–65. doi:10.4102ajip.v28i4.71

Martins, E. C., & Terblanche, F. (2003). Building organizational culture that stimulates creativity and innovation. *European Journal of Innovation Management*, *6*(1), 64–74. doi:10.1108/14601060310456337

Matos, M. H., & Machado, C. F. (2014). Institutional leaderships: The case of Crédito Agrícola in Portugal. *International Journal of Applied Management Sciences and Engineering*, *1*(1), 67–90. doi:10.4018/ijamse.2014010105

Matten, D., Crane, A., & Chapple, W. (2003). Behind the Mask: Revealing the True Face of Corporate Citizenship. *Journal of Business Ethics*, *45*(1/2), 109–120. doi:10.1023/A:1024128730308

Mayer, D., Kuenzi, M., Greenbaum, R., Bardes, M., & Salvador, R. B. (2009). How low does ethical leadership flow? Test of a trickle-down model. *Organizational Behavior and Human Decision Processes*, *108*(1), 1–13. doi:10.1016/j.obhdp.2008.04.002

Mclean, L. D. (2005). Organizational culture's influence on creativity and innovation: A review of the literature and implications for human resource development. *Advances in Developing Human Resources*, *7*(2), 226–246. doi:10.1177/1523422305274528

McLeary, C. N., & Cruise, P. A. (2015). A context-specific model of organizational trust – an examination of cognitive and socio-affective trust determinants in unique cultural settings. *Cross Cultural Management*, *22*(2), 297–302. doi:10.1108/CCM-11-2013-0180

McWilliams, A., & Siegel, D. (2001). Corporate social responsibility: A theory of the firm perspective. *Academy of Management Review*, *26*(1), 117–127. doi:10.5465/amr.2001.4011987

Milliman, J., Czaplewski, A. J., & Ferguson, J. (2003). Workplace spirituality and employee work attitudes: An exploratory empirical assessment. *Journal of Organizational Change Management*, *16*(4), 426–447. doi:10.1108/09534810310484172

Mintzberg, H., Simons, R., & Kunal, B. (2002). Beyond Selfishness. *MIT Sloan Management Review*, *44*(1), 67–74.

Mirvis, P., Herrera, M., Googins, B., & Albaredad, L. (2016). Corporate social innovation: How firms learn to innovate for the greater good. *Journal of Business Research*, *69*(11), 5014–5021. doi:10.1016/j.jbusres.2016.04.073

Moon, Y. S., & Franke, G. R. (2000). Cultural influences on agency practitioners' ethical perceptions: A comparison of Korea and the U.S. *Journal of Advertising*, *26*(1), 51–65. doi:10.1080/00913367.2000.10673603

Mulki, J. P., Jaramillo, J. F., & Locander, W. B. (2009). Critical role of leadership on ethical climate and salesperson performance. *Journal of Business Ethics*, *86*(2), 125–141. doi:10.100710551-008-9839-4

Mulyania, T., Wahyono, M., & Yanto, H. (2021). The Effect of Leadership on Performance Through Work Involvement Based on Engineering Process Values On Educational Quality Assurance School Team in Central Java Indonesia. *Turkish Journal of Computer and Mathematics Education*, *12*(13), 7037–7047.

Mustapha, N. D., Jali, M. N., & Muhamad, N. (2021). An Overview of Corporate Social Innovation (CSI). *International Journal of Academic Research in Business & Social Sciences*, *11*(7), 691–698. doi:10.6007/IJARBSS/v11-i7/10527

Nazariana, A., Atkinsonb, P., & Foroudi, P. (2017). Influence of national culture and balanced organizational culture on the hotel industry's performance. *International Journal of Hospitality Management*, *63*, 22–32. doi:10.1016/j.ijhm.2017.01.003

Nicholls, A., Simon, J., & Gabriel, M. (2015). Dimensions of Social Innovation. *New Frontiers in Social Innovation Research*, 1–26.

Patterson, M. G., West, M. A., Shackleton, V. J., Dawson, J. F., Lawthom, R., Maitlis, S., Robinson, D. L., & Wallace, A. M. (2005). Validating the organizational climate measure: Links to managerial practices, productivity and innovation. *Journal of Organizational Behavior, 26*(4), 379–408. doi:10.1002/job.312

Pereira, A. T., Santos, M. J., & Estevam, D. O. (2021). *Styles of leadership and perceptions of corporate social responsibility book: Sustainable Management for Managers and engineers*. John Wiley & Sons.

Peterson, D. K. (2004). The relationship between perceptions of corporate citizenship and organizational commitment. *Business & Society, 43*(3), 296–319. doi:10.1177/0007650304268065

Phillips, R., Freeman, R. E., & Wicks, A. C. (2003). What stakeholder theory is not. *Business Ethics Quarterly, 13*(4), 479–502. doi:10.5840/beq200313434

Porter, M. E., & Kramer, M. R. (2006). The link between competitive advantage and corporate social responsibility. *Harvard Business Review, 84*(12), 78–92. PMID:17183795

Porter, M. E., & Kramer, M. R. (2011). The big idea: Creating shared value. *Harvard Business Review, 89*, 2–17.

Pothukuchi, V., Damanpour, F., Choi, J., Chen, C. C., & Ho Park, S. (2002). National and Organizational Culture Differences and International Joint Venture Performance. *Journal of International Business Studies, 33*(2), 243–256. doi:10.1057/palgrave.jibs.8491015

Quélin, B. V., Cabral, S., Lazzarini, S., & Kivleniece, I. (2019). The private scope in public-private collaborations: An institutional and capability-based perspective. *Organization Science, 30*(4), 831–846. doi:10.1287/orsc.2018.1251

Ray, S., & Ray, I. (2012). Understanding Democratic Leadership: Some Key Issues And Perception. With Reference To India's Freedom Movement. *Afro Asian Journal of Social Sciences, 3*(1), 1-26.

Riketta, M. (2002). Attitudinal organizational commitment and job performance: A meta-analyis. *Journal of Organizational Behavior, 23*(3), 257–266. doi:10.1002/job.141

Rivera-Santos, M., Rufn, C., & Kolk, A. (2012). Bridging the institutional divide: Partnerships in subsistence markets. *Journal of Business Research, 65*(12), 1721–1727. doi:10.1016/j.jbusres.2012.02.013

Robertson, C. J., & Crittenden, W. F. (2003). Mapping Moral Philosophies: Strategic Implications for Multinational Firms. *Strategic Management Journal, 24*(4), 385–392. doi:10.1002mj.300

Rupp, D. E., & Mallory, D. B. (2015). Corporate social responsibility: Psychological, person-centric, and progressing. *Annual Review of Organizational Psychology and Organizational Behavior, 2*(1), 211–236. doi:10.1146/annurev-orgpsych-032414-111505

Russell, S. V., Evans, A., Fielding, K. S., & Hill, C. (2016). Turn it off: An action research study of top management influence on energy conservation in the workplace. *Frontiers in Psychology, 7*, 389. doi:10.3389/fpsyg.2016.00389 PMID:27047417

Saka-Helmhout, A., Chappin, M., & Rodrigues, S. (2021). Corporate Social Innovation in Developing Countries. *Journal of Business Ethics*. Advance online publication. doi:10.100710551-021-04933-x

Salman, Z. (2011). Leadership Styles and Employee Performance (A case Study of Shakarganj Mills Limited Jhang Pakistan). *Interdisciplinary of Contemporary Research in Business, 3*(6), 257–267.

Sánchez, F., Sandoval, A., Rodríguez-Pomeda, J., & Casani, F. (2020). Professional aspirations as indicators of responsible leadership style and corporate social responsibility. Are we training the responsible managers that business and society need? A cross-national study. *Journal of Work and Organizational Psychology, 36*(1), 49–61. doi:10.5093/jwop2020a5

Sanyal, R. (2005). Determinates of bribery in international business: The cultural and economic factors. *Journal of Business Ethics, 59*(1-2), 139–145. doi:10.100710551-005-3406-z

Saul, J. (2011). *Social innovation, inc.: 5 strategies for driving business growth through social change.* Wiley.

Schein, E. H. (1985). *Organizational culture and leadership.* Jessey Bess.

Schein, E. H. (2010). *Organizational culture and leadership* (4th ed.). Jossey-Bass.

Schminke, M., Ambrose, M. L., & Neubaum, D. O. (2005). The effect of leader moral development on ethical climate and employee attitudes. *Organizational Behavior and Human Decision Processes, 97*(2), 135–151. doi:10.1016/j.obhdp.2005.03.006

Scholtens, B., & Dam, L. (2007). Cultural values and international differences in business ethics. *Journal of Business Ethics, 75*(3), 273–284. doi:10.100710551-006-9252-9

Sergeant, A., & Frenkel, S. (2000). When do customer contact employees satisfy customers? *Journal of Service Research, 3*(1), 18–34. doi:10.1177/109467050031002

Sharkie, R. (2009). Trust in Leadership is vital for Employee Performance. *Management Research News, 32*(5), 491–498. doi:10.1108/01409170910952985

Sharma, L., & Singh, S. (2013). A Study on the Democratic Style of Leadership. *International Journal of Management & Information Technology, 3*(2), 54–57. doi:10.24297/ijmit.v3i2.1367

Smith, A., & Hume, E. C. (2005). Linking culture and ethics: A comparison of accountants' ethical belief systems in the individualism/collectivism and power distance contexts. *Journal of Business Ethics, 62*(3), 209–220. doi:10.100710551-005-4773-1

Soschinski, C., & Giordani, M., Klann, R., & Brizolla, M. (2021). Influence Of National Culture On Corporate Social Responsibility. *Cross-cultural Management Journal, 1*, 71–88.

Stoney, C., & Winstanley, D. (2001). Stakeholding: Confusion or utopia? Mapping the conceptual terrain. *Journal of Management Studies, 38*(5), 603–626. doi:10.1111/1467-6486.00251

Szutowski, D., & Ratajczak, P. (2016). The Relation between CSR and Innovation. Model Approach. *Journal of Entrepreneurship. Management and Innovation, 12*(2), 77–94.

Tabares, S. (2020). Insights from corporate social innovation: A research agenda. *Social Enterprise Journal, 16*(3), 317–338. doi:10.1108/SEJ-08-2019-0057

Thomas, A. S., & Mueller, S. L. (2000). A case for comparative entrepreneurship: Assessing the relevance of culture. *Journal of International Business Studies, 31*(2), 287–301. doi:10.1057/palgrave.jibs.8490906

Thome, L., & Saunders, B. (2002). The Socio-Cultural Embeddedness of Individuals' Ethical Reasoning in Organizations. *Journal of Business Ethics, 35*(1), 1–14. doi:10.1023/A:1012679026061

Torelli, C. J., Shavitt, S., Cho, Y. I., Holbrook, A. L., Johnson, T. P., & Weiner, S. (2015). Justice or compassion? Cultural differences in power norms affect consumer satisfaction with power-holders. *International Marketing Review, 32*(3/4), 279–306. doi:10.1108/IMR-09-2013-0222

Trompenaars, F., & Hampden-Turner, C. (1998). *Riding the Waves of Culture: Understanding Cultural Diversity in Global Business* (2nd ed.). McGraw-Hill.

Tuan, L. T., & Venkatesh, S. (2010). Organizational culture and technological innovation adoption in private hospitals. *International Business Research, 3*(3), 144–153. doi:10.5539/ibr.v3n3p144

Twati, J. M., & Gammack, J. G. (2006). The impact of organizational culture innovation on the adoption of IS/IT: The case of Libya. *Journal of Enterprise Information Management, 19*(1-2), 175–191. doi:10.1108/17410390610645076

Uzkurt, C., Kumar, R., & Ensari, N. (2013). Assessing organizational readiness for innovation: An exploratory study on organizational characteristics of innovativeness. *International Journal of Innovation and Technology Management, 10*(4), 1350018. doi:10.1142/S0219877013500181

Valencia, J. C. N., Valle, R. S., & Jiménez, D. J. (2010). Organizational culture as determinant of product innovation. *European Journal of Innovation Management, 13*(4), 466–480. doi:10.1108/14601061011086294

Van der Have, R., & Rubalcaba, L. (2016). Emerging Area of Innovation Studies? *Research Policy, 45*(9), 1923–1935. doi:10.1016/j.respol.2016.06.010

Van Marrewijk, M. (2003). Concepts and definitions of CSR and corporate sustainability: Between agency and communion. *Journal of Business Ethics, 44*(2–3), 95–105. doi:10.1023/A:1023331212247

Vargo, S. L., & Lusch, R. F. (2017). Service-dominant logic 2025. *International Journal of Research in Marketing, 34*(1), 46–67. doi:10.1016/j.ijresmar.2016.11.001

Verbekea, A., Kanob, L., & Yuanc, W. (2016). Inside the regional multinationals: A new value chain perspective on subsidiary capabilities. *International Business Review, 25*(3), 785–793. doi:10.1016/j.ibusrev.2016.01.019

Verleye, K., Perks, H., Gruber, T., & Voets, J. (2018). The long and winding road: Building legitimacy for complex social innovation in networks. *Journal of Product Innovation Management, 36*(6), 695–720. doi:10.1111/jpim.12506

Vos, J. (2003). Corporate social responsibility and the identification of stakeholders. *Corporate Social Responsibility and Environmental Management, 10*(3), 141–152. doi:10.1002/csr.39

Waddock, S., & Smith, N. (2000). Relationships: The real challenge of corporate global citizenship. *Business and Society Review, 105*(1), 47–62. doi:10.1111/0045-3609.00064

Waldman, D. A., & Siegel, D. S. (2008). Theoretical and practitioner letters: Defining the socially responsible leader. *The Leadership Quarterly*, *19*(1), 117–131. doi:10.1016/j.leaqua.2007.12.008

Walumbwa, F. O., & Lawler, J. J. (2003). Building effective organizations: Transformational leadership, collectivist orientation, work-related attitudes, and withdrawal behaviors in three emerging economies. *International Journal of Human Resource Management*, *14*(7), 1083–1101. doi:10.1080/0958519032000114219

Warhurst, A. (2001). Corporate citizenship and corporate social investment. *Journal of Corporate Citizenship*, *1*(1), 57–73. doi:10.9774/GLEAF.4700.2001.sp.00008

Wilson, M. (2003). Corporate sustainability: What is it and where does it come from? *Ivey Business Journal*, *67*(6), 1–5.

Wong, S. I., & Giessner, S. R. (2016). The thin line between empowering and laissez-faire leadership: An expectancy-match perspective. *Journal of Management*, *44*(2), 757–783. doi:10.1177/0149206315574597

Wood, J. D. (1991). Corporate Social Performance Revised. *Academy of Management Review*, *16*(4), 691–718. doi:10.2307/258977

Wood, J. D., & Logsdon, J. M. (2002). Business citizenship: From domestic to global level of analysis. *Business Ethics Quarterly*, *12*(2), 155–187. doi:10.2307/3857809

Wright, T. A., & Cropanzano, R. (2004). The role of psychological well-being in job performance: A fresh look at an age-old quest. *Organizational Dynamics*, *33*(4), 338–351. doi:10.1016/j.orgdyn.2004.09.002

Zairi, M., & Al-Mashari, M. (2005). Developing a sustainable culture of innovation management: A perspective approach. *Knowledge and Process Management*, *12*(3), 190–202. doi:10.1002/kpm.229

Zander, L., & Romani, L. (2004). When nationality matters: A study of departmental, hierarchical, professional, gender and age-based employee groupings' leadership preferences across 15 countries. *International Journal of Cross Cultural Management*, *4*(3), 291–315. doi:10.1177/1470595804047811

Zwetsloot, G. (2003). From management systems to corporate social responsibility. *Journal of Business Ethics*, *44*(2/3), 201–207. doi:10.1023/A:1023303917699

Chapter 9
Exergoeconomic Analysis as a Cost-Control Mechanism in Manufacturing Operations:
Focused on Industrial Beverage Carbonation Systems

Chukwuemeka Joshua Okereke
https://orcid.org/0000-0003-1739-2445
University of Ilorin, Nigeria

Olumuyiwa A. Lasode,
University of Ilorin, Nigeria

Idehai O. Ohijeagbon
University of Ilorin, Nigeria

ABSTRACT

The aim of the study is to demonstrate the application of exergoeconomic analysis as a cost control mechanism in manufacturing operations with a focus on industrial beverage carbonation system. Exergoeconomic analysis is a thermodynamic tool that can identify cost rate of exergy destruction (hidden cost) associated with a machine or a system that cannot be identified using typical cost management techniques applied in industries. Exergy analysis was performed to examine the performance parameters of each unit in the system using mass and energy balance. Cost rate of exergy destruction was performed using the thermoeconomic analysis of energy systems software. The study reviews that the total hidden cost in the carbonation system was $777.31/hr, while the total cost rate related to the investment and maintenance of the system was $45.13/hr. The study concluded that reduction of exergy destruction in the mixer and other subsystems within the carbonation system is very crucial to the improvement of the cost rate in beverage production.

DOI: 10.4018/978-1-6684-2339-4.ch009

INTRODUCTION

Owing to the fact that the rate of energy consumption in industrial processes is proportional to the cost of products and emissions, it is therefore imperative to manage energy in manufacturing plants in order to improve profit. Profit maximisation is the aim of any manufacturing organisation, who seeks to earn a maximum profit with low cost. Profit helps organisation to build strong economic base, improve performance and remains competitive in the business. In order to drive the operating cost downward, various organisation has developed and applied different cost management techniques such as process automation, line rationalisation, Kaizen system implementation, cost volume analysis, total quality management, standard cost analysis, budgeting control, just-in-time production system and so on. But none of these cost management techniques focus on energy consumed in the manufacturing process. They do not consider exergy destruction rate in the manufacturing process nor identify thermodynamic limitation of the machineries or process flow involved in the manufacturing operations. Ghannadzadeh (2013) state that the energy and fuel consumption of a machinery or process flow will be huge, if it has thermodynamic limitations, consequently operate at low efficiency with a stiff cost. The thermodynamic limitation of a machinery, or process flow, can be identified using exergy analysis and the cost implications by applying exergoeconomic analysis. Exergoeconomics is emerging thermodynamic tool, used to identify cost rate of exergy destroyed (hidden cost) in a machinery, process flow and systems (Okereke et al. 2020a). Exergoeconomics uses exergy, compared to energy, as the product of system conversion and the cogent reason for assigning the monetary costs to the system conversion. Nothing happens without conversion of energy and everything described as changes in time involves energy conversion, which may be large or small (Wall, 2009). According to the laws of thermodynamics, energy is conserved, while exergy is not conserved. Exergy is the component of energy that is convertible and useful, which has the ability to do work and economic value (Dincer & Cengel, 2001; Gundersen, 2011; Wall, 2009). Once the exergy is wasted, it can never be recovered (Dincer & Cengel, 2001; Gundersen, 2011; Wall, 2009). Exergy analysis is a method of accounting for the thermodynamic constraint of the actual system and improving the system efficiency (Ghannadzadeh, 2013). One of the major challenge in manufacturing operation is exergy wastage, which cannot be identified with regular cost management technique or conventional energy analysis method.

BACKGROUND

Carbonation refers to the process of dissolving carbon dioxide (CO_2) in a liquid at varied conditions depending on the application. (Thongrote, Wirjantoro, & Phianmongkhol, 2016). Carbonation has been applied to beverage production (Abdellatif, 2018; Newbold & Koppel, 2018), sugar refining processes (Sahu, 2018; Moodley, Schorn, Walthew, & Masinga, 2002), cement and concrete production (Šavija & Lukovic, 2016; Pham, 2013; Younsi, Turcry, Rozière, Aît-Mokhtar, & Loukili, 2011) and CO_2 capturing technologies for mineral carbonation (Giannoulakis, Volkart, & Bauer, 2014; Olajire, 2013). Temperature and pressure of a system are major factors which determine the solubility of CO_2 in a solution. In the solubility of CO_2, temperature is directly proportional to pressure and volume as postulated by Henry's gas law and Charles gas law (Steen, 2006), expressed in Equation (1). A solution of high temperature requires high pressure to maintain CO_2 in it, while high amount of CO_2 will be dissolved in a solution of low temperature.

$$Pv = MRT \quad (1)$$

where, P = pressure (bar), V =volume (M^3), M = molar mass (kg/mol), R = gas constant (kJ/kgK), and T = temperature (K)

The solubility of CO_2 increases with a decrease in the temperature (Wiebe & Gaddy, 1940). Consequently, beverages can be more easily carbonated at lower temperatures; however, low temperature processes correspond to a high energy usage and production cost. With the increasing competition, energy and production costs, beverage manufacturers are seeking techniques to deliver low-cost but high-quality products. Therefore, it is necessary to demonstrate the application of exergoeconomic analysis as a cost control mechanism in manufacturing operations, using industrial beverage carbonation system as a case study. The study identifies the areas of exergy destruction and corresponding cost rate due to exergy destruction in the carbonation process. This will enhance decision-making regarding cost control, reduce energy consumption in the carbonation system, increase efficiency and profitability.

From literatures on exergoeconomics, exergy and energy study in food and beverage manufacturing process, it shows that there is a gap in application of exergoeconomic analysis to carbonation process. Fadare, Nkpubre, Oni, Falana, Waheed, & Bamiro (2010) conducted the energy and exergy analysis for a malt drink production process, among the group of operations involved in the malt production process, the packaging house and brewhouse, corresponded to the highest energy consumption of 223.19 MJ/hr and 35.94 MJ/hr, respectively and the highest exergy destruction occurred at the pasteuriser. However, the study did not focus on the carbonation process of malt drinks and the cost rate due to exergy destruction. In a similar research, Waheed, Jekayinfa, Ojediran, & Imeokparia, (2008) examined the energy and exergy of the orange juice production process, which is a non-carbonated process. The adaptation method was employed to investigate the energy usage in the orange production process. The research showed that high exergy destruction occurred at pasteuriser compared to other components. The study did not consider the cost rate due to exergy destruction of the components, which can help enhance the decision making. Sulaiman, Oni, & Fadare, (2012) performed the exergy analysis of vegetable oil production and reported that the deodoriser process corresponded to the highest exergy destruction among all the processes in vegetable oil production. The authors recommended that exergy destroyed in the manufacturing process can be reduced by an increase in production capacity. However, the cost rate due to exergy destroyed was not considered, although it could help clarify the relationship between the production capacity and exergy destruction in the system, thereby enhancing the decision making to optimise the production process. In the analysis of quadruple-effect evaporator unit in a tomato paste factory, Sogut, Ilten, & Oktay, (2010) reported that, steam generation process has the highest exergy destruction, where, steam boiler accounting for 82% of the total value; while the crusher exhibited the lowest exergy efficiency (1.05%) compared to other components. Various techniques for system enhancement were highlighted; however, the cost rate due to exergy destruction was not determined, which could assist in the prioritisation of the improvement techniques. Singh, Singh, Tyagi, & Pandey (2019) performed the energy, exergy, and exergoeconomic analyses of the milk production process. The overall energy efficiency, exergy efficiency, and cost rate due to exergy destruction were noted to be 86.36%, 53.02% and 8143.90 €/hr. Nevertheless, the study focused on a non-carbonated beverage. Filippos, Francisco, Atze Jan Van Der Goot, & Remko, (2017) presented the analysis of food supply chain, using exergy indicators. The need for efficient food production system that uses fewer energy forms the authors' motivation. Among the indicators applied in the food supply chain analysis, exergetic efficiency rank first with 38% application rate, followed by exergy losses with 25%. Other indicators such as entropy generation, exergy destruction ratio, cumulative exergy consumption, sustainability index, cumulative exergy losses, exergetic factor, specific exergy destruction, relative irreversibility, exergetic renewability, eco-exergy, improvement potential, lack of productivity and renewability performance indicator has application rate between 1-19%. The authors highlighted that exergy analysis is a critical component in sustainability of industrial food supply chain, which is in line with this study. However, they did not review how exergy cost can help to make inform decisions about food supply chain.

From literature, there are many approach to exergoeconomic analysis, such as Specific exergy cost method, Exergy economics approach, Moran method, First exergoeconomics approach, Wonergy method, Structural analysis approach, Engineering functional analysis, Modified productive structure analysis, Thermoeconomic functional analysis, Last-in-first-out approach and Exergetic cost theory, (Meyer, Castillo, Buchgeister, & Tsatsaronis, 2009; Gorji-Bandpy & Ebrabimian, 2006; Haydargil & Abusoglu, 2018; Uysal & Keçebas, 2021; Sun, Wang, & Xie, 2021; Chan, et al., 2021). In the work of Haydargil & Abusoglu (2018), the analysis of biogas engine-powered cogeneration system, using commonly applied exergoeconomic analysis approach such as modified productive structure analysis, Exergetic cost theory, Specific exergy cost method, and Wonergy method. The authors observed a varied exergy cost of electricity production, then concludes that, compared to other approach, exergetic cost theory is more complete. Exergetic cost theory approach estimates the unit exergoeconomic cost of all flows in the system and in the subsystems. It does not separate the cost of exergy destruction in the subsystem from production cost. While, in the modified productive structure analysis approach, the exergy flow of the material streams is taken either as thermal exergy or mechanical exergy. Mechanical exergy is concerned with the pressure of the material stream flow, while thermal exergy is concerned with temperature. The unit exergoeconomic cost of all flows in the system is not calculated, a general cost balance equation applies to each subsystem. A new unit exergoeconomic cost is appointed according to the subsystem's principal product and the cost structure of the system, which is influenced by the entropy production of each component. However, wonergy, present energy and exergy combination depending on the prevailing assumptions and a general cost balance equation is established which is written to include a wonergetic unit cost, this cost balance equation is separated into three parts according to the unit classifications and the level of wonergy input, which plays the most important role in determining the product costs. Specific exergy cost method is a systematic and general methodology to calculate efficiencies and costs. The cost rates are obtained by using basic principles of business administration. The cost balance equation does not include exergy destruction costs in the subsystems. In this study, exergy cost analysis was evaluated with aid of thermoeconomic analysis of energy systems software (TAESS), which algorithm is based on Exergetic cost theory approach.

METHOD

This chapter describes the industrial beverage carbonation system, data collection, mathematical models of energy balance, mass balance, exergy destroyed, energy and exergy efficiencies of each component, and the overall system. The fundamental principles of performing exergoeconomic analysis of the equipment in the carbonation system using TAESS are presented.

Description of Beverage Carbonation System

The considered carbonation system involves two major subsystems: the industrial refrigeration system and mixer system, as illustrated in Figure 1. The mixer system is a water-to-syrup proportioning type, in which a mixture of the syrup and carbonated water is obtained at a target Brix of the beverage. Water is treated to a certain quality standard required for bottling operations, whereas syrup is prepared in a ready syrup tank. Syrup from the ready syrup tank and treated water from the treatment plant are transported to the syrup and water reservoirs of the mixer, respectively, with the aid of a pump. In the water reservoir of the mixer, deaeration of treated takes place to form carbonated water through diffusion. At the mixer, the carbonated water and syrup are proportioned through water and syrup orifices to prepare the beverage, based on the required Brix (sucrose concentration). The proportioned beverage flows to the flow mix reservoir, where it is transported to the carbonator with the aid of a flow mix pump. CO_2 from Carbo-trol is mixed with the beverage from the flow mix reservoir as it is being transferred to the carbonator at node point. A node, defined as non-returnable valve (NRV) that prevents backward pressure on the flow mix pump. The solubility of the CO_2 in the beverage to form carbonated beverage takes place inside the carbonator with the aid of an industrial refrigeration system, which maintains carbonation at a low temperature. An ammonia compressor drives the flow of the refrigerant through the refrigeration system. The air compressor system supplies compressed air for all pneumatic valves and level controls in the carbonation system. The CO_2 used by the carbonation system comes from a CO_2 storage tank.

Figure 1. Material flow streams in beverage carbonation system

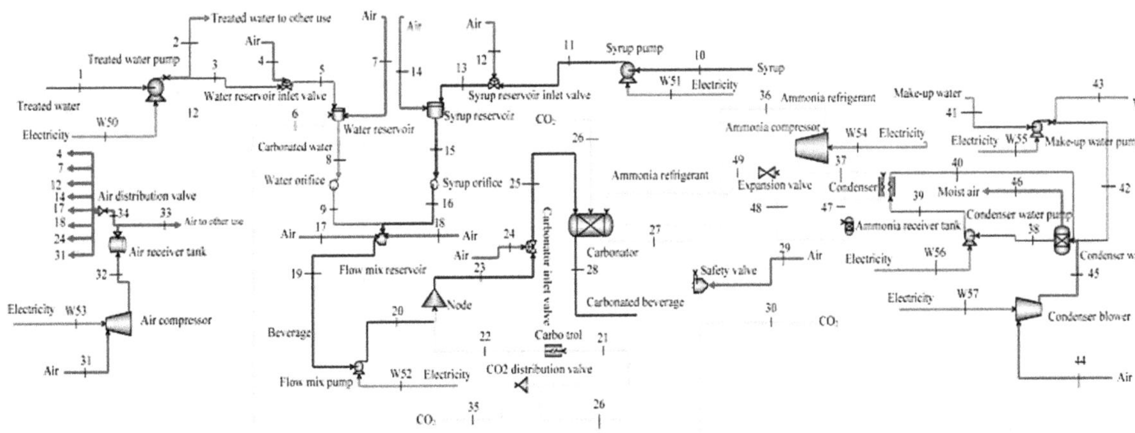

Data Collection

The operational thermodynamic variables of the carbonation system which are primary data (temperature, mass flow rate, pressure) and secondary data (energy, specific heat capacity, enthalpy, entropy, exergy), and data specific to the manufacturing process, such as brix, gas volume and calorific value, are obtained from the work of Okereke et al. (2019; 2020b) and presented in Table 1 to 2. The tools used in data collection and the calorimeter experiment procedure can be found in Okereke et al. (2019; 2020b)

Table 1. Operational thermodynamic variables of the carbonation system

State	Mass flow rate (kg/s)	Temperature (°C)	Pressure (bar)	Enthalpy (kJ/kg)	Entropy (kJ/kg K)	Energy (kW)	Physical exergy (kJ/kg)	Chemical exergy (kJ/kg)	Total exergy (kW)
1	2.50	27.00	1.70	2412.48	0.39374	6031.21	2295.08	50.00	5862.70
2	0.23	27.43	4.00	2414.27	0.39972	555.29	2295.09	50.00	539.37
3	2.27	27.43	4.00	2414.27	0.39972	5480.49	2295.09	50.00	5323.36
4	0.01	27.00	3.50	300.15	6.58186	2.70	106.07	0.00	0.95
5	2.27	27.50	4.00	2414.56	0.40069	5481.15	2295.10	50.00	5323.37
6	0.01	18.00	1.20	411.76	1.72440	4.12	4.12	451.59	4.56
7	0.01	29.00	3.50	302.15	6.58834	2.72	106.09	0.00	0.95
8	2.27	28.00	3.89	639.03	0.25163	1450.69	564.01	49.31	1392.24
9	2.27	28.00	3.89	639.03	0.25163	1450.68	564.01	49.31	1392.24
10	0.53	28.00	1.00	365.40	0.29397	193.66	277.75	8745.15	4782.14
11	0.53	30.00	5.00	371.42	0.31391	196.87	277.83	8745.15	4782.18
12	0.01	29.00	3.50	302.15	6.58834	2.72	106.09	0.00	0.95
13	0.53	31.00	5.00	374.43	0.32383	198.47	277.88	8745.15	4782.21
14	0.01	29.00	5.00	302.15	6.58834	2.42	136.61	0.00	1.09
15	0.53	32.00	5.00	377.45	0.33371	200.07	277.95	8431.31	4615.91
16	0.53	32.00	5.00	377.45	0.33371	200.06	277.95	8431.31	4615.91
17	0.01	29.00	3.50	302.15	6.58834	3.02	106.09	0.00	1.06
18.	0.01	29.00	3.50	302.15	6.58834	2.42	106.09	0.00	0.85
19	2.80	29.50	1.10	167.65	0.28981	469.44	81.24	1775.40	5198.61
20	2.80	30.10	4.20	169.35	0.29541	474.25	81.27	1775.40	5198.68
21	0.02	28.00	10.00	381.20	1.59990	9.15	129.00	451.59	13.93
22	0.02	28.00	3.40	381.20	1.59990	9.15	68.22	451.59	12.48
23	2.80	28.00	5.00	163.41	0.27577	457.67	81.27	1775.00	5197.34
24	0.02	29.00	3.50	302.15	6.58834	6.05	106.09	0.00	2.12
25	2.80	28.40	5.00	164.54	0.27952	460.83	81.27	1775.00	5197.37
26	0.13	27.00	10.80	386.39	1.61890	50.62	133.33	451.59	76.62
27	0.10	18.00	2.50	411.76	1.72440	41.18	50.95	451.59	50.25
28	2.82	10.00	4.20	17.53	0.07992	49.49	-6.30	1801.77	5063.24
29	0.01	29.00	3.50	302.15	6.58834	2.42	106.09	0.00	0.85
30	0.03	29.00	1.10	411.76	1.72440	12.35	4.65	451.59	13.69
31	0.16	26.00	1.00	299.15	6.86769	47.86	-1.13	0.00	-0.18
32	0.16	30.00	7.00	303.15	3.36730	48.52	165.42	0.00	26.47
33	0.08	29.00	6.80	302.15	6.34100	25.09	162.93	0.00	13.52
34	0.07	29.00	6.80	302.15	6.34100	21.16	162.95	0.00	11.41
35	0.23	28.00	12.00	381.20	1.59990	87.68	139.27	451.59	135.90
36	0.18	-10.00	2.60	1450.20	5.75500	261.04	-265.65	19876.47	3529.95

Continued on following page

Table 1. Continued

State	Mass flow rate (kg/s)	Temperature (°C)	Pressure (bar)	Enthalpy (kJ/kg)	Entropy (kJ/kg K)	Energy (kW)	Physical exergy (kJ/kg)	Chemical exergy (kJ/kg)	Total exergy (kW)
37	0.18	126.64	14.50	1717.78	5.82151	309.20	-17.90	19876.47	3574.54
38	8.50	34.00	1.80	142.62	0.49152	1212.42	-3.93	50.00	391.62
39	8.50	34.40	3.40	142.77	0.49147	1213.83	-3.76	50.00	393.02
40	8.44	40.00	1.00	167.62	0.57240	1415.08	-3.04	50.00	396.33
41	1.50	28.90	1.20	121.25	0.42156	181.88	-4.44	50.00	68.34
42	0.06	29.00	1.80	121.73	0.42293	7.31	-4.37	50.00	2.74
43	1.44	29.00	4.00	121.93	0.42286	175.65	-4.15	50.00	66.03
44	12.14	28.00	1.00	301.62	6.87572	3661.90	-13.54	0.00	-13.54
45	12.14	28.20	1.002	301.83	6.87614	3665.36	-11.44	0.00	-11.44
46	12.14	60.00	0.98	307.80	0.89660	3736.83	-1.05	86.04	1031.79
47	0.18	28.00	12.00	332.16	1.45657	59.79	-102.12	19876.47	3559.38
48	0.18	28.00	11.50	332.16	1.45657	59.79	-102.12	19876.47	3559.38
49	0.18	-4.00	3.60	332.13	1.50130	59.78	-115.48	19876.47	3556.98

Table 2. Electrical data of the power components of thee carbonation system

State	Components	Current (I)	Voltage (V)	Energy/ Exergy (kW)
W_{50}	Treated water pump	9.00	380.00	4.74
W_{51}	Syrup pump	9.50	374,00	4.92
W_{52}	Flow mix pump	14.00	370.00	7.18
W_{53}	Air compressor	70.00	371.00	35.99
W_{54}	Ammonia compressor	167.00	370.00	85.62
W_{55}	Make-up water pump	3.40	381.00	1.79
W_{56}	Condenser water pump	3.10	371.00	1.59
W_{57}	Condenser blower	7.40	373.00	3.82

Mass Balance, Exergy Destruction, and Energy and Exergy Efficiencies of the Carbonation System Components

The energy, exergy and mass balance for each component in the industrial beverage carbonation system presented in Figure 1 can be determined using Equations (2) to (4), respectively (Dincer & Rosen, 2007).

$$\mathrm{E} = Q - \mathrm{W} + \dot{m}_i\left(h_i + \frac{V_i^2}{2} + gz_i\right) - \dot{m}_e\left(h_e + \frac{V_e^2}{2} + gz_e\right) \tag{2}$$

$$Ex_d = \sum Ex_q - \dot{W} + \sum Ex_i - \sum Ex_e \quad (3)$$

$$\sum \dot{m}_i = \sum \dot{m}_e \quad (4)$$

where, ṁ = mass flow rate (kg/s), E= energy (kW), Q,q = heat (kW), W = work (kW), V = velocity (m/s), g = gravitational force (m/s^2), z = height (m), h = enthalpy (kJ/kg), Ex = exergy flow rate (kW), Ex_d = exergy destruction (kW), i = inlet flow to component and e = outlet flow from component

Treated Water Pump Control Volume Analysis

Treated water goes through the treated water pump control volume as presented in Figure1, with a change in the temperature and pressure at inlet state 1 and outlet states 2 and 3. The control volume is powered by electricity with an input power of W_{50}.
For treated water pump control volume:

$$\dot{m}_1 = \dot{m}_2 + \dot{m}_3 \quad (5)$$

$$\eta_{TWP} = \frac{E_2 + E_3 - E_1}{W_{50}} \times 100 \quad (6)$$

$$\eta_{\text{exTWP}} = \frac{Ex_2 + Ex_3 - Ex_1}{W_{50}} \times 100 \quad (7)$$

$$Ex_{d\,TWP} = (Ex_1 - Ex_2 + Ex_3) + W_{50} \quad (8)$$

where, η = energy efficiency (%), ηex_= exergy efficiency (%)

TWP, WRV….SYP = control volume name abbreviation, 1, 2, 3, 4 …57 = pipe/material flow stream number

Water Reservoir Inlet Valve Control Volume Analysis

The treated water and compressed air passing through the water reservoir inlet valve control volume as presented in Figure 1. The treated water enters at state 3 and leaves the control volume at state 5 with a change in the temperature. The compressed air used for pneumatic system control enters at state 4 and is dispersed to the environment at state 4'. There is no mixture of treated water and air in the control volume.
For water reservoir inlet valve control volume:

$$\dot{m}_3 = \dot{m}_5 \text{ (Treated water flow)} \tag{9}$$

$$\dot{m}_4 = \dot{m}_{4'} \text{ (Air flow)} \tag{10}$$

$$\eta_{WRV} = \frac{E_5}{E_3 + E_4} \times 100 \tag{11}$$

$$\eta_{\mathrm{exWRV}} = \frac{Ex_5}{Ex_3 + Ex_4} \times 100 \tag{12}$$

$$Ex_{d\,WRV} = Ex_4 + (Ex_3 - Ex_5) \tag{13}$$

Water Reservoir Control Volume Analysis

The treated water, CO_2, and compressed air passing through water reservoir control volume as presented in Figure 1. The treated water at state 5 and CO_2 at state 6 are mixed to form carbonated water at state 8. The compressed air, used for pneumatic system control, enters at state 7 and is dispersed to the environment at state 7'.

For water reservoir control volume:

$$\dot{m}_5 + \dot{m}_6 = \dot{m}_8 \text{ (Treated water and } CO_2 \text{ mixture)} \tag{14}$$

$$\dot{m}_7 = \dot{m}_{7'} \text{ (Air flow)} \tag{15}$$

$$\eta_{WR} = \frac{\mathrm{E}_8}{E_5 + E_6 + E_7} \times 100 \tag{16}$$

$$\eta_{exWR} = \frac{\mathrm{Ex}_8}{Ex_5 + Ex_6 + Ex_7} \times 100 \tag{17}$$

$$Ex_{d\,WR} = (Ex_5 + Ex_6 + Ex_7) - Ex_8 \tag{18}$$

Water Orifice Control Volume Analysis

Only carbonated water passes through the water orifice control volume as presented in Figure 1. Carbonated water enters at state 8 and leaves the control volume at state 9 with no change in the temperature and composition.
For water orifice control volume:

$$\dot{m}_8 = \dot{m}_9 \tag{19}$$

$$\eta_{WO} = \frac{E_9}{E_8} \times 100 \tag{20}$$

$$\eta_{ExWO} = \frac{Ex_9}{Ex_8} \times 100 \tag{21}$$

$$Ex_{d\,WO} = Ex_8 - Ex_9 \tag{22}$$

Syrup Pump Control Volume Analysis

Only syrup passes through the syrup pump control volume as presented in Figure1, with change in temperature and pressure at the inlet state 10 and outlet state 11. The control volume is powered by electricity with an input power of W_{51}.
For syrup pump control volume:

$$\dot{m}_{10} = \dot{m}_{11} \tag{23}$$

$$\eta_{SYP} = \frac{E_{11} - E_{10}}{W_{51}} \times 100 \tag{24}$$

$$\eta_{exSYP} = \frac{Ex_{11} - Ex_{10}}{W_{51}} \times 100 \tag{25}$$

$$Ex_{d\,SYP} = (Ex_{10} - Ex_{11}) + W_{51} \tag{26}$$

Syrup Reservoir Inlet Valve Control Volume

Syrup and compressed air pass through the syrup reservoir inlet valve control volume as presented in Figure 1. Syrup enters at state 11 and leaves the control volume at state 13 with a change in the temperature. The compressed air used for pneumatic system control enters at state 12 and is dispersed to the environment at state 12'. There is no mixture of the syrup and compressed air in the control volume. For syrup inlet valve control volume:

$$\dot{m}_{11}=\dot{m}_{13} \text{ (Syrup flow)} \tag{27}$$

$$\dot{m}_{12}=\dot{m}_{12'} \text{ (Air flow)} \tag{28}$$

$$\eta_{SRV}=\frac{E_{13}}{E_{11}+E_{12}}\times 100 \tag{29}$$

$$\eta_{\text{exSRV}}=\frac{Ex_{13}}{Ex_{11}+Ex_{12}}\times 100 \tag{30}$$

$$Ex_{d\,SRV}=Ex_{12}+(Ex_{11}-Ex_{13}) \tag{31}$$

Syrup Reservoir Control Volume Analysis

Syrup and compressed air pass through the syrup reservoir control volume as presented in Figure 1. Syrup enters at state 13 and leaves the control volume at state 15 with no changes in the temperature and pressure. The compressed air used for pneumatic system control enters at state 14 and is dispersed to the environment at state 14'. There is no mixture of the syrup and compressed air in the control volume. For syrup reservoir control volume:

$$\dot{m}_{13}=\dot{m}_{15} \text{ (Syrup flow)} \tag{32}$$

$$\dot{m}_{14}=\dot{m}_{14'} \text{ (Air flow)} \tag{33}$$

$$\eta_{SYR} = \frac{E_{15}}{E_{13} + E_{14}} \times 100 \tag{34}$$

$$\eta_{exSYR} = \frac{Ex_{15}}{Ex_{13} + Ex_{14}} \times 100 \tag{35}$$

$$Ex_{d\,SYR} = Ex_{14} + (Ex_{13} - Ex_{15}) \tag{36}$$

Syrup Orifice Control Volume Analysis

Only syrup passes through the syrup orifice control volume as presented in Figure 1. Specifically, the syrup enters at state 15 and leaves the control volume at state 16 with no change in the temperature and composition

For syrup orifice control volume:

$$\dot{m}_{15} = \dot{m}_{16} \tag{37}$$

$$\eta_{SYO} = \frac{E_{16}}{E_{15}} \times 100 \tag{38}$$

$$\eta_{ExSYO} = \frac{Ex_{16}}{Ex_{15}} \times 100 \tag{39}$$

$$Ex_{d\,SYO} = Ex_{15} - Ex_{16} \tag{40}$$

Flow Mix Reservoir Control Volume Analysis

Carbonated water, syrup, and compressed air pass through the flow mix reservoir control volume as presented in Figure 1. Carbonated water and syrup enter the control volume at states 9 and 16, respectively, are mixed to form the beverage. The mixture (beverage) is non-reactive and leaves the control volume at state 19. The compressed air used for level control and valve actuation enters the control volume at states 17 and 18, respectively, and is dispersed to the environment at states 17' and 18', respectively. For flow mix reservoir control volume:

$$\dot{m}_9+\dot{m}_{16} = \dot{m}_{19} \text{ (Syrup and carbonate water flow)} \quad (41)$$

$$\dot{m}_{17} = \dot{m}_{17'} \text{ (Air flow)} \quad (42)$$

$$\dot{m}_{18} = \dot{m}_{18'} \text{ (Air flow)} \quad (43)$$

$$\eta_{FMR} = \frac{\mathrm{E}_{19}}{E_9 + E_{16} + E_{17} + E_{18}} \times 100 \quad (44)$$

$$\eta_{exFMR} = \frac{\mathrm{Ex}_{19}}{Ex_9 + Ex_{16} + Ex_{17} + Ex_{18}} \times 100 \quad (45)$$

$$Ex_{d\,FMR} = (Ex_9 + Ex_{16} + Ex_{17} + Ex_{18}) - Ex_{19} \quad (46)$$

Flow Mix Pump Control Volume Analysis

The beverage passes through the flow mix pump control volume as presented in Figure 1, with a change in the temperature and pressure at inlet state 19 and outlet state 20. The control volume is powered by electricity with an input power of W_{52}.

For flow mix pump control volume:

$$\dot{m}_{19} = \dot{m}_{20} \quad (47)$$

$$\eta_{FMP} = \frac{E_{20} - E_{19}}{W_{52}} \times 100 \quad (48)$$

$$\eta_{\mathrm{exFMP}} = \frac{Ex_{20} - Ex_{19}}{W_{52}} \times 100 \quad (49)$$

$$Ex_{d\,FMP} = (Ex_{19} - Ex_{21}) + W_{52} \quad (50)$$

Node Control Volume Analysis

The beverage and CO_2 enter the node control volume as presented in Figure 1, at states 20 and 22, respectively; next, the fluids are mixed, and the mixture leaves the control volume at state 23
For node control volume:

$$\dot{m}_{20}+\dot{m}_{22}=\dot{m}_{23} \tag{51}$$

$$\eta_{NOD}=\frac{E_{23}}{E_{20}+E_{22}}\times 100 \tag{52}$$

$$\eta_{\mathrm{exNOD}}=\frac{Ex_{23}}{Ex_{20}+Ex_{22}}\times 100 \tag{53}$$

$$Ex_{d\,NOD}=(Ex_{20}+Ex_{22})-Ex_{23} \tag{54}$$

Carbonator Inlet Valve Control Volume Analysis

The beverage and compressed air pass through the carbonated inlet valve control volume as presented in Figure 1. The beverage enters at state 23 and leaves the control volume at state 25 with a change in the temperature. The compressed air used for valve actuation enters at state 24 and is dispersed to the environment at state 24'. There is no mixture of air and beverage in the control volume.
For carbonator inlet valve control volume:

$$\dot{m}_{23}=\dot{m}_{25}\ \text{(Beverage flow)} \tag{55}$$

$$\dot{m}_{24}=\dot{m}_{24'}\ \text{(Air flow)} \tag{56}$$

$$\eta_{CBV}=\frac{E_{25}}{E_{23}+E_{24}}\times 100 \tag{57}$$

$$\eta_{\mathrm{exCBV}}=\frac{Ex_{25}}{Ex_{23}+Ex_{24}}\times 100 \tag{58}$$

$$Ex_{d\,CBV} = Ex_{24} + (Ex_{25} - Ex_{23}) \tag{59}$$

Carbonator Control Volume Analysis

The beverage, $CO_{2,}$ and ammonia refrigerant pass through the carbonator control volume as presented in Figure 1, entering at states 25, 26, and 49, and exiting at states 27, 28, and 36 respectively, with changes in temperature and pressure. The beverage and CO_2 mixed during carbonation to form carbonated beverage at flow 28. The undissolved CO_2 leaves the carbonator control volume at flow 27.
For carbonator control volume:

$$\dot{m}_{25} + \dot{m}_{26} = \dot{m}_{27} + \dot{m}_{28} \text{ (Beverage and CO}_2\text{ flow)} \tag{60}$$

$$\dot{m}_{49} = \dot{m}_{36} \text{ (Refrigerant flow)} \tag{61}$$

$$\eta_{CBR} = \frac{\mathrm{E}_{27} + E_{28} + E_{36}}{E_{25} + E_{26} + E_{49}} \times 100 \tag{62}$$

$$\eta_{\mathrm{exCBR}} = \frac{\left(Ex_{27} + Ex_{28}\right)}{\left(Ex_{25} + Ex_{26}\right) + \left(Ex_{49} - Ex_{36}\right)} \times 100 \tag{63}$$

$$Ex_{d\,CBR} = (Ex_{25} + Ex_{26}) + (Ex_{49} - Ex_{36}) - (Ex_{27} + Ex_{28}) \text{ (64)}$$

Carbonator Safety Valve Control Volume Analysis

CO_2 from the carbonator enters the carbonator safety valve control volume as presented in Figure 1, at state 27 and leaves the control volume at states 6 and 30. The compressed air used for valve actuation enters at state 29 and is dispersed to the environment at state 29'
For carbonator safety valve control volume:

$$\dot{m}_{27} = \dot{m}_{6} + \dot{m}_{30} \text{ (CO}_2\text{ flow)} \tag{65}$$

$$\dot{m}_{29} = \dot{m}_{29'} \text{ (Air flow)} \tag{66}$$

$$\eta_{SV} = \frac{E_6 + E_{30}}{E_{27} + E_{29}} \times 100 \quad (67)$$

$$\eta_{exSV} = \frac{Ex_6}{(Ex_{27} - Ex_{30}) + Ex_{29}} \times 100 \quad (68)$$

$$Ex_{d\,SV} = (Ex_{27} + Ex_{30}) + Ex_{29} - Ex_6 \quad (69)$$

Carbo-Trol Control Volume Analysis

CO_2 enters the Carbo-Trol control volume as presented in Figure 1 at state 21 and leaves at state 22 with a change in pressure.
For carbo trol control volume:

$$\dot{m}_{21} = \dot{m}_{22} \quad (70)$$

$$\eta_{CBT} = \frac{E_{22}}{E_{21}} \times 100 \quad (71)$$

$$\eta_{ExCBT} = \frac{Ex_{22}}{Ex_{21}} \times 100 \quad (72)$$

$$Ex_{d\,CBT} = Ex_{21} - Ex_{22} \quad (73)$$

CO_2 Distribution Valve Control Volume Analysis

The CO_2 distribution valve control volume is shown in Figure 1. CO_2 from the storage tank enters the distribution valve at state 35 and leaves at states 21 and 26 with changes in temperature and pressure.
For CO_2 distribution valve control volume:

$$\dot{m}_{35} = \dot{m}_{21} + \dot{m}_{26} \quad (74)$$

$$\eta_{CDV} = \frac{E_{21} + E_{26}}{E_{35}} \times 100 \tag{75}$$

$$\eta_{\text{exCDV}} = \frac{Ex_{21} + Ex_{26}}{Ex_{35}} \times 100 \tag{76}$$

$$Ex_{d\,CDV} = Ex_{35} - (Ex_{21} + Ex_{26}) \tag{77}$$

Ammonia Compressor Control Volume Analysis

Ammonia goes through the ammonia compressor control volume as presented in Figure 1, with a change in the temperature and pressure with the inlet state 36 and outlet state 37. The control volume is powered by electricity with an input power of W_{54}.
For ammonia compressor control volume:

$$\dot{m}_{36} = \dot{m}_{37} \tag{78}$$

$$\eta_{AMC} = \frac{E_{37} - E_{36}}{W_{54}} \times 100 \tag{79}$$

$$\eta_{\text{exAMC}} = \frac{Ex_{37} - Ex_{36}}{W_{54}} \times 100 \tag{80}$$

$$Ex_{d\,AMC} = (Ex_{36} - Ex_{37}) + W_{54} \tag{81}$$

Make-Up Water Pump Control Volume Analysis

Water enters the make-up water pump control volume as presented in Figure 1, at state 41 and leaves at states 42 and 43 with changes in the temperature and pressure. The control volume is powered by electricity with an input power of W_{55}. Flow 42 is used to make-up for the loss due to evaporation in the condenser, whereas flow 43 is used for other utilities.
For make-up water pump control volume:

$$\dot{m}_{41} = \dot{m}_{42} + \dot{m}_{43} \tag{82}$$

$$\eta_{MWP}=\frac{E_{42}+E_{43}-E_{41}}{W_{55}}\times 100 \tag{83}$$

$$\eta_{\text{exMWP}}=\frac{Ex_{42}+Ex_{43}-Ex_{41}}{W_{55}}\times 100 \tag{84}$$

$$Ex_{d\,MWP}=(Ex_{41}-Ex_{42}+Ex_{43})+W_{55} \tag{85}$$

Condenser Blower Control Volume Analysis

Air from the environment passes through the condenser blower control volume as presented in Figure 1, to cool the water in the condenser tank. The air enters at state 44 and exits at state 45 with a change in the temperature and pressure. The control volume is powered by electricity with an input power of W_{57}. For condenser blower control volume:

$$\dot{m}_{44}=\dot{m}_{45} \tag{86}$$

$$\eta_{CDB}=\frac{E_{45}-E_{44}}{W_{57}}\times 100 \tag{87}$$

$$\eta_{\text{exCDB}}=\frac{Ex_{45}-Ex_{44}}{W_{57}}\times 100 \tag{88}$$

$$Ex_{d\,CDB}=(Ex_{44}-Ex_{45})+W_{57} \tag{89}$$

Condenser Water Tank Control Volume Analysis

The make-up water, recirculated water from the condenser, and air enter the condenser water tank control volume as presented in Figure 1, at states 40, 42, and 45, respectively. The make-up water is mixed with the recirculation water, and the mixture leaves the control volume for further recirculation at state 38 with a change in the temperature and pressure. The air leaves the control volume at state 46 after cooling the recirculation water, with a change in the temperature and pressure.

For condenser water tank control volume:

$$\dot{m}_{40} + \dot{m}_{42} = \dot{m}_{38} \text{ (Water flow)} \tag{90}$$

$$\dot{m}_{45} = \dot{m}_{46} \text{ (Air flow)} \tag{91}$$

$$\eta_{CWT} = \frac{E_{38} + E_{46}}{E_{40} + E_{42} + E_{45}} \times 100 \tag{92}$$

$$\eta_{exCWT} = \frac{Ex_{38}}{(Ex_{46} - Ex_{45}) + (Ex_{40} + Ex_{42})} \tag{93}$$

$$Ex_{d\,CWT} = (Ex_{46} - Ex_{45}) + (Ex_{40} + Ex_{42}) - Ex_{38} \tag{94}$$

Condenser Water Pump Control Volume Analysis

Recirculation water enters the condenser water pump control volume as presented in Figure 1, at state 38 and leaves at state 39. The control volume is powered by electricity with an input power of W_{56}. For condenser water pump control volume:

$$\dot{m}_{38} = \dot{m}_{39} \tag{95}$$

$$\eta_{CDP} = \frac{E_{39} - E_{38}}{W_{56}} \times 100 \tag{96}$$

$$\eta_{exCDP} = \frac{Ex_{39} - Ex_{38}}{W_{56}} \times 100 \tag{97}$$

$$Ex_{d\,CDP} = (Ex_{39} - Ex_{38}) + W_{56} \tag{98}$$

Condenser Control Volume Analysis

Ammonia and cooling water pass through the condenser control volume as presented in Figure 1. Ammonia enters at state 37 and exits at state 47, while the cooling water enters at state 39 and exists at state 40 with changes in the temperature and pressure. The cooling water at state 40 is recirculated.
For condenser control volume:

$$\dot{m}_{39} = \dot{m}_{40} \text{ (Water flow)} \quad (99)$$

$$\dot{m}_{37} = \dot{m}_{47} \text{ (Ammonia flow)} \quad (100)$$

$$\eta_{COND} = \frac{E_{40} - E_{39}}{E_{37} - E_{47}} \times 100 \quad (101)$$

$$\eta_{exCOND} = \frac{Ex_{40} - Ex_{39}}{Ex_{37} - Ex_{47}} \times 100 \quad (102)$$

$$Ex_{d\,COND} = (Ex_{37} - Ex_{47}) + (Ex_{40} - Ex_{39}) \quad (103)$$

Ammonia Receiver Tank Control Volume Analysis

Ammonia enters the ammonia receiver tank control volume as presented in Figure 1, at state 47 and leaves the control volume at state 48 with no change in the temperature and pressure.
For ammonia receiver tank control volume:

$$\dot{m}_{47} = \dot{m}_{48} \quad (104)$$

$$\eta_{ART} = \frac{E_{48}}{E_{47}} \times 100 \quad (105)$$

$$\eta_{ExART} = \frac{Ex_{48}}{Ex_{47}} \times 100 \quad (106)$$

$$Ex_{d\,ART} = Ex_{47} - Ex_{48} \tag{107}$$

Expansion Valve Control Volume Analysis

Ammonia enters the expansion valve control volume as presented in Figure 1, at state 48 and leaves at state 49 with a certain change in the temperature and pressure.
For expansion valve control volume:

$$\dot{m}_{48} = \dot{m}_{49} \tag{108}$$

$$\eta_{EPV} = \frac{\mathrm{E}_{49}}{\mathrm{E}_{48}} \times 100 \tag{109}$$

$$\eta_{\mathrm{ExEPV}} = \frac{Ex_{49}}{Ex_{48}} \times 100 \tag{110}$$

$$Ex_{d\,EPV} = Ex_{48} - Ex_{49} \tag{111}$$

Air Compressor Control Volume Analysis

Air from the environment enters the air compressor control volume as presented in Figure 1, at state 31 and exits at state 32 with a change in the temperature and pressure. The control volume is powered by electricity with an input power of W_{53}.
For air compressor control volume:

$$\dot{m}_{31} = \dot{m}_{32} \tag{112}$$

$$\eta_{ACP} = \frac{E_{32} - E_{31}}{W_{53}} \times 100 \tag{113}$$

$$\eta_{\mathrm{exACP}} = \frac{Ex_{32} - Ex_{31}}{W_{53}} \times 100 \tag{114}$$

$$Ex_{d\,ACP} = (Ex_{32} - Ex_{31}) + W_{53} \quad (115)$$

Air Receiver Tank Control Volume Analysis

Compressed air from the air compressor enters the air receiver tank control volume as presented in Figure 1, at state 32 and leaves at states 33 and 34 with changes in the temperature and pressure.
For the air receiver tank control volume:

$$\dot{m}_{32} = \dot{m}_{33} + \dot{m}_{34} \quad (116)$$

$$\eta_{ART} = \frac{E_{33} + E_{34}}{E_{32}} \times 100 \quad (117)$$

$$\eta_{\mathrm{exART}} = \frac{Ex_{33} + Ex_{34}}{Ex_{32}} \times 100 \quad (118)$$

$$Ex_{d\,ART} = Ex_{32} - (Ex_{33} + Ex_{34}) \quad (119)$$

Air Distribution Valve Analysis

Compressed air enters the air distribution valve control volume as present in Figure 1, at state 33 and leaves at states 4, 7, 12, 14, 17, 18, 24, and 29 with changes in the temperature and pressure.
For the air distribution control volume:

$$\dot{m}_{33} = \dot{m}_4 + \dot{m}_7 + \dot{m}_{12} + \dot{m}_{14} + \dot{m}_{17} + \dot{m}_{18} + \dot{m}_{24} + \dot{m}_{29} \quad (120)$$

$$\eta_{ADV} = \frac{E_4 + E_7 + E_{12} + E_{14} + E_{17} + E_{18} + E_{24} + E_{29}}{E_{33}} \times 100 \quad (121)$$

$$\eta_{\mathrm{exADV}} = \frac{Ex_4 + Ex_7 + Ex_{12} + Ex_{14} + Ex_{17} + Ex_{18} + Ex_{24} + Ex_{29}}{Ex_{33}} \times 100 \quad (122)$$

$$Ex_{d\,ADV} = Ex_{33} - (Ex_4 + Ex_7 + Ex_{12} + Ex_{14} + Ex_{17} + Ex_{18} + Ex_{24} + Ex_{29}) \tag{123}$$

Overall System Efficiencies

The overall energy efficiency of the carbonation system can be determined using Equation (124) (Woudstra, 2012)

$$\text{Overall energy efficiency} = \frac{\text{Total energy output from the system}}{\text{Total energy input to the system}} \tag{124}$$

$$\text{Total energy output from the system} = E_2 + E_4 + E_7 + E_{12} + E_{14} + E_{17} + E_{18} + E_{24} + E_{28} + E_{29} + E_{30} + E_{33} + (E_{37} - E_{36}) + (E_{40} - E_{38}) + E_{46} \tag{125}$$

$$\text{Total energy input to the system} = E_1 + E_{10} + E_{31} + E_{41} + E_{44} + E_{w50} + E_{w51} + E_{w52} + E_{w53} + E_{w54} + E_{w55} + E_{w56} + E_{w57} \tag{126}$$

The overall exergy efficiency of the system can be determined using Equation (127) (Woudstra, 2012)

$$\text{Overall exergy efficiency} = \frac{\text{Total exergy input to the system} - \text{Total exergy destroyed}}{\text{Total exergy input to the system}} \tag{127}$$

$$\text{Total exergy input to the system} = Ex_1 + Ex_{10} + Ex_{31} + Ex_{41} + Ex_{44} + Ex_{w50} + Ex_{w51} + Ex_{w52} + Ex_{w53} + Ex_{w54} + Ex_{w55} + Ex_{w56} + Ex_{w57} \tag{128}$$

$$\text{Total exergy destroyed} = \sum \text{exergy destroyed in each component of the system} \tag{129}$$

Exergoeconomics Analysis Calculation Process

The exergoeconomics can be examined by formulating a cost balance equation for each component in the system. The general equation for the exergoeconomics analysis is presented as Equation (130) (Valdimarsson, 2011; Valero & Torres, 2009; Igbong & Fakorede, 2014).

$$\Sigma \dot{C}_i + \dot{C}_w = \Sigma \dot{C}_e + \dot{C}_q + \dot{Z} \tag{130}$$

where,

$$\dot{C}_i = c_i Ex_i \tag{131}$$

$$\dot{C}_w = c_w Ex_w \tag{132}$$

$$\dot{C}_e = c_e Ex_e \tag{133}$$

$$\dot{C}_q = c_q Ex_q \tag{134}$$

where, $\dot{C}$ = cost rate due to exergy flow (\$/hr), c = unit exergy cost rate (\$/hr), $\dot{Z}$ = cost rate due to investment and maintenance (\$/hr), w= flow due to work, (kW) and q = flow due to heat (kW)

The investment cost can be expressed as in Equation (135) (Shamsi & Omidkhah, 2012, Atmaca, 2018, Querol et al., 2013)

$$\dot{Z} = \dot{Z}_{CL} + \dot{Z}_{OM} \tag{135}$$

where

$$\dot{Z}_{CL} = CFR(i,n) \times TCI \tag{136}$$

$$CFR = \frac{i(1+i)^n}{(1+i)^n - 1} \tag{137}$$

$$\dot{Z}_{OM} = f_{OM} \times TCI \tag{138}$$

$$\therefore \dot{Z} = \left(\frac{i(1+i)^n}{(1+i)^n - 1} + f_{OM} \right) \times TCI \tag{139}$$

and

$$\dot{Z}(\$/h)=\frac{\left(\frac{i(1+i)^{n}}{(1+i)^{n}-1}+f_{OM}\right)\times TCI}{N} \tag{140}$$

where, $\dot{Z}_{CL}$ = cost rate of investment (\$/hr), $\dot{Z}_{OM}$ =cost rate due to maintenance (\$/hr), CFR = capital recovering factor, I = interest rate, n =number of years operating the system, TCI =total cost of investment, f_{OM} = maintenance factor, and N = hours the system operated at full load in a year

In terms of the economic data related to the carbonation system, the total investment costs of the major equipment are listed in Table 3 as obtained from manufacturing plant.

Table 3. Total investment cost of equipment in the carbonation system

S/N	Parameter	Total investment cost ($)
1	Mixer	187,500.00
2	Ammonia compressor	15,540.69
3	Air compressor	10,818.01
4	Treated water pump	815.00
5	Syrup pump	875.00
6	Make-up water pump	792.50
7	Condenser	5,600.00

The manufacturing plant is considered to have been in operation for 7 years, and the electricity tariff for the manufacturing plant is 0.11/kWhr. Based on the cost of capital, available in plant record, the interest rate is considered as 20%

Three types of cost flows are associated with each component in the carbonation system, specifically,

i. Incoming exergy cost flow, also known as the fuel cost flow c_p,
ii. Component investment cost flow,
iii. Outgoing exergy cost flow, also known as the product exergy cost flow c_f.

The cost rate due to exergy destruction in a component is referred to as the hidden cost (Shamsi & Omidkhah, 2012), and it can be defined as in Equation (141).

$$\dot{C}_d = c_f \times \dot{E}x_d \tag{141}$$

Exergoeconomic factor measures the effect of the cost rate of destroyed exergy to investment cost of equipment. It is as defined in Equation (142) (Atmaca, 2018; Tsatsaronis, 2011).

$$f = \frac{\dot{Z}}{\dot{Z} + \dot{C}_d} \tag{142}$$

where, C_d = cost rate of exergy destroyed in a component (\$/hr), c_f =unit exergy cost of fuel (\$), c_p = unit exergy cost product (\$), and f = exergoeconomic factor

Exergoeconomic Analysis of the Equipment in the Carbonation System

Exergy-based methods have been established as the optimal techniques to obtain reliable information related to the efficiency, resource utilisation, and economic importance of any system or machine. The two main types of exergy-based analysis of energy system are life cycle analysis and exergoeconomic analysis. In this study, exergoeconomic analysis was performed. The key steps involved in exergoeconomic analysis are exergy analysis, economic analysis, exergoeconomic evaluation, and evaluation of exergoeconomic variables were adopted. The complete exergy cost analysis was performed with the aid of TAESS. Detailed explanation regarding the algorithm of TAESS is available in the work of Torres, Perez, & Valero (2018). In the TAESS environment, components/apparatus are classified into two major categories: productive devices and dissipative devices. Productive devices are those devices whose products are used for further processing or are the benefit of the system. In contrast, dissipative devices are those that reduce or eliminate the environmental impact of generated residue on other devices. In this study, all components of the investigated carbonation system were modelled as productive devices except the condenser, which was modelled as a dissipative device because it removed heat from the ammonia refrigerant. The following steps were followed in the use of TAESS in exergy cost analysis:

1. Develop the physical structure
2. Define the productive structure, by stating the numbers of productive devices, dissipative devices/ component and flows/streams. In this study, there 26 productive devices/components, 1 dissipative device/component and 57 flows/ streams.
3. Determine the fuel and product of each device
4. Input the device/component name, fuel and product streams of the device in the general data of the software interphase
5. Simulate the physical structure, to verify the correct input of fuel/product stream of each component. If it is not correct, the error signal will be indicated.
6. Load total exergy value of the flow stream as shown in Table 1 and 2. The exergy value is prepared in a separate file either .txt format or. EES format.
7. Run simulation to generate related matrixes and cost flows
8. Evaluate the exergoeconomic analysis using the matrixes and cost flow generated.

RESULTS AND DISCUSSION

In this chapter, the results obtained from the carbonation system data collection, exergy and exergoeconomic analysis are discussed and improvement techniques of the carbonation system presented.

Gas Volume of the Carbonated Product

The overall aim of beverage carbonation is to achieve a gas volume that will enhance product quality. Gas volume can be defined as the volume of dissolved gases divided by the volume of liquid in the container (Steen, 2006). The gas volume of the beverage is a pre-defined quality standard set by the control department of a beverage company. The gas volume of carbonated beverage varies from one product to another; it also depends on the beverage container type such as polyethylene terephthalate (PET) bottles, can, and glass containers. The thermodynamic variables of any carbonation system are aligned to achieve the pre-defined gas volume, considering other quality parameters. From the study, beverages were carbonated at a temperature of 10°C, pressure of 4.2 bar, mass flow rate of 2.82 kg/s, and gas volume of 3.6 ml. The operational thermodynamic variables presented in Table 1 to 2 achieved a carbonated product of gas volume of 3.6 ml.

Energy and Exergy Efficiencies of the Carbonation System Components

Figure 2 presents the results of the energy and exergy efficiencies of the carbonation system components. The overall energy and exergy efficiencies of the carbonation system were 47.07% and 42.27%, respectively. All the pumps in the system exhibited relatively high energy efficiencies (ranging from 64% to 98%), compared to the corresponding exergy efficiencies. Apart from the condenser pump with a high exergy efficiency of 78.09%, other pumps exergy efficiencies are low, ranging from 0.6% to 27.00%.

Figure 2. Energy and exergy efficiencies of the carbonation system

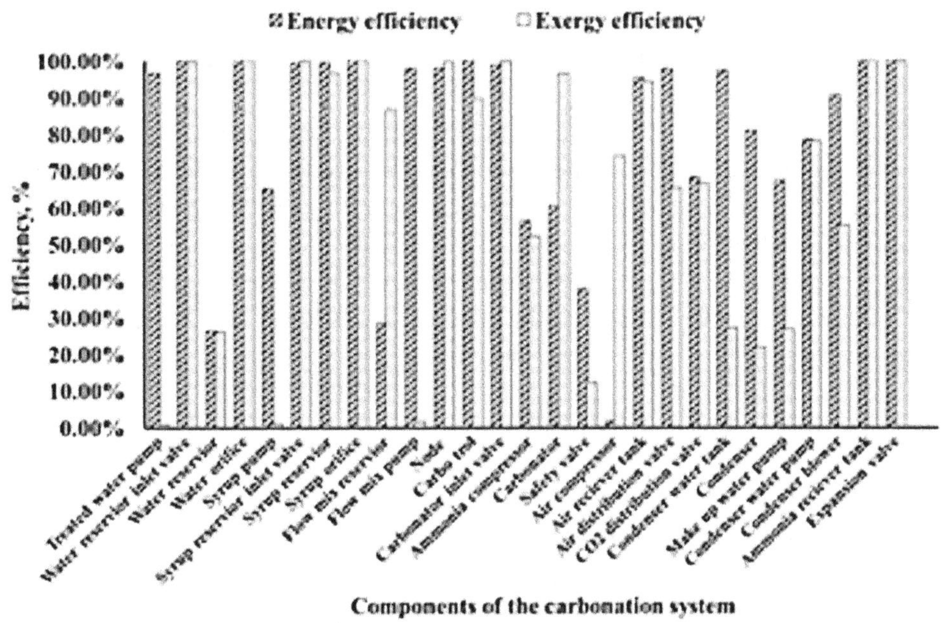

Kaya, Kayabasi, Kilinc, Eyidogan, Selimli, & Ozkaymak, (2013) reported on the high variations in the energy efficiencies (42.17–82.56%) and low exergy efficiencies of industrial pumps, which could be due to motor selection and poor pipe network. The authors recommended that the industrial pump efficiencies could be enhanced by optimising the pump system piping network and pump motor selection. The energy and exergy efficiencies of the inlet valves of the water reservoir, syrup reservoir and carbonator and expansion valves ranged from 97% to 100%. The air distribution valve exhibited a higher energy efficiency (98.68%) than that of the CO_2 distribution valve (68.17%), although the exergy efficiency of both valves ranged from 64% to 70%. Compared to the other valves, the safety valve exhibited considerably low energy and exergy efficiencies of 37.78% and 12.18%, respectively. This phenomenon could be attributed to the release in the excess pressure build-up in the carbonator. Only a few literatures focused on the energy and exergy analysis of valve, although valve is a major component in every fluid transport system and control design. Mrzljak, Orovic, & Lorencin, (2019) reported that the exergy efficiency of a valve is approximately 90%. The results from Mrzljak et al. (2019) are in accordance with certain results obtained in this research, even though a different valve type (steam valve) and application (power plant) were considered. The condenser blower energy efficiency (90.60%) was found higher than that of the ammonia compressor (56.25%), and the air compressor energy efficiency (1.83%) was lower than that of the other components. The exergy efficiency of the air compressor (74.05%) was found to be higher than that of ammonia compressor and condenser blower, which ranges between 52–55%. Nevertheless, among the components of the carbonation system, in addition to the air compressor, the flow mix reservoir, node, and carbonator exhibited a higher exergy efficiency compared to the energy efficiency. Wayne and Kissock (2000) noted that at various operating conditions, the exergy efficiency of an industrial air compressor is higher than its energy efficiency. The authors demonstrated that the energy efficiency cannot be used as the true efficiency for energy conversion devices. Furthermore, Audrius (2013) stated that although the energy efficiency can help evaluate the performance of energy conversion devices, the obtained results may be misleading owing to the accepted definition of the energy efficiency or method of calculation, as several methods do not consider the quality aspect of the energy. The energy efficiencies of other components such as condenser, Carbo-trol, air receiver tank, ammonia receiver tank, condenser water tank, and syrup ranged from 80% to 100%. The energy efficiency of the water reservoir (26.43%) was found lower than that of the other components. In general, the exergy and energy efficiencies of the water reservoir are correlated owing to the deaeration process that occurs in this reservoir. The exergy efficiency of the syrup reservoir, air receiver tank, ammonia receiver tank, node, and Carbo-trol ranged from 79% to 100%. The condenser water tank exhibited a lower exergy efficiency (27.15%) compared to the energy efficiency owing to the heat loss from the cooling water. Similarly, the exergy efficiency (21.83%) of the condenser was lower than the corresponding energy efficiency, likely because of the heat loss from the ammonia refrigerant.

Exergy Destruction in the Carbonation System

The total exergy destruction in the system was computed to be 6339.69 kW. The highest exergy destruction occurred in the water reservoir (62.10%), followed by that in condenser water tank (16.57%) and flow mix reservoir (12.80%), as presented in Figure 3. Deaeration process may be responsible for the high exergy destruction in the water reservoir. The remaining components of the carbonation system contributed 8.53% to the total exergy destroyed, among which the syrup reservoir and carbonator corresponded to 30.82% and 34.93%, respectively.

Figure 3. Exergy destroyed in the carbonation system

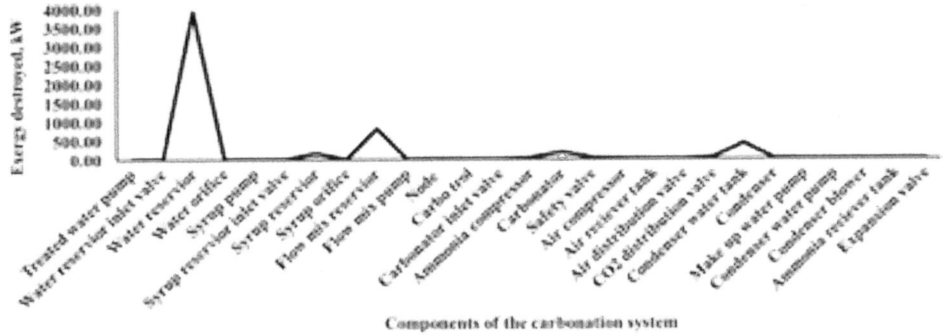

Cost Rate Due to Exergy Destruction (Hidden Cost)

From the study, the overall cost rate of the operating carbonation system corresponding to the exergy destruction was determined to be $777.47/hr, and the total cost rate related to the investment and maintenance of the system equalled $45.30/hr as presented in Table 4. Reducing the exergy destruction in the carbonation system will reduce the overall cost rate of the system operation. The mixer and condenser exhibited the lowest exergoeconomic factors of 5.60% and 1.39%, respectively, among all the equipment in the carbonation system, while the cost rate of the exergy destruction in the mixer ($644.60/hr) was higher than that of the condenser $80.84/hr.

Table 4. Cost rate due to exergy destruction in the carbonation system equipment

Machine/Equipment	TCI ($)	$_{Cd}$ ($/hr)	$_{Z}$ ($/hr)	$_{C}$ ($/hr)	f
Treated water pump	815.00	0.50	0.17	0.67	24.95%
Syrup pump	875.00	0.52	0.18	0.70	25.62%
Ammonia compressor	15,540.69	4.36	3.17	7.53	42.11%
Air compressor	10,818.01	1.23	2.21	3.44	64.15%
Mixer	187,500.00	644.60	38.27	682.87	5.60%
Make up water pump	792.50	0.12	0.16	0.28	56.75%
Condenser	5,600.00	80.84	1.14	81.98	1.39%

By optimisation, a reduction of mass flow rate of beverage at the carbonator inlet from 2.82 to 2 kg/s led to an increase in the overall system energy efficiency from 47.07% to 59.15%, while exergy efficiency increases from 42.27% to 46.46% and a total decrease in exergy destruction by 14.14% with a corresponding decrease of 12.26% in the cost rate due to exergy destruction. The water reservoir exhibited the highest exergy destruction of 3936.64 kW, as shown in Figure 3, and thus corresponded to the highest cost rate due to exergy destroyed in the mixer. The deaeration process in the water reservoir is responsible for this high exergy destruction. This shows that alternative deaeration technology should be sought for water deaeration process and the water treatment procedure must be optimised to reduce oxygen and other dissolved gases/elements.

CONCLUSION AND RECOMMENDATIONS

Conclusion

A significant amount of energy is involved in beverage manufacturing, which led to high cost of production. The high amount of energy usage is primarily owed to the manufacturing processes involved, such as carbonation system, boiler and steam generation system, clean-in-place (sanitation) system, and heavy machine operation, which formed the drive for this study. The reduction or elimination of cost rate due to exergy destruction will reduce the operating cost of beverage manufacturing process and hence increase profit margin. The study has demonstrated the application of exergoeconomic analysis in manufacturing operations to reduce operating cost and improve machine efficiency with a focus on beverage carbonation process. It was found that beverages were carbonated at a temperature of 10.00 °C, pressure of 4.20 bar, mass flow rate of 2.82 kg/s, and gas volume of 3.6 ml. The overall exergy destruction in the carbonation system was found to be 6,339.69 kW and the overall energy and exergy efficiencies for the carbonation system was found to be 47.07% and 42.27%, respectively. The lowest exergoeconomic factors of 1.39% and 5.60%, were found in the condenser and mixer, respectively. The cost rate due to exergy destruction in the mixer ($682.87/hr) was higher compared to that of the condenser ($82.10/hr). The mixer subsystem has highest cost rate due to exergy destruction because of high exergy destruction, occurred in the following mixer components- carbonator, flow mix reservoir and water reservoir. The overall cost rate of operating the carbonation system corresponding to the exergy destruction was determined to be $777.31/hr, while the overall cost rate related to the investment and maintenance of the system was $45.13/hr. Reduction of mass flow rate at the carbonator inlet led to a reduction of total exergy destruction by 14.14% and a decrease in the total cost rate of exergy destruction by 12.26%. The study also shows that water reservoir has the highest exergy destruction, which suggests alternative deaeration technology and water treatment procedure should be optimisation to reduce oxygen and other dissolved gases/elements.

Juxtapose with the frequently used cost management techniques in manufacturing operations, such as process automation, line rationalisation, Kaizen system implementation, cost volume analysis, total quality management, standard cost analysis, budgeting control, just-in-time production system, none of the techniques can identify the hidden cost, which is cost rate due to exergy destruction in the carbonation system which led to more energy consumption and high cost of production. This research concludes that, that reduction of exergy destruction in the mixer and other subsystems within the carbonation system is very crucial to the improvement of the cost rate in beverage production. The study, therefore, recommended that exergy destruction reduction techniques should be applied to beverage carbonation system to enhance its performance and exergoeconomic analysis should be adopted in the manufacturing operations as a cost control mechanism.

Recommendation for Further Studies

Further studies on this work should include: Energy, exergy and exergoeconomic analysis of a complete beverage manufacturing process; a comparison of exergy destructions in refrigeration carbonation system, gas bubble carbonation system and in-line carbonation system; cost saving on the application of exergoeconomic analysis in manufacturing operations.

REFERENCES

Abdellatif, A. S. (2018). The beverages. *Agricultural Research & Technology: Open Access Journal, 14*(5), 1–9.

Atmaca, A. (2018). Energy, exergy and exergoeconomic assessment of a dry type rotary kiln. *Anadolu University of Sciences & Technology A. Applied Sciences and Engineering, 19*(1), 192–205.

Audrius, B. (2013, September 16). *UKERC Blog: The Current Event.* Retrieved from UK Energy Research Centre: https://ukerc.wordpress.com/2013/09/16/a-few-thoughts-on-energy-and-exergy-efficiences/

Chan, W., Li, H., Li, X., Chang, F., Wang, L., & Feng, Z. (2021). Exergoeconomic analysis and optimization of the Allam cycle with liquefied natural gas cold exergy utilization. *Energy Conversion and Management, 235*, 113972. doi:10.1016/j.enconman.2021.113972

Dincer, I., & Cengel, Y. A. (2001). Energy, entropy and exergy concepts and their roles in thermal engineering. *Entropy (Basel, Switzerland), 3*(3), 116–149. doi:10.3390/e3030116

Dincer, I., & Rosen, M. A. (2007). *Exergy, energy, environment and sustainable development.* Elsevier.

Fadare, D. A., Nkpubre, D. O., Oni, A. O., Falana, A., Waheed, M. A., & Bamiro, O. A. (2010). Energy and exergy analysis of malt drink production in Nigeria. *Energy, 35*(12), 5336–5346. doi:10.1016/j.energy.2010.07.026

Filippos, K. Z., & Francisco, J. R.-M. (2017). The use of exergetic indicators in the food industry – A review. *Critical Reviews in Food Science and Nutrition, 57*(1), 197–211. doi:10.1080/10408398.2014.975335 PMID:26066523

Ghannadzadeh, A. (2013). *Exergetic balances and analysis in a process simulator: A way to enhance process energy integration.* Retrieved 3 1, 2014, from www.ethesis.inp-toulouse.fr/archive/00002116/01/ghannazadeh.pdf

Giannoulakis, S., Volkart, K., & Bauer, C. (2014). Life cycle and cost assessment of mineral carbonation for carbon capture and storage in European power generation. *International Journal of Greenhouse Gas Control, 21*, 140–157. doi:10.1016/j.ijggc.2013.12.002

Gorji-Bandpy, M., & Ebrabimian, V. (2006). Exergoeconomic analysis of gas turbine power plants. *International Energy Journal, 7*(1), 35–41.

Gundersen, T. (2011). *An introduction to the concept of exergy and energy quality.* Retrieved 7 26, 2014, from www.ivt.ntnu.no/ept/fag/.../ Exergy%20Light%20 Version%204.pdf

Haydargil, D., & Abusoglu, A. (2018). A comparative thermoeconomic cost accounting analysis and evaluation of biogas engine-powered cogeneration. *Energy*, *159*, 97–114. doi:10.1016/j.energy.2018.06.102

Igbong, D. I., & Fakorede, D. O. (2014). Exergoeconomic analysis of a 100 MW unit GE Frame 9 gas turbine plant in Ughelli, Nigeria. *IACSIT International Journal of Engineering and Technology*, *4*(8), 463–468.

Kaya, D., Kayabasi, E., Kilinc, E., Eyidogan, M., Selimli, S., & Ozkaymak, M. (2013). Energy and exergy efficiencies in industrial pumps. *International Journal of Renewable Energy Resources*, *3*(1), 27–42.

Meyer, L., Castillo, R., Buchgeister, J., & Tsatsaronis, G. (2009). Application of exergoeconomicand exergoenvironmental analysis to an SOFC system with an allothermal biomass gasifier. *International Journal of Thermodynamics*, *12*(4), 177–186.

Moodley, M., Schorn, P. M., Walthew, D. C., & Masinga, P. (2002). Optimising the carbonation process. *Proc S Afr Sug Technol Ass*, *76*, 469–476.

Mrzljak, V., Oroviæ, J., Poljak, I., & Lorencin, I. (2019). Exergy analysis of steam turbine governing valve from a super critical thermal power plant. In *XXVII International Scientific Conference Trans & Motauto*'19 (pp. 99-102). Scientific-Technical Union of Mechanical Engineering, Industry-4.0.

Newbold, D., & Koppel, K. (2018). Carbonated dairy beverages: Challenges and opportunities. *Beverages*, *4*(3), 66. doi:10.3390/beverages4030066

Okereke, C. J., Lasode, O. A., & Ohijeagbon, I. O. (2020a). Exergoeconomic analysis of an industrial beverage mixer system. *Heliyon, 6*. 10.2020.e04402 doi:.1016/j.heliyon

Okereke, C. J., Lasode, O. A., & Ohijeagbon, I. O. (2020b). Exergoeconomic analysis of an industrial beverage mixer system: Process data. *Data in Brief*, *32*, 106125. doi:10.1016/j.dib.2020.106125 PMID:32904336

Okereke, C. J., Ohijeagbon, I. O., & Lasode, O. A. (2019). Energy and exergy analysis of vapor compression refrigeration system with flooded evaporator. *International Journal of Air-Conditioning and Refrigeration*, *27*(04), 1950041. doi:10.1142/S201013251950041X

Olajire, A. A. (2013). A review of mineral carbonation technology in sequestration of CO_2. *Journal of Petroleum Science Engineering*, *109*, 364–392. doi:10.1016/j.petrol.2013.03.013

Pham, S. T. (2013). Experimental Investigation and Modelling of Carbonation Process in Cement Materials. *The Open Civil Engineering Journal*, *7*(1), 116–125. doi:10.2174/1874149520130815006

Querol, E., Gonzalez-Regueral, B., & Perez-Benedito, J. L. (2013). *Exergy concept and determination. In Practical Approach to Exergy and Thermoeconomic Analysis of Industrial Process*. Springer. doi:10.1007/978-1-4471-4622-3_1

Sahu, O. (2018). Assessment of sugarcane industry: Suitability for production, consumption, and utilization. *Annals of Agrarian Science*, *16*(4), 389–395. doi:10.1016/j.aasci.2018.08.001

Šavija, B., & Lukovic, M. (2016). Carbonation of cement paste: Understanding, challenges, and opportunities. *Construction & Building Materials*, *117*, 285–301. doi:10.1016/j.conbuildmat.2016.04.138

Shamsi, S., & Omidkhah, M. R. (2012). Optimization of steam pressure levels in a total site using a thermoeconomic method. *Energies*, *5*(3), 702–717. doi:10.3390/en5030702

Singh, G., Singh, P. J., Tyagi, V., & Pandey, A. K. (2019). Thermal and exergoeconomic analysis of a dairy food processing plant. *Journal of Thermal Analysis and Calorimetry*, *136*(3), 1365–1382. doi:10.100710973-018-7781-y

Sogut, Z., Ilten, N., & Oktay, Z. (2010). Energetic and exergetic performance evaluation of the quadruple effect evaporator unit in tomatoes paste production. *Energy*, *35*(9), 3821–3826. doi:10.1016/j.energy.2010.05.035

Steen, D. P. (2006). Carbon dioxide, carbonation and the principles of filling technology. In D. P. Steen & P. R. Ashurst (Eds.), *Carbonated soft drinks: formulation and manufacture* (pp. 112–116). Blackwell Publishing Ltd. doi:10.1002/9780470996034

Sulaiman, M. A., Oni, A. O., & Fadare, D. A. (2012). Energy and exergy analysis of a vegetable oil Refinery. *Energy and Power Engineering*, *4*(5), 358–364. doi:10.4236/epe.2012.45047

Sun, L., Wang, D., & Xie, Y. (2021). Energy, exergy and exergoeconomic analysis of two supercritical CO2 cycles for waste heat recovery of gas turbine. *Applied Thermal Engineering*, *196*, 117337. doi:10.1016/j.applthermaleng.2021.117337

Thongrote, C., Wirjantoro, T. I., & Phianmongkhol, A. (2016). Effect of carbonation sources and its addition levels on carbonated mango juice. *International Food Research Journal*, *23*(5), 2159–2165.

Torres, C., Perez, E., & Valero, A. (2018). A. Guidelines to develop software for thermoeconomic analysis of energy systems. In *Proceedings of ECOS 2018: the 31st International Conference on Efficiency, Cost, Optimization, Simulation and Environmental Impact of Energy Systems* (pp. 435–453). Academic Press.

Uysal, C., & Keçebaş, A. (2021). Advanced exergoeconomic analysis with using modified productive structure analysis: An application for a real gas turbine cycle. *Energy*, *223*, 120085. doi:10.1016/j.energy.2021.120085

Valdimarsson, P. (2011). Basic concepts of thermo-economics. In *Short Course on Geothermal Drilling, Resource Development and Power Plants, organized by UNU-GTP and LaGeo* (pp. 1-7). Retrieved 9 1, 2014, from http://www.os.is/gogn/unu-gtp-sc/UNU-GTP-SC-12-41.pdf

Valero, A., & Torres, C. (2009). Thermoeconomic Analysis. In C. A. Frangopoulos (Ed.), *Exergy, Energy System Analysis and Optimization–Volume II: Thermoeconomic Analysis Modeling, Simulation and Optimization in Energy Systems* (pp. 1–35). EOLSS Publications.

Waheed, M. A., Jekayinfa, S. O., Ojediran, J. O., & Imeokparia, O. E. (2008). Energetic analysis of fruit juice processing operations in Nigeria. *Energy*, *33*(1), 35–45. doi:10.1016/j.energy.2007.09.001

Wall, G. (2009). *Exergetics*. Retrieved October 17, 2013, from http://www.exergy.se /ftp/exergetics.pdf

Wayne, T. B., & Kissock, K. J. (2000). Exergy analysis of industrial air compression. *Proceedings from the Twenty-Second National Industrial Energy Technology Conference*, 89–98.

Wiebe, R., & Gaddy, V. (1940). The Solubility of carbon dioxide in water at various temperatures from 12 to 40° and at pressures to 500 atmospheres. Critical Phenomena. *Journal of the American Chemical Society*, *64*(4), 815–817. doi:10.1021/ja01861a033

Woudstra, N. (2012). *Sustainable energy systems- limitations and challenges based on exergy analysis*. Retrieved September 19, 2014, from http://repository.tudelft.nl/ view/ir/uuid%3A112039e0-89ee-44fa-8d52-17eca3ec41f0/

Younsi, A., Turcry, P., Rozière, E., Aît-Mokhtar, A., & Loukili, A. (2011). Performance-based design and carbonation of concrete with high fly ash content. *Cement and Concrete Composites*, *33*(10), 993–1000. doi:10.1016/j.cemconcomp.2011.07.005

ADDITIONAL READING

Okereke, C. J., Lasode, O. A., & Ohijeagbon, I. O. (2020b). Exergoeconomic analysis of an industrial beverage mixer system: Process data. *Data in Brief*, *32*, 106125. doi:10.1016/j.dib.2020.106125 PMID:32904336

Okereke, C. J., Ohijeagbon, I. O., & Lasode, O. A. (2019). Energy and exergy analysis of vapor compression refrigeration system with flooded evaporator. *International Journal of Air-Conditioning and Refrigeration*, *27*(04), 1950041. doi:10.1142/S201013251950041X

Singh, G., Singh, P. J., Tyagi, V., & Pandey, A. K. (2019). Thermal and exergoeconomic analysis of a dairy food processing plant. *Journal of Thermal Analysis and Calorimetry*, *136*(3), 1365–1382. doi:10.100710973-018-7781-y

Sogut, Z., Ilten, N., & Oktay, Z. (2010). Energetic and exergetic performance evaluation of the quadruple effect evaporator unit in tomatoes paste production. *Energy*, *35*(9), 3821–3826. doi:10.1016/j.energy.2010.05.035

Steen, D. P. (2006). Carbon dioxide, carbonation and the principles of filling technology. In D. P. Steen & P. R. Ashurst (Eds.), *Carbonated soft drinks: formulation and manufacture* (pp. 112–116). Blackwell Publishing Ltd. doi:10.1002/9780470996034

KEY TERMS AND DEFINITIONS

Carbonation: Refers to the process of dissolving carbon dioxide in a liquid at a varied conditions depending on the application.

Control Volume: Is a space/ boundary where mass, momentum and energy are allowed to cross.

Cost Control Mechanism: Is a method or procedure used to assess elements of cost in operation, system, or environment.

Exergoeconomic Analysis: Is a method of determining and assigning economic values to the exergy flow.

Exergy: Is the component of energy that is convertible and useful.

Exergy Analysis: Is a method of determining exergy value of a system, process, or device.
Exergy Destruction: Is a measure of energy component of a device that are not useful.
Exergy Efficiency: Is a measures of the quality of exergy a device, process or system has harness.

Chapter 10
To Build a Future Abroad:
The Strength of Women as an Example of Tenacity and Achievement of Goals

Alessandra Panzella
School of Economics and Management, University of Minho, Portugal

Carolina Feliciana Machado
https://orcid.org/0000-0002-9685-1576
School of Economics and Management, University of Minho, Portugal & Interdisciplinary Centre of Social Sciences (CICS), University of Minho, Portugal

ABSTRACT

Conscious of the challenges and changes that we are facing, this chapter aims to describe the characteristics of a work experience abroad from the point of view of a woman. In an organizational context, the value resulting from international assignments is getting more decisive. Overseas tasks represent a great occasion for workers to improve their professional path, but it also has positive outcomes on their personal life. Its growing importance is due also to the increased number of people accepting job in a foreign country. Among others, there will be highlighted the reasons why women should consider going and what they can gain from it as well as the obstacles that prevent them from accepting an international assignment or the difficulties that they could eventually find.

INTRODUCTION

"It has become imperative for international human resource experts to attract, select, develop and retain employees who can live and work effectively outside their national borders." (Adler and Bartholomew, 1992)

Globalization has challenged companies to confront each other on international areas; it has changed the economy and the markets, which have become global, considering a much larger scale, with different demands from and to customers, suppliers, investors, among others.

DOI: 10.4018/978-1-6684-2339-4.ch010

Globalization has led all companies to deal with foreign markets. It was necessary to be open to the world, establish relationships and create international and intercontinental connections. In recent decades, it has affected not only national and international economies but also social, cultural, political, and technological areas. Opportunities have multiplied in economic terms but also in terms of thinking, approaches and work development. In some way, worldwide business growth is built upon international trade and experiences. In this context, organizations could benefit greatly from interculturality, such as a greater labour market, exports, and a more homogeneous world.

This chapter aims to describe the characteristics of a work experience abroad from the point of view of a woman. In an organizational context, the value resulting from international assignments is getting more and more decisive. Overseas tasks represent a great occasion for workers to improve their professional path, but it also has positive outcomes on their personal life. Its growing importance is due also to the increased number of people accepting job in a foreign country.

The theme will be first introduced generally, then it will be elaborated from the women perspective. There will be outlined the reasons why they should consider going and what they can gain from it, but also which are the obstacles that prevent them from accepting an international assignment, or the difficulties that they could eventually find. There will be identified in which countries women would rather go. For example, countries in which it is spoken a language they already know or countries that have more to offer, compared to their home one.

In conclusion, asking for more innovative and creative types of management, this chapter intends to describe the objectives achievable through the experience abroad, in terms of promotions and career advancements.

1. WOMEN ABROAD

An international experience has a strategic importance for those who want to improve their skills, acquire new ones, know more about the context in which their company operates. As shown from a BCG study (Abouzahr, Hoteit, Krentz, and Tracey, 2017) *"Women on the Move: Shaping Leaders Through Overseas Postings"*, women are increasingly interested in living an experience abroad and it seems to be considered a priority in their careers.

International assignments have a critical importance for both the organizations and those who will experience it. People, who are taking advantage of overseas job opportunities, are taking a chance to prove themselves in a different work context. This gives them the possibility to acquire cross-cultural skills that will differentiate them from their colleagues and provide them a significant added value. On the other side, having employees abroad helps the organizations to maintain international negotiations, acknowledge and experiment new technologies, organizational and productive systems. It also contributes and allow that the organization have a better knowledge and understanding of the multicultural environment in which it is involved, offering, in this way, more innovative and creative products and services, at the same time that it is possible to obtain and develop new markets.

As we can imagine, it is a difficult path, from the selection that organizations conduct, to the experience itself for those who decide to live it. It is necessary that people, who find themselves in a foreign country, learn to work cross-culturally, facing more challenges than the ones they would normally have. These difficulties are more evident when the employee in question is a woman. For example, one of the most common and encountered obstacle is "being female in cultures where women are treated differently" (The Telegraph, 2011).

Despite the challenges that they must face, they are generally satisfied of their experiences abroad and there is an increasing number of women interested in this experience, as it is also demonstrated by the available literature on this topic (Culpan and Wright, 2002; Adler, 1984). On the other side, women have the possibility to prove themselves, gain visibility and the opportunity to create stronger and better interpersonal relationship.

2. POSITIVE SIDES OF EXPATRIATION

Having a global career allow to achieve a wide range of benefits. Through an international assignment, women can seize upon challenges and experiences that will help them grow and that may not be available in their home country. These constructive experiences allow people to obtain information, knowledge and competencies which contribute to the training of those entering the world of work.

This section pretends to describe the motives that encourage women in considering an international experience, as well as the resulting advantages. Additionally, it identifies the necessary skills that will favour the experience abroad, but also those competencies that could be developed throughout it.

2.1. Motives and Benefits

There are several reasons why women are predisposed to take on international missions.

First, we can say that women are more motivated. Indeed, since ever they need to prove their value, abilities and competencies in the labor world, deeply masculine, which has always forced them to try harder to get noticed in the work context.

This experience can accelerate their careers and improving their occupational prospects. Undoubtedly, it represents an opportunity to expand knowledge and competencies related to their specific role, because it gives the possibility to explore and understand a different way of work and specific practices.

It also allows to establish better connections and expand the professional network. Despite all the advantages offered by internet in shortening distances, meeting people personally is more satisfying, and the direct contact allows to develop greater trust.

Women can get the so called "added value", supported by the development of new and essential skills through the experience abroad. All these characteristics will contribute to women success, leading them to become better managers and leaders in future. For example, learning to collaborate and work with multicultural teams, and, by doing so, also understand how to eventually manage them in the future.

A further motive to take on international assignments could be an economic incentive with the intention of improving the standard of living of the expatriate.

Regarding personal characteristics and people behaviour in general, staying abroad is considered to have a positive influence on personal growth.

People develop a global mindset and broaden their horizons; they become more inclined to accept new challenges and adapt to them easily. The tenacity in pursuing a goal represents one of the fundamental levers to be successful also in the work area.

Going abroad means giving up the comforts of your own country and the habits of your own culture and being aware of having to face difficulties that will require greater sacrifices. It will be necessary to adopt behaviours and life dynamics distant from their everyday routine. Women can also experience the local cultures and improve their cross-cultural competence. If comprehend the local culture means improving your own knowledge and enriching the personal baggage, this is even more true for a woman who must also overcome the prejudices present in every country in the world regarding 'being a woman'.

The desire for professional growth in a specific role and area arises from the impossibility of being able to do it sometimes in one's own country, because the organization does not give the possibility. This explains many choices of orienting towards foreign companies. CV are sent to firms that run similar activities, with the ambition to perform roles not available in the current organization in the home country.

2.2. Necessary Skills to have Success Abroad

According to Ines Wichert, "Women are well-suited for international assignments due to their strong interpersonal skills which can make an important difference when a person has to show cultural sensitivity and build relationship with new colleagues and customers once abroad" (Torfx, 2016).

It is necessary to consider that women have interpersonal skills, such as empathy, which allow them to get attention more easily from foreign clients and it is easier to talk to a woman about different topics.

Certainly, women with social and open personalities will be more suitable for an experience overseas. It is an important value to be able to easily live new contexts.

Expatriate women with specific personal characteristics and skills are more likely to be successful and work well in international contexts. Among these particular features there can be included emotional stability and self-confidence. It is also important to have flexibility and adaptability, not to be ethnocentric, but also have cultural empathy and interest in the host culture. Competencies that can facilitate the stay abroad are knowing and be fluent in one (or more) foreign language(s), intercultural competence, good focus, and communication skills. Some of the skills could also be acquired and developed or improved during the experience in the foreign country (Girdauskiene and Eyvazzade, 2015).

All these competencies and characteristics will help in overcoming the difficulties of working in a foreign country with different cultures. For example, a woman who knows the local language will communicate and better understand others. This way, she can earn the trust of co-workers and improve relationship with people in the host country. In the workplace, language skills can be useful to complete tasks and develop better communication.

This skill also helped women to cope with discrimination at workplace, they can use the everyday vocabulary to end the prejudices about them.

2.3. Where Women Prefer to Go

A further important factor is represented by the location, taken as the country in which women are going to carry out overseas tasks. Usually, women would rather go in English speaking country, due to the internationalization of the language, or Western European countries. At the same time, women could usually be more open to any destinations if they have already had a previous international assignment.

According to the article *Women say working abroad improves their career* prospects (The Telegraph, 2011), women perceived a greater situation of equality in Canada, New Zealand, the US, and Australia.

In accordance with Business Insider (2017), some countries considered the best for women to work abroad are:

- New Zealand, Ecuador, Hungary and Australia for the good balance between work and personal life and the possibility of spending their time off work in the best way possible;
- Germany, due to the greater presence of expatriate women in managerial and responsible roles;
- Taiwan, because of the good career prospects within the country.

3. NEGATIVE SIDES OF EXPATRIATION

Several barriers and obstacles prevent women from experiencing a period of work abroad or make it more difficult than what it should be. Complications regarding the job itself or the cultural environment of the new country that may affect their psychological status, but also their work outcomes.

The aim of this section is to describe the reasons why this happens. Personal motives have been identified, because, for example, having children will influence the decision of accepting or refusing an international assignment. However, it is not always about children and family or about a husband who prevent the professional growth of his spouse. Sometimes, as a matter of fact, women cannot choose, considering that it is not given to them the opportunity to take on international assignments, because they are not even selected. Sometimes, this is caused by gender discrimination, still present within the workplace, that ensure that women do not have the same opportunities as men. Companies managers tend to think that women would not accept, in case of married women and/or with children, or that they would not be accepted in the foreign country because of the different culture, or even that they are not qualified and would not be effective.

3.1. Personal Reasons

Regarding the personal reasons, women may not be interested in being relocated or assigned to an international task. For example, they could not be interested in the available position.

In other cases, as anticipated, managers do not offer this possibility to married women because they think that the husbands could not be interested in relocation, therefore, women would not accept. Or, for example, a married woman would be available in going abroad if there is a vacant position or a job for their spouse too. These are questions that influence women' decision of leaving or not. In addition, most of the times, it happens that managers do not even ask if they are interested in international positions assuming that they would not want to leave their children.

Women are considered as the "family's primary caregiver" (The Telegraph, 2019a) and it is because of this idea that "children" are considered the first obstacles for a woman career.

Jessica Bruce helps reverse the situation by saying that "Having children does not slow down female leaders, it simply changes the way we work and manage our time" (The Telegraph, 2019b). Women could excel as leaders and be great mothers at the same time, having children will simply turn them more determined in showing their value and proving themselves.

3.2. Selection Process

Companies, specifically HR experts, have an important role since they will choose the employees who are going abroad to, in some way, represent the firm. Therefore, they should be professional figures extremely qualified. The recruitment for overseas allocation is a complex and difficult process. It should be ensured that men and women have the same opportunities, and that the selection process is managed in an unbiased way, but this does not always happen. Managers often believe that women are not interested in pursuing a career overseas and because of this, they are not taken into consideration in the selection process. Most of times, women express their willingness to participate to this experience. Companies should not assume about what women are disposed to do or what they want for their careers (Adler, 1984).

3.3. Gender Inequality

The gender bias is still present within the work context, as it is also demonstrated by a study conducted by Yale Scientific (Midura, 2013), *John Vs. Jennifer: A Battle of the Sexes*. This study is based on an experiment in which professors from six North American research universities had to evaluate the application submitted by two fictitious students (*John and Jennifer*), to become laboratory supervisor. The peculiarity of the experiment lies in the fact that the documentation presented is the same, with identical characteristics and experiences, but the name and gender of the candidate changes. *Jennifer* scored significantly lower than *John* in all the assessed aspects and would have deserved a lower pay as well.

Another obstacle woman may encounter is the prejudice that exist since forever that has always seen women far from top positions in the workplace. In organizational context, it is still present inequality between women and men, due to myths that persist in cultures and within companies themselves, especially for some specific professional figures, seen as more suitable for men. Women are considered unqualified and tend to have less opportunities than men.

As a matter of fact, it is widespread the concept of glass-ceiling when talking about gender differences. It refers to all the barriers that women face and that prevent them from achieving certain positions within organizations, just because of their gender.

3.4. Culture of the Host Country

Host country culture play an important role, since it has the capacity to influence employees' behaviour and therefore their work performance. Women working in a foreign country will have to deal with different reactions from the local people. For example, one of the difficulties is the prejudice about them, going to countries where women are present in lower percentage within the business context or considered as not suitable for some specific roles. Women face multiples challenges and need to make greater efforts to be considered.

Because of the resistance and challenges they will probably have to face in countries with a "traditionally male-dominated culture" (Culpan and Wright, 2002), organizations may be reluctant in sending women abroad. However, foreign women are considered in a different way, because some culture accept their contributions and work and, therefore, recognize their value.

To face this new challenge, it will be critical for women to believe in themselves and "use their female 'difference'" (The Telegraph, 2011) to emerge. Furthermore, it could be helpful a local support, chosen among the new colleagues, or a 'cultural adjustment training' (Chen, 2019).

3.5. Repatriation

Going back to their country, women will face the so-called 'reverse culture shock', that is the psychological discomfort due to returning home after a period abroad.

Despite the advantages described, international experience may not guarantee a promotion, career advancement or a salary increase when returning to the home company, thus generating feelings of anxiety and worry in the minds of employees. Women often perceive that the skills and knowledge acquired abroad are not recognized. Organizations should value the work done abroad and entrust tasks like those carried out in the other country. Otherwise, employees will feel less satisfied and not motivated; this will also negatively affect their performance.

In addition, they may feel distant from their country of origin, their original culture, and ways of doing things; the more time spent abroad the greater this feeling of alienation.

4. FEMALE LEADERSHIP

International experience is a great opportunity for those who want to grow within the company proceeding toward leadership positions.

With a female leader there are greater chances of fostering a management style that is more attentive to differences, which therefore allows you to enhance the contribution that each person can bring to the achievement of organization's objectives (Career Addict, 2017; Grant, 1988).

As a leader, a woman distinguish herself from the male colleagues for distinctive characteristics such as flexibility, empathy, and tenacity. Also, women are more likely to mentor and motivate others by converting their own goal into the company's goal. They tend to produce situations in which people can feel good about their work and themselves, encouraging them, therefore they will accomplish their tasks the best way possible. Women are also able to gain easily trust and confidence of their co-workers. Characteristics as cooperation and communication facilitate a more democratic decision-making process, which also means a higher commitment from members of the organization (Deaconu and Raşcã, 2015).

As a matter of fact, in organizations where gender enhancement is applied there are evident and measurable economic improvements: motivation and engagement are improved, conflicts are reduced, the positive dynamics of human resources increase, growth of the individual and the entire workforce is favoured.

A female leader is a sign of creativity, innovation and introduces different points of view that can best meet the needs of a diversified organization.

Around the world, more women are reaching leadership positions passing through the glass ceiling. Throughout time, there are many examples of female talents: women leaders who have left their mark in history, in medicine, in politics where at the head of large organizations or communities they demonstrate to be great leaders acting in the most effective way and behaving to overcome the challenges of work and the future.

5. FINAL REMARKS

Expatriation is a crucial experience, because of its opportunities and benefits. It is 'complete' in which circumstances are created for people to leave their comfort zone and prove themselves in a wider work contest. It also encloses aspects of private life regarding the knowledge and awareness of oneself, one's abilities, one's strengths and weaknesses.

Women shows an incredible courage when making the choice to live an experience abroad, despite the prejudice that dominates the female world. The audacity and tenacity in facing the problems that may arise in terms of culture and modus vivendi, together with the hunger for more specialized skills, fits effectively into a world that is increasingly moving towards internationalization and the abstract abolition of borders between countries. Despite all the obstacles and potential barriers, what can be gained from expatriation is so satisfying that repays all the hardships experienced.

REFERENCES

Abouzahr, K., Hoteit, L., Krentz, M., & Tracey, C. (2017). *Women on the Move: Shaping Leaders Through Overseas Postings*. Boston Consulting Group. Retrieved from https://www.bcg.com/publications/2017/people-organization-leadership-talent-women-on-the-move

Adler, N. (1984). Women Do Not Want International Careers: And Other Myths About International Management. *Organizational Dynamics*, *13*(2), 66–79. doi:10.1016/0090-2616(84)90019-6

Adler, N., & Bartholomew, S. (1992). Managing Globally Competent People. *The Academy of Management Executive*, *6*(3), 52–62.

Business Insider. (2017). *The 9 best countries in the world for working women, as voted by expats*. Retrieved from https://www.businessinsider.com/internations-best-countries-for-women-working-abroad-2017-3?IR=T

Career Addict. (2017). *Top 10 Benefits of Working Abroad*. Retrieved from https://www.careeraddict.com/working-abroad-benefit-career

Chen, M. (2019). The Impact of Expatriates' Cross-Cultural Adjustment on Work Stress and Job Involvement in the High-Tech Industry. *Frontiers in Psychology*, *10*, 2228. doi:10.3389/fpsyg.2019.02228 PMID:31649581

Culpan, O., & Wright, G. (2002). Women abroad: Getting the best results from women managers. *International Journal of Human Resource Management*, *13*(5), 784–801. doi:10.1080/09585190210125921

Deaconu, A., & Raşcă, L. (2015). Women's Role in Leading Organizations. *Revista de Management Comparat Internaţional.*, *16*(4), 439–450.

Girdauskiene, L., & Eyvazzade, F. (2015). The Profile of an Effective Female Leadership in Multicultural Context. *Procedia: Social and Behavioral Sciences*, *210*, 11–20. doi:10.1016/j.sbspro.2015.11.323

Grant, J. (1988). Women as managers: What they can offer to organizations. *Organizational Dynamics*, *16*(3), 56–63. doi:10.1016/0090-2616(88)90036-8

Midura, M. (2013). John Vs. Jennifer: A Battle of the Sexes. *Yale Scientific*. Retrieved from https://www.yalescientific.org/2013/02/john-vs-jennifer-a-battle-of-the-sexes/

The Telegraph. (2011, July 22). *Women say working abroad improves their career prospects*. Retrieved from https://www.telegraph.co.uk/finance/personalfinance/expat-money/8651745/Women-say-working-abroad-improves-their-career-prospects.html

The Telegraph. (2019a, March 25). *The five barriers to women in business*. Retrieved from https://www.telegraph.co.uk/business/women-entrepreneurs/five-barriers/

The Telegraph. (2019b, March 25). *Ten myths about women in business*. Retrieved from https://www.telegraph.co.uk/business/women-entrepreneurs/ten-myths/

Torfx. (2016). *Why Women are choosing to Work Abroad to Advance their Careers*. Retrieved from https://blog.torfx.com/emigrating/why-women-are-choosing-to-work-abroad-to-advance-their-careers/

Compilation of References

Abdallah, J., & Jibai, S. (2020). Women in leadership: Gender personality traits and skills. *Business Excellence and Management*, *10*(1), 5–15. doi:10.24818/beman/2020.10.1-01

Abdellatif, A. S. (2018). The beverages. *Agricultural Research & Technology: Open Access Journal*, *14*(5), 1–9.

Abouzahr, K., Hoteit, L., Krentz, M., & Tracey, C. (2017). *Women on the Move: Shaping Leaders Through Overseas Postings*. Boston Consulting Group. Retrieved from https://www.bcg.com/publications/2017/people-organization-leadership-talent-women-on-the-move

Abramowitz, M., & David, P. (1996). Technological change and the rise of intangible investments: The US economy's growth path in the Twentieth Century. In D. Foray & B.-Å. Lundvall (Eds.), Employment and growth in the knowledge-based economy. Paris: OECD.

Abreu, F. de S. (1997). QFD - desdobramento da função qualidade - estruturando a satisfação do cliente. *Revista de Administração de Empresas*, *37*(2), 47–55. doi:10.1590/S0034-75901997000200005

Abu-Jarad, I. Y., Yusof, N., & Nikbin, D. (2010). A review paper on organizational culture and organizational performance. *International Journal of Business and Social Science*, *1*(3), 26–46.

Ackfeldt, A., & Wong, V. (2006). The antecedents of prosocial service behaviors: An empirical investigation. *Service Industries Journal*, *26*(7), 727–745. doi:10.1080/02642060600898237

Adler, N. (1984). Women Do Not Want International Careers: And Other Myths About International Management. *Organizational Dynamics*, *13*(2), 66–79. doi:10.1016/0090-2616(84)90019-6

Adler, N., & Bartholomew, S. (1992). Managing Globally Competent People. *The Academy of Management Executive*, *6*(3), 52–62.

Aga, D. (2016). Transactional Leadership and Project Success: The Moderating Role of Goal Clarity. *Procedia Computer Science*, *100*, 517–525. doi:10.1016/j.procs.2016.09.190

Agars, M. D., Baer, J., & Kaufman, J. C. (2005). The many creativities of business and the APT Model of creativity. *The International Journal of Creativity & Problem Solving*, *15*(2), 133–141.

Ahmadabadi, H. Z., Zamzam, F., Meybodi, F. R., Streimikiene, D., & Mardanim, A. (2018). Developent of a new sesame product using QFD and DOE methods: A case study of sesame product in yazd. *Montenegrin Journal of Economics*, *14*(1), 27–44. doi:10.14254/1800-5845/2018.14-1.2

Ahmed Khan, Z., Nawaz, A., & Khan, I. (2016). Leadership theories and styles: A literature review. *Journal of Resources Development and Management*, *16*, 1–7.

Ahuja, R. K., & Orlin, J. B. (1997). Developing fitter genetic algorithms. *INFORMS Journal on Computing*, *9*(3), 251–253. doi:10.1287/ijoc.9.3.251

Akao, Y. (1997). Past, Present, and Future. *International Symposium on QFD '97 – Linköping*, (2), 1–12.

Akao, Y., & Mazur, G. (2003). The leading edge in QFD: Past, present and future. *International Journal of Quality & Reliability Management*, *20*(1), 20–35. doi:10.1108/02656710310453791

Akinola, M., Kapadia, C., Lu, J. G., & Mason, M. F. (2019). Incorporating physiology into creativity research and practice: The effects of bodily stress responses on creativity in organizations. *The Academy of Management Perspectives*, *33*(2), 163–184. doi:10.5465/amp.2017.0094

Akram, T., Lei, S., Hussain, S. T., Haider, M. J., & Akram, M. W. (2016). Does relational leadership generate organizational social capital? A case of exploring the effect of relational leadership on organizational social capital in China. *Future Business Journal*, *2*(2), 116–126. doi:10.1016/j.fbj.2016.06.001

Al-Ali, A. A., Singh, S. K., Al-Nahyan, M., & Sohal, A. S. (2017). Change management through leadership: The mediating role of organizational culture. *The International Journal of Organizational Analysis*, *25*(4), 723–739. doi:10.1108/IJOA-01-2017-1117

Alba, F. E., de La Nieta, A. G. S., Casas, A. G. S., González, A. M., Cendõn, J. A., & Limas, M. C. (2014). A new device for dosing additives in the food industry using quality function deployment. *Journal of Food Process Engineering*, *37*(4), 387–395. doi:10.1111/jfpe.12094

Alencar, E. M. (1995). *Criatividade* (2nd ed.). Editora da Universidade de Brasília.

Alencar, E. M. (1998). Promovendo um ambiente favorável à criatividade nas organizações. *Revista de Administração de Empresas*, *38*(2), 18–25. doi:10.1590/S0034-75901998000200003

Alexandridis, G., Hoepner, A., Huang, Z., & Oikonomou, I. (2022). Corporate social responsibility culture and international M&As. *The British Accounting Review*, *54*(1), 101035. doi:10.1016/j.bar.2021.101035

Alon, I., Farrell, M., & Li, S. (2020). Regime type and COVID-19 response. *FIIB Business Review*, *9*(3), 152–160. doi:10.1177/2319714520928884

Alonso-Almeida, M., Perramon, J., & Bagur-Femenias, L. (2017). Leadership styles and corporate social responsibility management: Analysis from a gender perspective. *Business Ethics (Oxford, England)*, *26*(2), 147–161. doi:10.1111/beer.12139

Altuna, N., Contri, A. M., Dell'Era, C., Frattini, F., & Maccarrone, P. (2015). Managing social innovation in for-profit organizations: The case of Intesa Sanpaolo. *European Journal of Innovation Management*, *18*(2), 258–280. doi:10.1108/EJIM-06-2014-0058

Alvesson, M., Kärreman, D., & Ybema, S. (2017). Studying culture in organizations: Not taking for granted the taken-for-granted. In A. Wilkinson, S. Armstrong, & M. Lounsbury (Eds.), *The Oxford Handbook of Management* (pp. 103–126). Oxford University Press.

Amabile, T. M. (1988). A model of creativity and innovation in organizations. *Research in Organizational Behavior*, *10*(1), 123–167.

Amabile, T. M. (1995). Discovering the unknowable, managing their unmanageable. In C. M. Ford & D. A. Gioia (Eds.), *Creative action in organizations*. Sage Publications. doi:10.4135/9781452243535.n7

Amabile, T. M., Conti, R., Coon, H., Lazenby, J., & Herron, M. (1996). Assessing the work environment for creativity. *Academy of Management Journal, 39*(5), 1154–1184.

Amalric, F. (2005). *The Equator Principles: a step towards sustainability.* Center for Corporate Responsibility and Sustainability. Working Paper No. 01/05.

Ancelotti, C., Brady, C., & Forde, M. (2017). *Quiet leadership: Winning hearts, minds and matches.* Penquin Books.

Andrei, J. V., Panait, M., & Voica, C. (2018). Challenges and approaches for the corporate social responsibility and human resource management in the financial sector. *Economics. Management and Financial Markets, 13*(3), 415–431.

Ansah, M., & Louw, L. (2019). The influence of national culture on organizational culture of multinational companies. *Cogent Social Sciences, 5*(1), 1–15.

ANTAD. (2019). *INDICANTAD, 2018. Datos a septiembre de 2019.* https://antad.net/indicadores/indicantad/

Antonakis, J., Bastardoz, N., Jacquart, P., & Shamir, B. (2016). Charisma: An ill-defined and ill-measured gift. *Annual Review of Organizational Psychology and Organizational Behavior, 3*(1), 293–319. doi:10.1146/annurev-orgpsych-041015-062305

Apekey, T. A., McSorley, G., Tilling, M., & Siriwardena, A. N. (2011). Room for improvement? Leadership, innovation culture and uptake of quality improvement methods in general practice. *Journal of Evaluation in Clinical Practice, 17*(2), 311–318. doi:10.1111/j.1365-2753.2010.01447.x PMID:20438607

Arruda, Â. F., Chrisóstomo, E., & Rios, S. S. (2010). A importância da liderança nas organizações. *Revista Razão Contábil & Finanças, 1*(1).

Arvey, R., Zhang, Z., Avolio, B., & Krueger, R. (2007). Developmental and genetic determinants of leadership role occupancy among females. *The Journal of Applied Psychology, 92*(3), 693–706. doi:10.1037/0021-9010.92.3.693 PMID:17484551

Ashford, S. J., & Sitkin, S. B. (2019). From problems to progress: A dialogue on prevailing issues in leadership research. *The Leadership Quarterly, 30*(4), 454–460. doi:10.1016/j.leaqua.2019.01.003

Ashmos, D. P., & Duchon, D. (2000). Spirituality at work: A conceptualization and measure. *Journal of Management Inquiry, 9*(2), 134–145. doi:10.1177/105649260092008

Atkinson, L., & Kim, Y. (2014). I drink it anyway and I know I shouldn't": Understanding green consumers' positive evaluations of norm-violating non-green products and misleading green advertising. *Environmental Communication, 9*(1), 37–57. doi:10.1080/17524032.2014.932817

Atmaca, A. (2018). Energy, exergy and exergoeconomic assessment of a dry type rotary kiln. *Anadolu University of Sciences & Technology A. Applied Sciences and Engineering, 19*(1), 192–205.

Audrius, B. (2013, September 16). *UKERC Blog: The Current Event.* Retrieved from UK Energy Research Centre: https://ukerc.wordpress.com/2013/09/16/a-few-thoughts-on-energy-and-exergy-efficiences/

Babu, M. M., Dey, B. L., Rahman, M., Roy, S. K., Alwi, S. F. S., & Kamal, M. M. (2020). Value co-creation through social innovation: A study of sustainable strategic alliance in telecommunication and financial services sectors in Bangladesh. *Industrial Marketing Management, 89*, 13–27. doi:10.1016/j.indmarman.2020.06.003

Babu, N., De Roeck, K., & Raineri, N. (2020). Hypocritical organizations: Implications for employee social responsibility. *Journal of Business Research, 114*, 376–384. doi:10.1016/j.jbusres.2019.07.034

Back, T. (1992), The interaction of mutation rate, selection, and self-adaptation within a genetic algorithm. *Proceedings of the 2nd International Conference on Parallel Problem Solving From Nature*, 85-94.

Baer, J., & Kaufman, J. C. (2005). Bridging generality and specificity: The amusement park theoretical (APT) model of creativity. *Roeper Review*, *27*(3), 158–163. doi:10.1080/02783190509554310

Bagherzad, A., Chavosh, A., & Hosseinikhah, S. (2011). The influence of relationship marketing tactics on customer loyalty in B2C relationships–the role of communication and personalization. *European Journal of Economics, Finance and Administrative Science*, (31), 49–56.

Bagnoli, M., & Watts, S. G. (2003). Selling to socially responsible consumers: Competition and the private provision of public goods. *Journal of Economics & Management Strategy*, *12*(3), 419–445. doi:10.1162/105864003322309536

Baierlein, R. (2001). The elusive chemical potential. *American Journal of Physics*, *69*(4), 423–434. doi:10.1119/1.1336839

Baker, K. R. (1974). *Introduction to Sequencing and Scheduling*. John Wiley & Sons.

Baker, S., & Mehmood, A. (2015). Social innovation and the governance of sustainable places. *Local Environment*, *20*(3), 321–334. doi:10.1080/13549839.2013.842964

Baker, W. E., & Sinkula, J. M. (2009). Complementary effects of market orientation and entrepreneurial orientation on profitability in small businesses. *Journal of Small Business Management*, *47*(4), 443–464. doi:10.1111/j.1540-627X.2009.00278.x

Bakker, H., Boersma, K., & Oreel, S. (2006). Creativity (ideas) management in industrial R&D organizations: A crea-political process model and an empirical illustration of Corus RD&T. *Creativity and Innovation Management*, *15*(3), 296–309. doi:10.1111/j.1467-8691.2006.00397.x

Balasubramanian, S. K., Karrh, J. A., & Patwardhan, H. (2006). Audience response to product placements: An integrative framework and future research agenda. *Journal of Advertising*, *35*(3), 115–141. doi:10.2753/JOA0091-3367350308

Balmer, J. M., Fukukawa, T., & Gray, E. R. (2007). The Nature and Management of Ethical Corporate Identity: A Commentary on Corporate Identity, Corporate Social Responsibility and Ethics. *Journal of Business Ethics*, *76*(1), 7–15. doi:10.100710551-006-9278-z

Bang-Jensen, J., Gutin, G., & Yeo, A. (2004). When the greedy algorithm fails. *Discrete Optimization*, *1*(2), 121–127. doi:10.1016/j.disopt.2004.03.007

Bank Act. (1991). *Bank Canada Act*. Retrieved from https://laws-lois.justice.gc.ca/PDF/B-1.01.pdf

Bank of Mexico. (2017). *Balanza Comercial de Mercancías de México*. SNIEG. Información de Interés Nacional. Información disponible en: https://www.banxico.org.mx/SieInternet/consultarDirectorioInternetAction.do?accion=consultarCuadro&idCuadro=CE37&locale=es

Barabasz, M. (1991). Hypnotizability in bulimia. *International Journal of Eating Disorders*, *10*(1), 117–120. doi:10.1002/1098-108X(199101)10:1<117::AID-EAT2260100113>3.0.CO;2-T

Barbot, B., & Reiter-Palmon, R. (2019). Creativity assessment: Pitfalls, solutions, and standards. *Psychology of Aesthetics, Creativity, and the Arts*, *13*(2), 131–132. doi:10.1037/aca0000251

Barney, J. B. (1986). Organizational culture: Can it be a source of sustained sompetitive sdvantage? *Academy of Management Review*, *11*(3), 656–665. doi:10.5465/amr.1986.4306261

Barrantes, R. J., & Eaton, A. A. (2018). Sexual orientation and leadership suitability: How being a gay man affects perceptions of fit in gender-stereotyped positions. *Sex Roles*, *79*(9), 549–564. doi:10.100711199-018-0894-8

Barry, N. (2002). The stakeholder concept of corporate control is illogical and impractical. *The Independent Review*, *6*(4), 541-554.

Barry, C. B., & Brown, S. J. (1985). Differential information and security market equilibrium. *Journal of Financial and Quantitative Analysis*, *20*(4), 407–422. doi:10.2307/2330758

Bass, B. M. (1985). *Leadership and performance beyond expectations*. Free Press.

Bass, B. M. (1998). *Transformational leadership: industrial, military, and educational impact*. Lawrence Erlbaum Associates Publishers.

Beasley, J. E. (1990). An algorithm for set covering problems. *European Journal of Operational Research*, *31*(1), 85–93. doi:10.1016/0377-2217(87)90141-X

Bech, A. C., Hansen, M., & Wienberg, L. (1997). Application of house of quality in translation of consumer needs into sensory attributes mea- surable by descriptive sensory analysis. *Food Quality and Preference*, *8*(5-6), 329–348. doi:10.1016/S0950-3293(97)00062-1

Bedi, T., Coudouel, A., & Simler, K. (Eds.). (2007). *More than a pretty picture: using poverty maps to design better policies and interventions*. World Bank Publications.

Beghetto, R. A., & Kaufman, J. C. (2007). The genesis of creative greatness: Mini-c and the expert performance approach. *High Ability Studies*, *18*(1), 59–61. doi:10.1080/13598130701350668

Beghetto, R. A., & Plucker, J. A. (2006). The Relationship Among Schooling, Learning, and Creativity: "All Roads Lead to Creativity" or "You Can't Get There from Here"? In J. C. Kaufman & J. Baer (Eds.), *Creativity and reason in cognitive development* (pp. 316–332). Cambridge University Press. doi:10.1017/CBO9780511606915.019

Bejan, A., Errera, M. R., & Gunes, U. (2020). Energy theory of periodic economic growth. *International Journal of Energy Research*, *44*(7), 5231–5242. doi:10.1002/er.5267

Belfiore, N. P., Esposito, A., & Vincent, T. L. (1998). A theoretical and experimental study of crossover operators of genetic algorithms. *Journal of Optimization Theory and Applications*, *99*(2), 271–302. doi:10.1023/A:1021766025497

Benner, M., Geerts, R. F. R., Linnemann, A. R., Jongen, W. M. F., Folstar, P., & Cnossen, H. J. (2003b). A chain information model for structured knowledge management: Towards effective and efficient food product improvement. *Trends in Food Science & Technology*, *14*(11), 469–477. doi:10.1016/S0924-2244(03)00154-7

Benner, M., Linnemann, A. R., Jongen, W. M. F., & Folstar, P. A. (2003a). Quality Function Deployment (QFD) - Can it be used to develop food products? *Food Quality and Preference*, *14*(4), 327–339. doi:10.1016/S0950-3293(02)00129-5

Berg, M. E., & Karlsen, J. T. (2016). A study of coaching leadership style practice in projects. *Management Research Review*, *39*(9), 1122–1142. doi:10.1108/MRR-07-2015-0157

Berg, M., & Karlsen, J. (2020). Coaching leadership style: A learning process. *International Journal of Knowledge and Learning*, *13*(4), 356. doi:10.1504/IJKL.2020.10033158

Besemer, S. P., & Treffinger, D. J. (1981). Analysis of creative products: Review and synthesis. *The Journal of Creative Behavior*, *15*(3), 158–178. doi:10.1002/j.2162-6057.1981.tb00287.x

Bevilacqua, M., Ciarapica, F. E., & Marchetti, B. (2012). Development and test of a new fuzzy-QFD approach for characterizing customers rating of extra virgin olive oil. *Food Quality and Preference*, *24*(1), 75–84. doi:10.1016/j.foodqual.2011.09.005

Bigné, E. & Curras, R. (2008). ¿Influye la imagen de Responsabilidad Social en la Intención de compra? El papel de la identificación del consumidor con la empresa. *Universia Business Review*, *19*, 10-23.

Bigné, J. E., & Andreu, L. (2004). Modelo cognitivo-afectivo de la satisfacción en servicios de ocio y turismo. *Cuadernos de Economía y Dirección de Empresa*, *21*, 89–120.

Binnewies, C., Ohly, S., & Niessen, C. (2008). Age and creativity at work: The interplay between job resources, age and idea creativity. *Journal of Managerial Psychology*, *23*(4), 438–457. doi:10.1108/02683940810869042

Bishop, W. R. J. (1984). Competitive Intelligence. *Progressive Grocer*, (March), 19–20.

Black, B. (2005). Comparative industrial relations theory: The role of national culture. *International Journal of Human Resource Management*, *16*(7), 1137–1158. doi:10.1080/09585190500143980

Blauth, M., Mauer, R., & Brettel, M. (2014). Fostering creativity in new product development through entrepreneurial decision making. *Creativity and Innovation Management*, *23*(4), 495–509. doi:10.1111/caim.12094

Bock, A. J., Opsahl, T., George, G., & Gann, D. M. (2012). The effects of culture and structure on strategic flexibility during business model innovation. *Journal of Management Studies*, *49*(2), 279–305. doi:10.1111/j.1467-6486.2011.01030.x

Boschetti, M., & Maniezzo, V. (2015). A set covering based matheuristic for a real-world city logistics problem. *International Transactions in Operational Research*, *22*(1), 169–195. doi:10.1111/itor.12110

Botosan, C. A. (2006). Disclosure and the cost of capital: What do we know? *Accounting and Business Research, 36*(sup1), 31-40.

Bravo-Ibarra, E. R., & Herrera, L. (2009). Capacidad de innovación y configuración de recursos organizativos. *Intagible Capital*, *5*(3), 301–320.

Brettel, M., & Cleven, N. J. (2011). Innovation culture, collaboration with external partners and NPD performance. *Creativity and Innovation Management*, *20*(4), 253–272. doi:10.1111/j.1467-8691.2011.00617.x

Brin, P., & Nehme, M. N. (2019). Corporate social responsibility: Analysis of theories and models. *Eureka: Social and Humanities*, *5*, 22–30. doi:10.21303/2504-5571.2019.001007

Brodeur, A., Clark, A. E., Fleche, S., & Powdthavee, N. (2021). COVID-19, lockdowns and well-being: Evidence from Google Trends. *Journal of Public Economics*, *193*, 104–346. doi:10.1016/j.jpubeco.2020.104346 PMID:33281237

Brown, M., & Treviño, L. (2006). Ethical leadership: A review and future directions. *The Leadership Quarterly*, *17*(6), 595–616. doi:10.1016/j.leaqua.2006.10.004

Brucker, P. (2001). *Scheduling Algorithms* (3rd ed.). Springer-Verlag. doi:10.1007/978-3-662-04550-3

Bruno-Faria, M. F., & Alencar, E. M. (1996). Estímulos e obstáculos à criatividade no ambiente de trabalho. *Revista ADM*, *31*(2), 50–61.

Bruno-Faria, M., & Fonseca, M. (2014). Culture of Innovation: Concepts and Theoretical Models. *RAC. Revista de Administração Contemporânea*, *18*(4), 372–396. doi:10.1590/1982-7849rac20141025

Brunsø, K., Fjord, T. A., & Grunert, K. G. (2002). Consumers' Food Choice and Quality Perception. *Aarhus School of Business*, *77*, 1–60.

Bryant, J. (1982). A thermodynamic approach to economics. *Energy Economics*, *4*(1), 36–50. doi:10.1016/0140-9883(82)90042-1

Bryant, J. (2007). A thermodynamic theory of economics. *International Journal of Exergy*, *4*(3), 302–337. doi:10.1504/IJEX.2007.013396

Bryman, A., Gillingwater, D., & McGuinness, I. (1992). Leadership and organisational transformation. *Management Research News*, *15*(5/6), 26–27. doi:10.1108/eb028221

Business Insider. (2017). *The 9 best countries in the world for working women, as voted by expats*. Retrieved from https://www.businessinsider.com/internations-best-countries-for-women-working-abroad-2017-3?IR=T

Çakar, N. D., & Ertürk, A. (2010). Comparing innovation capability of small and medium-sized enterprises: Examining the effects of organizational culture and empowerment. *Journal of Small Business Management*, *48*(3), 325–359. doi:10.1111/j.1540-627X.2010.00297.x

Cameron, K. S., Bright, D., & Caza, A. (2004). Exploring the relationships between organizational virtuousness and performance. *The American Behavioral Scientist*, *47*(6), 766–790. doi:10.1177/0002764203260209

Cameron, K. S., & Quinn, R. E. (2011). *Diagnosing and changing organizational culture: Based on the competing values framework*. John Wiley & Sons.

Camilleri, M. A. (2017). Corporate sustainability and responsibility: Creating value for business, society and the environment. *Asian Journal of Sustainability and Social Responsibility*, *2*(1), 59–74. doi:10.118641180-017-0016-5

Campbell, H. G., Dudek, R. A., & Smith, M. L. (1970). A heuristic algorithm for the n-job m-machine sequencing problem. *Management Science*, *16*(10), B630–B637. doi:10.1287/mnsc.16.10.B630

Cappelli, A., Parretti, C., Cini, E., & Citti, P. (2019). Development of a new washing machine in olive oil extraction plant: A first application of usability-based approach. *Journal of Agricultural Engineering*, *50*(3), 134–142. doi:10.4081/jae.2019.949

Cardinal, L. B., Alessandri, T. M., & Turner, S. F. (2001). Knowledge codifiability, resources, and science-based innovation. *Journal of Knowledge Management*, *5*(2), 195–204. doi:10.1108/13673270110393266

Cardoso, J. de F., Casarotto Filho, N., & Cauchick Miguel, P. A. (2015). Application of Quality Function Deployment for the development of an organic product. *Food Quality and Preference*, *40*, 180–190. doi:10.1016/j.foodqual.2014.09.012

Career Addict. (2017). *Top 10 Benefits of Working Abroad*. Retrieved from https://www.careeraddict.com/working-abroad-benefit-career

Carroll, A. B. (1980). Social Responsibility as an Objective of Business: Evolving Toward a Model of Corporate Social Performance In Business Policy and Strategic Management (pp. 62-70). McGraw-Hill.

Carroll, A. B. (1991). The Pyramid of Corporate Social Responsibility: Toward the Moral Management of Corporate Stakeholders. *Business Horizons*, *34*(4), 39–48. doi:10.1016/0007-6813(91)90005-G

Carroll, A. B. (1998). The four faces of corporate citizenship. *Business & Society*, *100*(1), 1–7.

Carroll, A. B. (1999). Corporate social responsibility: Evolution of a definitional construct. *Business & Society*, *38*(3), 268–295. doi:10.1177/000765039903800303

Carroll, A. B., & Schwartz, M. (2003). Corporate Social Responsibility: A Three-Domain Approach. *Business Ethics Quarterly*, *13*(4), 503–530. doi:10.5840/beq200313435

Carvalhal, C. (2014). *Entre Linhas* [Between Lines]. Prime Books.

Cauchick Miguel, P. A. (2005). Evidence of QFD best practices for product development: A multiple case study. *International Journal of Quality & Reliability Management*, *22*(1), 72–82. doi:10.1108/02656710510573002

Cavusgil, S. T., Calantone, R. J., & Zhao, Y. (2003). Tacit knowledge transfer and firm innovation capability. *Journal of Business and Industrial Marketing*, *18*(1), 6–21. doi:10.1108/08858620310458615

Cavusgil, S. T., Knight, G., & Riesenberger, J. R. (2014). *International Business – The New Realities* (3rd ed.). Pearson Education.

Cela, E., Deineko, V. G., & Woeginger, G. J. (2014). Linearizable special cases of the QAP. *Journal of Combinatorial Optimization*, 1–11.

Cerni, T., Curtis, G., & Colmar, S. (2013). Information-processing and leadership: An introductory review and the cognitive leadership model (CLM). *Proceedings of the International Conference on Management, Leadership and Governance*.

Chakrabarti, B. K., Chakraborti, A., & Chatterjee, A. (Eds.). (2006). *Econophysics and sociophysics: trends and perspectives*. John Wiley & Sons. doi:10.1002/9783527610006

ChamizoA.Novales CincaA. (2015). *Econometric models of credit spreads*. Available at SSRN 2707644.

Changar, M., & Atan, T. (2021). The role of transformational and transactional leadership approaches on environmental and ethical aspects of CSR. *Sustainability*, *13*(3), 1411. doi:10.3390u13031411

Chan, L. K., & Wu, M. L. (2002). Quality function deployment: A literature review. *European Journal of Operational Research*, *143*(3), 463–497. doi:10.1016/S0377-2217(02)00178-9

Chan, L. K., & Wu, M. L. (2004). A systematic approach to quality function deployment with a full illustrative example. *Omega*, *33*(2), 119–139. doi:10.1016/j.omega.2004.03.010

Chan, W., Li, H., Li, X., Chang, F., Wang, L., & Feng, Z. (2021). Exergoeconomic analysis and optimization of the Allam cycle with liquefied natural gas cold exergy utilization. *Energy Conversion and Management*, *235*, 113972. doi:10.1016/j.enconman.2021.113972

Chapela G. (2012). Problemas y Oportunidades en el mercado para las empresas Sociales Forestales en México. Agencia de los Estados Unidos para el Desarrollo Internacional (USAID). *Proyecto Competitividad de las Empresas Sociales Forestales*.

Chaston, I., & Scott, G. J. (2012). Entrepreneurship and open innovation in an emerging economy. *Management Decision*, *50*(7), 1161–1177. doi:10.1108/00251741211246941

Chatman, J. A., & O'Reilly, C. A. (2016). Paradigm lost: Reinvigorating the study of organizational culture. *Research in Organizational Behavior*, *36*, 199–224. doi:10.1016/j.riob.2016.11.004

Chatterjee, S., Carrera, C., & Lynch, L. A. (1996). Genetic algorithms and traveling salesman problems. *European Journal of Operational Research*, *93*(3), 490–510. doi:10.1016/0377-2217(95)00077-1

Chaudhry, A., Vidyarthi, P. R., Liden, R. C., & Wayne, S. J. (2021). Two to tango? Implications of alignment and misalignment in leader and follower perceptions of LMX. *Journal of Business and Psychology*, *36*(3), 383–399. doi:10.100710869-020-09690-8

Chen, C. (2022). *The mediating effect of corporate culture on the relationship between business model innovation and corporate social responsibility: A perspective from small- and medium-sized enterprises*. Asia Pacific Management Review. doi:10.1016/j.apmrv.2022.01.001

Chen, C. L., Vempati, V. S., & Aljaber, N. (1995). An application of genetic algorithms for flow shop problems. *European Journal of Operational Research*, *80*(2), 389–396. doi:10.1016/0377-2217(93)E0228-P

Cheng, R., & Gen, M. (1994). Crossover on intensive search and traveling salesman problem. *Computers & Industrial Engineering*, *27*(1-4), 485–488. doi:10.1016/0360-8352(94)90340-9

Chen, J., Lim, S., Nam, H., & Phillips, J. (2020). Local culture as a corporate social responsibility multiplier: Confucian values' mediation between firm policies and employees' attitude. *Asia-Pacific Journal of Business Administration*, *12*(3-4), 387–407. doi:10.1108/APJBA-04-2019-0088

Chen, M. (2019). The Impact of Expatriates' Cross-Cultural Adjustment on Work Stress and Job Involvement in the High-Tech Industry. *Frontiers in Psychology*, *10*, 2228. doi:10.3389/fpsyg.2019.02228 PMID:31649581

Chen, N., Huang, H., & Lin, C. (2018). Equator principles and bank liquidity. *International Review of Economics & Finance*, *55*, 185–202. doi:10.1016/j.iref.2017.07.020

Chen, Y. S., Lin, M. J. J., & Chang, C. H. (2009). The positive effects of relationship learning and absorptive capacity on innovation performance and competitive advantage in industrial markets. *Industrial Marketing Management*, *38*(2), 152–158. doi:10.1016/j.indmarman.2008.12.003

Chesbrough, H. (2010). Business model innovation: Opportunities and barriers. *Long Range Planning*, *43*(2), 354–363. doi:10.1016/j.lrp.2009.07.010

Chih, H., Chih, H., & Chen, T. (2010). On the determinants of corporate social responsibility: International evidence on the financial industry. *Journal of Business Ethics*, *93*(1), 115–135. doi:10.100710551-009-0186-x

Chin, J. L. (2013). Diversity leadership: Influence of ethnicity, gender, and minority status. *Open Journal of Leadership*, *2*(01), 1–10. doi:10.4236/ojl.2013.21001

Chiu, C.-Y., Balkundi, P., & Weinberg, F. J. (2017). When managers become leaders: The role of manager network centralities, social power, and followers' perception of leadership. *The Leadership Quarterly*, *28*(2), 334–348.

Chong, M., Shang, Y., Richards, M., & Zhu, X. (2018). Two sides of the same coin? Leadership and organizational culture. *Leadership and Organization Development Journal*, *39*(8), 975–994. doi:10.1108/LODJ-05-2017-0122

Christopher, A. N., Victoria Kuo, S., Abraham, K. M., Noel, L. W., & Linz, H. E. (2004). Materialism and affective well being: The role of social support. *Personality and Individual Differences*, *37*(3), 463–470. doi:10.1016/j.paid.2003.09.015

Ciutacu, C., Chivu, L., & Preda, D. (2005). Company's Social Responsibility-A Challenge for Contemporary World. *Romanian Journal of Economics*, *20*(1 (29)), 79–94.

Clark, R. A., Hartline, M. D., & Jones, K. C. (2009). The effects of leadership style on hotel employees' commitment to service quality. *Cornell Hospitality Quarterly*, *50*(2), 209–231. doi:10.1177/1938965508315371

Coelho, L. C., & Laporte, G. (2015). Classification, models and exact algorithms for multi-compartment delivery problems. *European Journal of Operational Research*, *242*(3), 854–864. doi:10.1016/j.ejor.2014.10.059

Coelho, O. (2004). *Pedagogia do desporto: Contributos para uma compreensão do desporto juvenil* [Pedagogy of sport: Contributions to an understanding of youth sport]. Livros Horizonte.

Coleman, M. (2003). Gender and the orthodoxies of leadership. *School Leadership & Management*, *23*(3), 325–339. doi:10.1080/1363243032000112810

COM 2001/366, GREEN PAPER Promoting a European framework for Corporate Social Responsibility, retrieved Microsoft Word - en 366-5.doc (europa.eu)

COM 2010/284, Green Paper - Corporate governance in financial institutions and remuneration policies, retrieved COM(2010)284/F1 - EN (europa.eu)

Conforto, E. C., Amaral, D. C., & Silva, S. L. (2011). Roadmap for systematic literature review: application in product development and project management (in portuguese). *Congresso Brasileiro de Gestão e Desenvolvimento de Produto,* (8), 1-12.

Contreras, G., Bos, J. W. B., & Kleimeier, S. (2019). Self-regulation in sustainable finance: The adoption of the equator principles. *World Development, 122*, 306–324. doi:10.1016/j.worlddev.2019.05.030

Cook, S. A. (1971). The complexity of theorem-proving procedures. *Proceedings of the 3rd Annual ACM Symposium on Theory of Computing*, 151-158. 10.1145/800157.805047

Cook, W. J., Cunningham, W. H., Pulleyblank, W. R., & Schrijver, A. (1998). *Combinatorial Optimization*. John Wiley & Sons.

Cooter, R., & Ulen, T. (1988). *Law and economics*. Academic Press.

Cormen, T., Leiserson, C. E., & Rivest, R. L. (1990). *Introduction to Algorithms*. The MIT Press.

Costa, A. I. A., Dekker, M., & Jongen, W. M. F. (2001). Quality function deployment in the food industry: A review. *Trends in Food Science & Technology, 11*(9–10), 306–314.

Costa, A., Cappadonna, F. A., & Fichera, S. (2015). A hybrid genetic algorithm for minimizing makespan in a flow-shop sequence-dependent group scheduling problem. *Journal of Intelligent Manufacturing*, 1–15.

Cotter, K. N., Silvia, P. J., & Fayn, K. (2018). What does feeling like crying when listening to music feel like? *Psychology of Aesthetics, Creativity, and the Arts, 12*(2), 216–227. doi:10.1037/aca0000108

Counsell, A., Furtado, M., Iorio, C., Anand, L., Canzonieri, A., Fine, A., Fotinos, K., Epstein, I., & Katzman, M. A. (2017). Intolerance of uncertainty, social anxiety, and generalized anxiety: Differences by diagnosis and symptoms. *Psychiatry Research, 252*, 63–69. doi:10.1016/j.psychres.2017.02.046 PMID:28254577

Cromwell, J. R., Amabile, T. M., & Harvey, J.-F. (2018). An integrated model of dynamic problem solving within organizational constraints. In R. Reiter-Palmon, V. L. Kennel, & J. C. Kaufman (Eds.), *Individual creativity in the workplace* (pp. 53–81). Academic Press. doi:10.1016/B978-0-12-813238-8.00003-6

Cropley, A. (2006). Creativity: A social approach. *Roeper Review, 28*(3), 125–130. doi:10.1080/02783190609554351

Cropley, D. H. (2015). Promoting creativity and innovation in engineering education. *Psychology of Aesthetics, Creativity, and the Arts, 9*(2), 161–171. doi:10.1037/aca0000008

Crust, L., & Lawrence, I. (2006). A review of leadership in sport: Implications for football management. *Athletic Insight, 8*(4), 28–48.

Csikszentmihalyi, N. M. (1988). Society, culture, and person: A systems view of creativity. In R. Sternberg (Ed.), *The nature of creativity*. Cambridge University Press.

Culpan, O., & Wright, G. (2002). Women abroad: Getting the best results from women managers. *International Journal of Human Resource Management, 13*(5), 784–801. doi:10.1080/09585190210125921

Cunha, M. P., Rego, A., Cunha, R. C., Cabral-Cardoso, C., & Neves, P. (2016). *Manual de Comportamento Organizacional e Gestão* [Handbook of Organizational Behavior and Management]. RH Editora.

Curral, L., Gomes, C., Marques-Quinteiro, P., & Lind, P. (2016). *Caos e Complexidade: Novos Conceitos para a Gestão das Organizações* [Chaos and Complexity: New Concepts for Managing Organizations]. RH Editora.

Curtis, S. A. (2003). The classification of greedy algorithms. *Science of Computer Programming*, *49*(1-3), 125–157. doi:10.1016/j.scico.2003.09.001

Dannenbring, D. G. (1977). An evaluation of flow shop sequencing heuristics. *Management Science*, *23*(11), 1174–1182. doi:10.1287/mnsc.23.11.1174

Dantzig, G. B., Fulkerson, D. R., & Johnson, S. M. (1954). Solution of a large scale traveling salesman problem. *Operations Research*, *2*, 393–410.

Darroch, J., & McNaughton, R. (2002). Examining the link between knowledge management practices and types of innovation. *Journal of Intellectual Capital*, *3*(3), 210–222. doi:10.1108/14691930210435570

Das, G. S. (2003). Preparedness for innovation: An Indian perspective. *Global Business Review*, *4*(1), 27–39. doi:10.1177/097215090300400103

Davenport, T. H., De Long, D. W., & Beers, M. C. (1998). Successful knowledge management projects. *Sloan Management Review*, (Winter), 43–57.

Davies, G., & Olmedo-Cifuentes, I. (2016). Corporate misconduct and the loss of trust. *European Journal of Marketing*, *50*(7/8), 1426–1447. doi:10.1108/EJM-11-2014-0729

Davis, L. (1985). Applying adaptive algorithms to epistatic domains. *Proceedings of the 9th International Joint Conference on Artificial Intelligence*, 162–164.

Day, D., & Antonakis, J. (2012). Leadership: Past, present, and future. In J. Antonakis & D. V. Day (Eds.), *The nature of leadership* (pp. 3–25). Sage Publications, Inc. doi:10.1093/oxfordhb/9780199928309.013.0022

Day, G. S. (1994). The capabilities of market-driven organizations. *Journal of Marketing*, *58*(4), 37–52. doi:10.1177/002224299405800404

de Jong, M. D., Huluba, G., & Beldad, A. D. (2020). Different shades of greenwashing: Consumers' reactions to environmental lies, half-lies, and organizations taking credit for following legal obligations. *Journal of Business and Technical Communication*, *34*(1), 38–76. doi:10.1177/1050651919874105

De Koker, L., & Jentzsch, N. (2013). Financial inclusion and financial integrity: Aligned incentives? *World Development*, *44*, 267–280. doi:10.1016/j.worlddev.2012.11.002

De León AriasA. (2010). Teaching the Solow Through the Building of Economic Growth Scenarios: An Application to China. Available at SSRN 1749699. doi:10.2139/ssrn.1749699

De Pelsmaeker, S., Gellynck, X., Delbaere, C., Declercq, N., & Dewettinck, K. (2014). Consumer-driven product development and improvement combined with sensory analysis: A case-study for European filled chocolates. *Food Quality and Preference, 41*, 20–29.

De Silva, M., Al-Tabbaa, O., & Khan, Z. (2019). Business model innovation by international social purpose organizations: The role of dynamic capabilities. *Journal of Business Research, 125*, 733–749. doi:10.1016/j.jbusres.2019.12.030

Deaconu, A., & Raşcă, L. (2015). Women's Role in Leading Organizations. *Revista de Management Comparat Internaţional.*, *16*(4), 439–450.

Deephouse, D. L., & Suchman, M. (2008). Legitimacy in organizational institutionalism. In R. Greenwood, C. Oliver, K. Sahlin, & R. Suddaby (Eds.), *The Sage Handbook of Organizational Institutionalism* (pp. 49–77). London Sage Publications. doi:10.4135/9781849200387.n2

Delmas, M. A., & Burbano, V. C. (2011). The drivers of greenwashing. *California Management Review*, *54*(1), 64–87. doi:10.1525/cmr.2011.54.1.64

DeLone, W. H., & McLean, E. R. (2004). Measuring e-Commerce Success: Applying the DeLone & McLean Information Systems Success Model. *International Journal of Electronic Commerce*, *9*(1), 31–47. doi:10.1080/10864415.2004.11044317

Demil, B., & Lecocq, X. (2010). Business model evolution: In search of dynamic consistency. *Long Range Planning*, *43*(2/3), 227–246. doi:10.1016/j.lrp.2010.02.004

Deshpandé, R., & Farley, J. U. (2004). Organizational culture, market orientation, innovativeness, and firm performance: An international research odyssey. *International Journal of Research in Marketing*, *21*(1), 3–22. doi:10.1016/j.ijresmar.2003.04.002

Deshpandé, R., Farley, J. U., & Webster, F. E. Jr. (1993). Corporate culture, customer orientation, and innovativeness in Japanese firms: A quadrad analysis. *Journal of Marketing*, *57*(1), 23–37. doi:10.1177/002224299305700102

Deshpandé, R., & Webster, F. E. Jr. (1989). Organizational culture and marketing: Defining the research agenda. *Journal of Marketing*, *53*(1), 3–15. doi:10.1177/002224298905300102

Devaraj, S., Fan, M., & Kohli, R. (2002). Antecedents of B2C Channel Satisfaction and Preference: Validating e-Commerce Metrics. *Information Systems Research*, *13*(3), 316–333. doi:10.1287/isre.13.3.316.77

Dewhirst, H. (2020). *Bank of England accused of Greenwashing*. Positive Money Europe. https://positivemoney.org/2020/06/bank-of-england-governor-accused-of-greenwashing/

Dias Júnior, A. F., Andrade, C. R., Milan, M., Brito, J. O., Andrade, A. M., & de Souza, N. D. (2020). Quality function deployment (QFD) reveals appropriate quality of charcoal used in barbecues. *Scientia Agrícola*, *77*(6), 1–11. doi:10.1590/1678-992x-2019-0021

Dickson, M. W., Castaño, N., Magomaeva, A., & Den Hartog, D. N. (2012). Conceptualizing leadership across cultures. *Journal of World Business*, *47*(4), 483–492. doi:10.1016/j.jwb.2012.01.002

Dijkstra, E. W. (1959). A note on two problems in connexion with graphs. *Numerische Mathematik*, *1*(1), 269–271. doi:10.1007/BF01386390

Dill, M. D., Revillion, J. P. P., Barcellos, J. O. J., Dias, E. A., Mércio, T. Z., & de Oliveira, T. E. (2014). Procedural priorities of the pork loin supply chain. *Journal of Technology Management & Innovation*, *9*(1), 84–92. doi:10.4067/S0718-27242014000100007

Dimitrova, Y. (2019). Corporate Culture Change Management. *Economic Alternatives*, *2*, 296–312.

Dincer, I., & Cengel, Y. A. (2001). Energy, entropy and exergy concepts and their roles in thermal engineering. *Entropy (Basel, Switzerland)*, *3*(3), 116–149. doi:10.3390/e3030116

Dincer, I., & Rosen, M. A. (2007). *Exergy, energy, environment and sustainable development*. Elsevier.

Dinh, J. E., Lord, R. G., Gardner, W. L., Meuser, J. D., Liden, R. C., & Hu, J. (2014). Leadership theory and research in the new millennium: Current theoretical trends and changing perspectives. *The Leadership Quarterly*, *25*(1), 36–62. doi:10.1016/j.leaqua.2013.11.005

Dionisio, M., & Vargas, E. (2020). Corporate social innovation: A systematic literature review. *International Business Review*, *29*(2), 101641. doi:10.1016/j.ibusrev.2019.101641

Djekic, I., Radivojevic, D., & Milivojevic, J. (2019). Quality perception throughout the apple fruit chain. *Journal of Food Measurement and Characterization*, *13*(4), 3106–3118. doi:10.100711694-019-00233-1

Djekic, I., Skunca, D., Nastasijevic, I., Tomovic, V., & Tomasevic, I. (2018). Transformation of quality aspects throughout the chicken meat supply chain. *British Food Journal*, *120*(5), 1132–1150. doi:10.1108/BFJ-08-2017-0432

Dnishev, F., & Alzhanova, F. (2016). Globalization of technological development and opportunities for national innovation systems of developing countries. *The Journal of Asian Finance, Economics, and Business*, *3*(4), 67–79. doi:10.13106/jafeb.2016.vol3.no4.67

Dobni, C. B. (2008). Measuring innovation culture in organizations: The development of a generalized innovation culture construct using exploratory factor analysis. *European Journal of Innovation Management*, *11*(4), 539–559. doi:10.1108/14601060810911156

Doherty, B., Haugh, H., & Lyon, F. (2014). Social enterprises as hybrid organizations: A review and research agenda. *International Journal of Management Reviews*, *16*(4), 417–436. doi:10.1111/ijmr.12028

Doh, J., Rodrigues, S., Saka-Helmhout, A., & Makhija, M. (2017). International business responses to institutional voids. *Journal of International Business Studies*, *48*(3), 293–307. doi:10.105741267-017-0074-z

Dolgun, L. E., & Köksal, G. (2018). Effective use of quality function deployment and Kansei engineering for product planning with sensory customer requirements: A plain yogurt case. *Quality Engineering*, *30*(4), 569–582. doi:10.1080/08982112.2017.1366511

Dombrowski, C., Kim, J. Y., Desouza, K. C., Brabanza, A., Papagari, S., Baloh, P., & Jha, S. (2007). Elements of innovative cultures. *Knowledge and Process Management*, *14*(3), 190–202. doi:10.1002/kpm.279

Donaldson, T., & Preston, L. (1995). The Stakeholder theory of the corporation: Concepts, evidence and implications. *Academy of Management Review*, *20*(1), 65–91. doi:10.2307/258887

Dora, Y. M. (2021). Adaptability, Creativity And Innovation To Be Survive. Study On Arts And Culture Tourism Bandung. *Turkish Journal of Computer and Mathematics Education*, *12*(8), 973–979.

Doyle, M. (1984). New Ways of Measuring Value. *Progressive Grocer-Value, Executive Report*, 15-19.

Drucker, P. F. (1985). The discipline of innovation. *Harvard Business Review*, *63*(3), 67–72. PMID:10272260

Drucker, P. F. (1993). *Sociedade pós-capitalista*. Pioneira.

Drucker, P. F. (2004). A disciplina da inovação. *Harvard Business Review*, *82*(8), 80–85.

du Plessis, M., & Boon, J. A. (2004). Knowledge management in eBusiness and customer relationship management: South African case study findings. *International Journal of Information Management*, *24*(1), 73–86. doi:10.1016/j.ijinfomgt.2003.10.002

Du, D. Z., & Pardalos, P. M. (Eds.). (2005). *Handbook of Combinatorial Optimization*. Springer-Verlag. doi:10.1007/b102533

Dzafic, J., & Petersson, A. (2016). *Greenwashing in CSR reports-A case study of two entities*. http://www.diva-portal.org/smash/get/diva2:932386/FULLTEXT01.pdf

Edmonds, J. (1971). Matroids and the greedy algorithm. *Mathematical Programming*, *1*(1), 126–136. doi:10.1007/BF01584082

Edvinsson, L. (1997). Developing intellectual capital at Skandia. *Long Range Planning*, *30*(3), 366–373. doi:10.1016/S0024-6301(97)90248-X

Ehrhart, M., Schneider, B., & Macey, W. (2014). *Organizational climate and culture: An introduction to theory, research, and practice*. Routledge.

Eisenbeiss, S. A., Knippenberg, D., & Boerner, S. (2008). Transformational leadership and team innovation: Integrating team climate principles. *The Journal of Applied Psychology*, *93*(6), 1438–1446. doi:10.1037/a0012716 PMID:19025260

Eisenberger, R., Armeli, S., Rexwinkel, B., Lynch, P. D., & Rhoades, L. (2001). Reciprocation of perceived organizational support. *The Journal of Applied Psychology*, *86*(1), 42–51. doi:10.1037/0021-9010.86.1.42 PMID:11302232

Ekawati, Y., & Bazarado, M. (2016). Designing food products based on carrots using the product design phase of Quality Function Deployment. *Journal of Engineering and Applied Sciences (Asian Research Publishing Network)*, *11*(5), 3109–3116.

Elleuch, H., Dafaoui, E., El Mhamedi, A., & Chabchou, H. (2016). A Quality Function Deployment approach for Production Resilience improvement in Supply Chain: Case of Agrifood Industry. *IFAC-PapersOnLine*, *49*(31), 125–130. doi:10.1016/j.ifacol.2016.12.173

Ene, C., & Panait, M. (2017). The financial education-Part of corporate social responsibility for employees and customers. *Revista Romana de Economie*, *44*(1), 145–154.

Engelen, A., Gupta, V., Strenger, L., & Brettel, M. (2015). Entrepreneurial orientation, firm performance, and the moderating role of transformational leadership behaviors. *Journal of Management*, *41*(4), 1069–1097. doi:10.1177/0149206312455244

Equator Principles. (2020). *The Equator Principles July 2020 A financial industry benchmark for determining, assessing and managing environmental and social risk in projects.* Author.

Erlichson, H. (1984). Internal energy in the first law of thermodynamics. *American Journal of Physics*, *52*(7), 623–625. doi:10.1119/1.13601

Esen, A., & Maden-Eyiusta, C. (2019). Delineating the concept of corporate social innovation: Toward a multidimensional model. *International Journal of Entrepreneurship and Innovation Management*, *23*(1), 23–45. doi:10.1504/IJEIM.2019.096497

Espasandín-Bustelo, F., Ganaza-Vargas, J., & Diaz-Carrion, R. (2021). Employee happiness and corporate social responsibility: The role of organizational culture. *Employee Relations*, *43*(3), 609–629. doi:10.1108/ER-07-2020-0343

Etiler, O., Toklu, B., Atak, M., & Wilson, J. M. (2004). A genetic algorithm for flow shop scheduling problems. *The Journal of the Operational Research Society*, *55*(8), 830–835. doi:10.1057/palgrave.jors.2601766

Evan, W. M., & Freeman, R. E. (1993). A stakeholder theory of the modern corporation: Kantian capitalism. In Ethical theory and business (pp. 75-84). Prentice-Hall.

Eva, N., Robin, M., Sendjaya, S., van Dierendonck, D., & Liden, R. C. (2019). Servant Leadership: A systematic review and call for future research. *The Leadership Quarterly*, *30*(1), 111–132. doi:10.1016/j.leaqua.2018.07.004

Facchinetti, G., Ricci, R. G., & Muzzioli, S. (1998). Note on ranking fuzzy triangular numbers. *International Journal of Intelligent Systems*, *13*(7), 613–622. doi:10.1002/(SICI)1098-111X(199807)13:7<613::AID-INT2>3.0.CO;2-N

Fadare, D. A., Nkpubre, D. O., Oni, A. O., Falana, A., Waheed, M. A., & Bamiro, O. A. (2010). Energy and exergy analysis of malt drink production in Nigeria. *Energy*, *35*(12), 5336–5346. doi:10.1016/j.energy.2010.07.026

Falkenauer, E., & Bouffouix, S. (1991). A genetic algorithm for job shop. *Proceedings of the IEEE International Conference on Robotics and Automation*, 824–829.

Farmer, R., & Sundberg, N. D. (1986). Boredom proneness - the development and correlates of a new scale. *Journal of Personality Assessment, 50*(1), 4–17. doi:10.120715327752jpa5001_2 PMID:3723312

Felder, G. N., & Felder, K. M. (2015). *Mathematical methods in engineering and physics*. John Wiley & Sons.

Felfe, J., & Petersen, L.-E. (2007). Romance of leadership and management decision making. *European Journal of Work and Organizational Psychology, 16*(1), 1–24. doi:10.1080/13594320600873076

Feo, T. A., & Resende, M. G. C. (1995). Greedy randomized adaptive search procedures. *Journal of Global Optimization, 6*(2), 109–133. doi:10.1007/BF01096763

Fernandes, J. M. R., & Rebelato, M. G. (2006). Proposta de um método para integração entre QFD e FMEA. *Gestão & Produção, 13*(2), 245–259. doi:10.1590/S0104-530X2006000200007

Fiedler, F. E. (1978). The Contingency Model and the Dynamics of the Leadership Process. In L. Berkowitz (Ed.), Advances in Experimental Social Psychology (Vol. 11, pp. 59–112). Academic Press. doi:10.1016/S0065-2601(08)60005-2

Fietz, B., & Günther, E. (2021). Changing organizational culture to establish sustainability. *Control Management Review, 65*(3), 32–40. doi:10.100712176-021-0379-4

Filippos, K. Z., & Francisco, J. R.-M. (2017). The use of exergetic indicators in the food industry – A review. *Critical Reviews in Food Science and Nutrition, 57*(1), 197–211. doi:10.1080/10408398.2014.975335 PMID:26066523

Financial Times. (2019). *Bankers push back against 'greenwashing' criticism*. https://www.ft.com/content/c4cefa40-f4ab-11e9-b018-3ef8794b17c6

Financial Times. (2021). *Greenwashing in finance: Europe's push to police ESG investing*. https://www.ft.com/content/74888921-368d-42e1-91cd-c3c8ce64a05e

Fiordelisi, F., Renneboog, L., Ricci, O., & Lopes, S. S. (2019). Creative corporate culture and innovation. *Journal of International Financial Markets, Institutions and Money, 63*, 101–137. doi:10.1016/j.intfin.2019.101137

Fisher, C. M., & Barrett, F. J. (2019). The experience of improvising in organizations: A creative process perspective. *The Academy of Management Perspectives, 33*(2), 148–162. doi:10.5465/amp.2017.0100

Fleury, M. T., & Oliveira, M. M., Jr. (Orgs.). (2001). Gestão estratégica do conhecimento. São Paulo: Atlas.

Flint, S. W., Plumley, D. J., & Wilson, R. J. (2014). You don't know what you're doing! The impact of managerial change on club performance in the English Premier League. *Managing Leisure, 19*(6), 390–399. doi:10.1080/13606719.2014.910000

Fong, P. S., Men, C., Luo, J., & Jia, R. (2018). Knowledge hiding and team creativity: The contingent role of task interdependence. *Management Decision, 56*(2), 329–343. doi:10.1108/MD-11-2016-0778

Ford, C. M. (2000). Creative developments in creativity theory. Academy of Management. *Academy of Management Review, 25*(2), 284–285. doi:10.5465/amr.2000.26079331

Foss, N., & Saebi, T. (2017). Business models and business model innovation: Between wicked and paradigmatic problems. *Long Range Planning, 51*(1), 9–21. doi:10.1016/j.lrp.2017.07.006

Franke, G. R., & Nadler, S. S. (2008). Culture, economic development and national ethical attitudes. *Journal of Business Research, 61*(3), 254–264. doi:10.1016/j.jbusres.2007.06.005

Freeman, E. (1984). *Strategic Management: A Stakeholder Approach*. Pitman Publishing.

Freeman, E., & Reed, D. L. (1983). Stockholders and stakeholders: A new perspective on corporate governance. *California Management Review*, *25*(3), 88–106. doi:10.2307/41165018

French, S. (1982). *Sequencing and Scheduling: An Introduction to the Mathematics of the Job-Shop*. Ellis Horwood.

Friedman, M. (1970). The Social Responsibility of Business is to Increase its Profits. *New York Times Magazine, 13*, 122-126.

Friedman, A. L., & Miles, S. (2001). Socially responsible investment and corporate social and environmental reporting in the UK: An exploratory study. *The British Accounting Review*, *33*(4), 523–548. doi:10.1006/bare.2001.0172

Friedman, M. (1962). *Capitalism and Freedom*. The University Chicago Press.

Frontiera, J. (2010). Leadership and organizational culture transformation in professional sport. *Journal of Leadership & Organizational Studies*, *17*(1), 71–86. doi:10.1177/1548051809345253

Frost, P. (1991). *Reframing Organizational Culture*. Sage Publications.

Galanaki, E. (2004). Are children able to distinguish among the concepts of aloneness, loneliness, and solitude? *International Journal of Behavioral Development*, *28*(5), 435–443. doi:10.1080/01650250444000153

Gamble, A., & Kelly, G. (2001). Shareholder Value and the Stakeholder Debate in the UK. *Corporate Governance*, *9*(2), 110–117. doi:10.1111/1467-8683.00235

García-Sánchez, I. M., Gómez-Miranda, M. E., David, F., & Rodríguez-Ariza, L. (2019). Analyst coverage and forecast accuracy when CSR reports improve stakeholder engagement: The Global Reporting Initiative-International Finance Corporation disclosure strategy. *Corporate Social Responsibility and Environmental Management*, *26*(6), 1392–1406.

Garey, M. R., & Johnson, D. S. (1979). *Computers and Intractability a Guide to the Theory of NP-Completeness*. W.H. Freeman and Company.

Garey, M. R., Johnson, D. S., & Sethi, R. (1976). The complexity of flowshop and jobshop scheduling. *Mathematics of Operations Research*, *1*(2), 117–129. doi:10.1287/moor.1.2.117

Garganta, J. (2004). Atrás do palco. Nas oficinas do Futebol [Behind the stage. At the Football Workshops]. In Futebol de muitas cores e sabores. Reflexões em torno do desporto mais popular do mundo [Football of many colors and flavors. Reflections around the most popular sport in the world] (pp. 227–234). Campo das Letras.

Garvin, D. A. (1983). Can industry self-regulation work? *California Management Review*, *25*(4), 37–52. doi:10.2307/41165031

Gasques, J. G., de Rezende, G. C., Verde, C. M. V., Salerno, M. S., da Conceição, J. C. P. R., & Carvalho, J. C. de S. (2004). *Desempenho e Crescimento do Agronegócio no Brasil*. Instituto de Pesquisa Econômica Aplicada – IPEA.

Gatti, L., Seele, P., & Rademacher, L. (2019). Grey zone in–greenwash out. A review of greenwashing research and implications for the voluntary-mandatory transition of CSR. *International Journal of Corporate Social Responsibility*, *4*(1), 1–15. doi:10.118640991-019-0044-9

Gavin, J. H., & Mason, R. O. (2004). The virtuous organization: The value of happiness in the workplace. *Organizational Dynamics*, *33*(4), 379–392. doi:10.1016/j.orgdyn.2004.09.005

Gaylin, W. (1979). *Feelings: Our Vital Signs*. Harper and Row.

Genest, C. M. (2005). Cultures, organizations and philanthropy. *Corporate Communications, 10*(4), 315–327. doi:10.1108/13563280510630115

George, G., Lakhani, K. R., & Puranam, P. (2020). What has changed? The impact of Covid pandemic on the technology and innovation management research agenda. *Journal of Management Studies, 57*(8), 1754–1758. doi:10.1111/joms.12634

George, G., & Zahra, S. A. (2002). *Being entrepreneurial and being market-driven – Exploring the interaction effects of entrepreneurial and market orientation on firm performance. In Frontiers of Entrepreneurship Research.* Babson College.

Georgescu-Roegen, N. (1971). *The Entropy Law and the Economic Process.* Harvard University Press.

Gerhart, B., & Fang, M. (2005). National culture and human resource management: Assumptions and evidence. *International Journal of Human Resource Management, 16*(6), 971–986. doi:10.1080/09585190500120772

Gesteland, R. R. (1999). *Cross-Cultural Business Behavior.* Copenhagen Business School Press.

Ghannadzadeh, A. (2013). *Exergetic balances and analysis in a process simulator: A way to enhance process energy integration.* Retrieved 3 1, 2014, from www.ethesis.inp-toulouse.fr/archive/00002116/01/ ghannazadeh.pdf

Ghasabeh, M. S., Reaiche, C., & Soosay, C. (2015). The emerging role of transformational leadership. *Journal of Developing Areas, 49*(6), 459–467. doi:10.1353/jda.2015.0090

Giannoulakis, S., Volkart, K., & Bauer, C. (2014). Life cycle and cost assessment of mineral carbonation for carbon capture and storage in European power generation. *International Journal of Greenhouse Gas Control, 21*, 140–157. doi:10.1016/j.ijggc.2013.12.002

Gigauri, I. (2021). Corporate Social Responsibility and COVID-19 Pandemic Crisis: Evidence From Georgia. *International Journal of Sustainable Entrepreneurship and Corporate Social Responsibility, 6*(1), 30–47. doi:10.4018/IJSECSR.2021010103

Gigauri, I., & Damenia, N. (2020). Cooperation between Social Entrepreneurs and Government to Develop Solutions to Social Problems. *Business and Economic Review, 10*(3), 116–136. doi:10.5296/ber.v10i3.17383

Gill, A. S., & Mathur, N. (2007). Improving employee dedication and pro-social behavior. *International Journal of Contemporary Hospitality Management, 19*(4), 328–334. doi:10.1108/09596110710747661

Girdauskiene, L., & Eyvazzade, F. (2015). The Profile of an Effective Female Leadership in Multicultural Context. *Procedia: Social and Behavioral Sciences, 210*, 11–20. doi:10.1016/j.sbspro.2015.11.323

Glavas, A. (2016). Corporate social responsibility and employee engagement: Enabling employees to employ more of their whole selves at work-. *Frontiers in Psychology, 7*, 796. doi:10.3389/fpsyg.2016.00796 PMID:27303352

Gloet, M., & Terziovski, M. (2004). Exploring the relationship between knowledge management practices and innovation performance. *Journal of Manufacturing Technology Management, 15*(5), 402–409. doi:10.1108/17410380410540390

Glozer, S., & Morsing, M. (2020). Helpful hypocrisy? Investigating 'double-talk' and irony in CSR marketing communications. *Journal of Business Research, 114*, 363–375. doi:10.1016/j.jbusres.2019.08.048

Goldberg, D. E., & Lingle, R. (1985). Alleles, loci and the travelling salesman problem. In *Proceedings of the 2nd International Conference on Genetic Algorithms.* Lawrence Erlbaum.

Gond, J.-P., Akremi, A. E., Swaen, V., & Babu, N. (2017). The psychological microfoundations of corporate social responsibility: A person-centric systematic review. *Journal of Organizational Behavior, 38*(2), 225–246. doi:10.1002/job.2170

Gonzalez, T., & Sahni, S. (1978). Flowshop and Jobshop Schedules: Complexity and Approximation. *Operations Research, 26*(1), 36–52. doi:10.1287/opre.26.1.36

Gorji-Bandpy, M., & Ebrabimian, V. (2006). Exergoeconomic analysis of gas turbine power plants. *International Energy Journal, 7*(1), 35–41.

Gorton, G. B., & Zentefis, A. K. (2020). *Corporate culture as a theory of the firm.* National Bureau of Economic Research Working Paper Series, No. 27353. doi:10.3386/w27353

Govers, C. P. M. (1996). What and how about quality function deployment (QFD). *International Journal of Production Economics, 46–47*(95), 575–585. doi:10.1016/0925-5273(95)00113-1

Grabowski, J., & Wodecki, M. (2004). A very fast tabu search algorithm for the permutation flow shop problem with makespan criterion. *Computers & Operations Research, 31*(11), 1891–1909. doi:10.1016/S0305-0548(03)00145-X

Graen, G. B., & Uhl-Bien, M. (1995). Relationship-based approach to leadership: Development of leader-member exchange (LMX) theory of leadership over 25 years: Applying a multi-level multi-domain perspective. *The Leadership Quarterly, 6*(2), 219–247. doi:10.1016/1048-9843(95)90036-5

Grant, J. (1988). Women as managers: What they can offer to organizations. *Organizational Dynamics, 16*(3), 56–63. doi:10.1016/0090-2616(88)90036-8

Grefenstette, J. J., Gopal, R., Rosmaita, B., & Van Gucht, D. (1985), Genetic algorithms for the traveling salesman problem. *Proceedings the 2nd International Conference on Genetic Algorithms*, 160-168.

Greiner, W., & Müller, B. (2012). *Quantum mechanics: symmetries.* Springer Science & Business Media.

Grigorescu, A., Cerchia, A. E., Oachesu, M. M., & Udroiu, F. (2017). Enhancing Internet Banking-Solutions for Customer Relationship Management. *Saudi J. Bus. Manag. Stud, 2*, 38–43.

Grimaldi, R. P. (1989). *Discrete and Combinatorial Mathematics.* Addison-Wesley.

Grinstein, A. (2008). The relationships between market orientation and alternative strategic orientations: A meta-analysis. *European Journal of Marketing, 42*(1/2), 115–134. doi:10.1108/03090560810840934

Groysberg, B., Lee, J., Price, J., & Cheng, J. (2018). The leader's guide to corporate culture. *Harvard Business Review, 96*(1), 44–52.

Grunert, K. G. (2005). Food quality and safety: Consumer perception and demand. *European Review of Agriculture Economics, 32*(3), 369–39. doi:10.1093/eurrag/jbi011

Guellec, D., & Van Pottelsberghe de la Potterie, B. (2004). From R&D to productivity growth: Do the institutional settings and the source of funds of R&D matter? *Oxford Bulletin of Economics and Statistics, 66*(3), 353–378. doi:10.1111/j.1468-0084.2004.00083.x

Guilford, J. P. (1950). Creativity. *The American Psychologist, 4*(9), 444–454. doi:10.1037/h0063487 PMID:14771441

Gundersen, T. (2011). *An introduction to the concept of exergy and energy quality.* Retrieved 7 26, 2014, from www.ivt.ntnu.no/ept/fag/.../ Exergy%20Light%20 Version%204.pdf

Gutin, G. (2013). Section 4.6 Traveling Salesman Problems. Handbook of Graph Theory, 336.

Gutin, G., & Punnen, A. P. (2002). *The Traveling Salesman Problem and Its Variations.* Kluwer Academic Publishers.

Gutin, G., & Yeo, A. (2002). Anti-matroids. *Operations Research Letters, 30*(2), 97–99. doi:10.1016/S0167-6377(02)00117-7

Gutin, G., Yeo, A., & Zverovich, A. (2002). Traveling salesman should not be greedy: Domination analysis of greedy-type heuristics for the TSP. *Discrete Applied Mathematics, 117*(1-3), 81–86. doi:10.1016/S0166-218X(01)00195-0

Hajipour, B., Rahimi, F., & Hooshmand, M. (2013). Market orientation: Review and new research agenda. *Journal of Management Research, 5*(1), 92–111.

Haller, M., & Hadler, M. (2006). How social relations and structures can produce happiness and unhappiness: An international comparative analysis. *Social Indicators Research, 75*(2), 169–216. doi:10.100711205-004-6297-y

Hamilton, R., & Chernev, A. (2013). Low prices are just the beginning: Price image in retail management. *Journal of Marketing, 77*(6), 1–20. doi:10.1509/jm.08.0204

Hampden-Turner, C. (1992). *Creating corporate culture: from discord to harmony.* Addison-Wesley Publishing Company Inc.

Ha, Q. M., Deville, Y., Pham, Q. D., & Hà, M. H. (2020). A hybrid genetic algorithm for the traveling salesman problem with drone. *Journal of Heuristics, 26*(2), 219–247. doi:10.100710732-019-09431-y

Harkema, S. (2003). A complex adaptive perspective on learning within innovation projects. *The Learning Organization, 10*(6), 340–346. doi:10.1108/09696470310497177

Hartley, J. (2014). New development: Eight and a half propositions to stimulate frugal innovation. *Public Money & Management, 34*(3), 227–232. doi:10.1080/09540962.2014.908034

Hauser, J. R., & Clausing, D. (1988). The House of Quality. *Harvard Business Review, 66*(5/6), 63–73.

Haydargil, D., & Abusoglu, A. (2018). A comparative thermoeconomic cost accounting analysis and evaluation of biogas engine-powered cogeneration. *Energy, 159*, 97–114. doi:10.1016/j.energy.2018.06.102

Hemingway, C. A., & Maclagan, P. W. (2004). Managers' personal values as drivers of corporate social responsibility. *Journal of Business Ethics, 50*(1), 33–44. doi:10.1023/B:BUSI.0000020964.80208.c9

Hemphill, T. (2004). Monitoring global corporate citizenship: Industry self-regulation at a crossroads. *Journal of Corporate Citizenship, 14*(14), 81–95. doi:10.9774/GLEAF.4700.2004.su.00009

Hendry, J. (2003). Environmental NGOs and business: A grounded theory of assessment, targeting, and influencing. *Business & Society, 42*(2), 267–276. doi:10.1177/0007650303042002005

Hernández, R., Fernández, C., & Baptista, P. (2018). Metodología de la investigación. McGraw-Hill.

Herrbach, O., & Mignonac, K. (2004). How organizational image affects employee attitudes. *Human Resource Management Journal, 14*(4), 76–88. doi:10.1111/j.1748-8583.2004.tb00134.x

Herrera, M. E. B. (2015). Creating competitive advantage by institutionalizing corporate social innovation. *Journal of Business Research, 68*(7), 1468–1474. doi:10.1016/j.jbusres.2015.01.036

Hersey, P., & Blanchard, K. H. (1988). *Management and Organizational Behavior.* Prentice-Hall.

Hockerts, K. (2007). Social Entrepreneurship. In W. Visser, D. Matten, M. Pohl, & N. Tolhurst (Eds.), *The A to Z of Corporate Social Responsibility: A Complete Guide to Concepts, Codes and Organisations.* John Wiley.

Hofestede, G. (1980). Culture's consequences: International differences in work-related values. *Sage (Atlanta, Ga.).*

Hofestede, G. (1994). Management scientists are human. *Management Science, 40*(1), 4–13. doi:10.1287/mnsc.40.1.4

Hofestede, G. (1997). *Cultures and Organizations: Software of the minds.* MacGraw-Hill.

Hofestede, G. (2001). *Culture's consequences: Comparing values, behaviors, institutions and organizations across nations* (2nd ed.). Sage.

Hofmeister, K. (1991). Quality Function Deployment: market success through customer driven products. New York. In E. Graf & I. S. Saguy (Eds.), *Food product development: from concept to the market place* (pp. 189–210). doi:10.1007/978-1-4613-0675-7_9

Hofstede, G. (1984). Cultural dimensions in management and planning. *Asia Pacific Journal of Management, 1*(2), 81–99. doi:10.1007/BF01733682

Hofstede, G. (1991). Cultural Routes of Economic Performance: A Research Note. *Strategic Management Journal, 12*(S1), 165–173. doi:10.1002mj.4250120912

Hofstede, G., & Bond, M. H. (1988). The Confucius Connection: From cultural roots to economic growth. *Organizational Dynamics, 16*(4), 5–21. doi:10.1016/0090-2616(88)90009-5

Ho, J. C. (1995). Flowshop sequencing with mean flowtime objective. *European Journal of Operational Research, 81*(3), 571–578. doi:10.1016/0377-2217(93)E0353-Y

Holland, J. H. (1975). *Adaptation in natural and artificial systems: an introductory analysis with applications to biology, control, and artificial intelligence*. University of Michigan Press.

Horowitz, E., & Sahni, S. (1978). *Fundamentals of Computer Algorithms*. Computer Science Press.

House, R. J. (1996). Path-goal theory of leadership: Lessons, legacy, and a reformulated theory. *The Leadership Quarterly, 7*(3), 323–352. doi:10.1016/S1048-9843(96)90024-7

House, R. J. (1999). Cultural influences on Leadership and Organizations: Project Globe. *Advances in Global Leadership, 1*, 171–233.

House, R. J. (2004). Culture, leadership and organizations: The Globe study of 62 societies. *Sage (Atlanta, Ga.)*.

House, R. J., & Aditya, R. N. (1997). The social scientific study of leadership: Quo Vadis? *Journal of Management, 23*(3), 409–473. doi:10.1177/014920639702300306

House, R. J., Javidan, M., Hanges, P., & Dorfman, P. (2002). Understandig cultures and implicit leadership theories across the globe: An introduction to project Globe. *Journal of World Business, 37*(1), 3–10. doi:10.1016/S1090-9516(01)00069-4

Hsia, T. C., Liao, R. C., & Huang, S. C. (2015). Enhancing the quality of rice milling by systematic innovation techniques. *International Journal of Systematic Innovation, 3*(4), 27–36.

Hua, W., & Wei, P. (2016). National culture, population age, and other country factors in volume–Price volatility relationship. *Global Finance Journal, 32*, 1044–1283.

Huey Yiing, L., & Zaman Bin Ahmad, K. (2009). The moderating effects of organizational culture on the relationships between leadership behaviour and organizational commitment and between organizational commitment and job satisfaction and performance. *Leadership and Organization Development Journal, 30*(1), 53–86. doi:10.1108/01437730910927106

Hughes, D. J., Lee, A., Tian, A. W., Newman, A., & Legood, A. (2018). Leadership, creativity, and innovation: A critical review and practical recommendations. *The Leadership Quarterly, 29*(5), 549–569. doi:10.1016/j.leaqua.2018.03.001

Hysa, E. (2011). Corruption and human development correlation in Western Balkan countries. *Euro Economica, 30*(04), 148–157.

Hysa, E., Kruja, A., Rehman, N. U., & Laurenti, R. (2020). Circular economy innovation and environmental sustainability impact on economic growth: An integrated model for sustainable development. *Sustainability, 12*(12), 4831. doi:10.3390u12124831

Iacovoiu, V. B. (2018). An Empirical Analysis of Some Factors Influencing Financial Literacy. *Economic Insights - Trends and Challenges, 70*(2).

Igbong, D. I., & Fakorede, D. O. (2014). Exergoeconomic analysis of a 100 MW unit GE Frame 9 gas turbine plant in Ughelli, Nigeria. *IACSIT International Journal of Engineering and Technology, 4*(8), 463–468.

Ilies, R., Gerhardt, M. W., & Le, H. (2004). Individual differences in leadership emergence: Integrating meta-analytic findings and behavioral genetics estimates. *International Journal of Selection and Assessment 12*(3), 207–219. doi:10.1111/j.0965-075X.2004.00275.x

INEGI DENUE. (2018). *Directorio Estadístico Nacional de Unidades Económicas 2018, Información para la actualización e incorporación de unidades económicas al DENUE; datos a abril de 2018.* http://www3.inegi.org.mx/rnm/index.php/catalog/477

INEGI. (2013). Instituto Nacional de Estadística y Geografía. *Encuesta Nacional de Gastos de los Hogares*. https://www.inegi.org.mx/est/contenidos/proyectos/encuestas/hogares/regulares/engasto/2012/doc/engasto_resultados.pdf

INEGI. (2018). Instituto Nacional de Estadística y Geografía. *Directorio Estadístico Nacional de Unidades Económicas.* http://www.beta.inegi.org.mx/app/mapa/denue/

Ismail, W. K. W., & Abdmajid, R. (2007). Framework of the culture of innovation: A revisit. *Journal Kemanusiaan, 9*, 38–49.

Iyer, S. K., & Saxena, B. (2004). Improved genetic algorithm for the permutation flowshop scheduling problem. *Computers & Operations Research, 31*(4), 593–606. doi:10.1016/S0305-0548(03)00016-9

Jaakson, K., Vadi, M., & Tamm, K. (2009). Organizational culture and CSR: An exploratory study of Estonian service organizations. *Social Responsibility Journal, 5*(1), 6–18. doi:10.1108/17471110910939962

Jacobs, D., & Snijders, H. (2008). *Innovation Routine: How Managers Can Support Repeated Innovation // Stichting Management Studies.* Van Gorcum.

Jacoby, J., & Olson, J. C. (1985). *Perceived Quality: How Consumers View Stores and Merchandise*. Lexington Books.

Jakimowicz, A. (2020). The role of entropy in the development of economics. *Entropy (Basel, Switzerland), 22*(4), 452. doi:10.3390/e22040452 PMID:33286226

Jali, M., Abss, Z., & Ariffin, A. (2017). Corporate Social Responsibility and Corporate Social Innovation: A Conceptual Understanding. *SHS Web of Conferences, 34*, 1-9.

Janiunaite, B., & Petraite, M. (2010). The relationship between organizational innovative culture and knowledge sharing in organization: The case of technological innovation implementation in a telecommunication organization. *Socialiniai Mokslai, 3*(69), 14–23.

Jaskyte, K., & Dressler, W. W. (2004). Studying culture as an integral aggregate variable: Organizational culture and innovation in a group of nonprofit organizations. *Field Methods, 16*(3), 265–284. doi:10.1177/1525822X03262281

Javidan, M., & Carl, D. E. (2004). East meets west: A cross-cultural comparison of charismatic leadership among Canadian and Iranian executives. *Journal of Management Studies, 41*(4), 665–691. doi:10.1111/j.1467-6486.2004.00449.x

Jepson, D. (2009). Studying leadership at cross-country level: A critical analysis. *Leadership, 5*(1), 61–80. doi:10.1177/1742715008098310

Johannessen, J.-A., & Olsen, B. (2011). Projects as communicating systems: Creating a culture of innovation and performance. *International Journal of Information Management, 31*(1), 30–37. doi:10.1016/j.ijinfomgt.2010.04.006

Johnson, S. M. (1954). Optimal Two-and-Three-Stage Production Schedules with Set-up Times Included. *Naval Research Logistics, 1*(1), 61–68. doi:10.1002/nav.3800010110

Johnstone, L. (2018). Theorising and modelling social control in environmental management accounting research. *Social and Environmental Accountability Journal, 38*(1), 30–48. doi:10.1080/0969160X.2017.1422778

Jones, C., Motta, J., & Alderete, M. V. (2016). Gestión estratégica de tecnologías de información y comunicación y adopción del comercio electrónico en Mipymes de Córdoba, Argentina. *Estudios Gerenciales, 32*(138), 4–13. doi:10.1016/j.estger.2015.12.003

Jones, D. A., Willness, C. R., & Glavas, A. (2017). When corporate social responsibility (CSR) meets organizational psychology: New frontiers in micro-CSR research, and fulfilling a quid pro quo through multilevel insights. *Frontiers in Psychology, 8*, 520. doi:10.3389/fpsyg.2017.00520 PMID:28439247

Kaler, J. (2003). Differentiating stakeholder theories. *Journal of Business Ethics, 46*(1), 71–83. doi:10.1023/A:1024794710899

Kaletnik, H., Lutsiak, V., Melnichuk, O., Dovhan, Y., & Malicki, M. (2019). Organizational basis of the development of innovative functional food products by the Ukrainian enterprises of deep walnut processing. *Ukrainian Food Journal, 2*(1), 1–148.

Kapoor, H., & Kaufman, J. C. (2020). Meaning-making through creativity during COVID-19. *Frontiers in Psychology, 11*, 1–9. doi:10.3389/fpsyg.2020.595990 PMID:33391115

Kass, S. L., & McCarroll, J. M. (2006). The revised equator principles. *New York Law Journal, 236*(44), 3.

Kaufman, J. C., & Beghetto, R. A. (2009). Beyond big and little: The four c model of creativity. *Review of General Psychology, 13*(1), 1–12. doi:10.1037/a0013688

Kaufman, J. C., & Beghetto, R. A. (2013). Do people recognize the four Cs? Examining layperson conceptions of creativity. *Psychology of Aesthetics, Creativity, and the Arts, 7*(3), 229–236. doi:10.1037/a0033295

Kaya, D., Kayabasi, E., Kilinc, E., Eyidogan, M., Selimli, S., & Ozkaymak, M. (2013). Energy and exergy efficiencies in industrial pumps. *International Journal of Renewable Energy Resources, 3*(1), 27–42.

Kaya, F. B., Sahin, G. G., & Gurson, P. (2010). Intellectual capital in organizations. *Problems and Perspectives in Management, 8*(1), 153–160.

Kelly, S. (2008). Understanding the role of the football manager in britain and ireland: A weberian approach. *European Sport Management Quarterly, 8*(4), 399–419. doi:10.1080/16184740802461652

Kernbach, S., & Schutte, N. S. (2005). The impact of service provider emotional intelligence on customer satisfaction. *Journal of Services Marketing, 19*(7), 438–444. doi:10.1108/08876040510625945

Kets de Vries, M. R. (2001). Creating authentizotic organizations: Well-functioning individuals in vibrant companies. *Human Relations, 54*(1), 101–111. doi:10.1177/0018726701541013

Kets de Vries, M. R., & Florent-Treacy, E. (2002). Creating high commitment organizations. *Organizational Dynamics, 30*(4), 295–309. doi:10.1016/S0090-2616(02)00067-0

Kim, J.-Y., & Miner, A. S. (2007). Vicarious learning from the failures and near-failures of others: Evidence from the U.S. commercial banking industry. *Academy of Management Journal, 50*(3), 687–714. doi:10.5465/amj.2007.25529755

Kim, M. K., Park, M. C., & Jeong, D. H. (2004). The effects of customer satisfaction and switching barrier on customer loyalty in Korean mobile telecommunication services. *Telecommunications Policy, 28*(2), 145–159. doi:10.1016/j.telpol.2003.12.003

Kim, M. S., & Thapa, B. (2018). Relationship of ethical leadership, corporate social responsibility and organizational performance. *Sustainability, 10*(2), 447. doi:10.3390u10020447

Kim, M., Yin, X., & Lee, G. (2020). The effect of CSR on corporate image, customer citizenship behaviors, and customers' long-term relationship orientation. *International Journal of Hospitality Management, 88*, 1–8. doi:10.1016/j.ijhm.2020.102520

Kirk, C. P., & Rifkin, L. S. (2020). I'll trade you diamonds for toilet paper: Consumer reacting, coping and adapting behaviors in the COVID-19 pandemic. *Journal of Business Research, 117*, 124–131. doi:10.1016/j.jbusres.2020.05.028 PMID:32834208

Kirkpatrick, S., & Locke, E. (1991). Leadership: Do Traits Matter? *The Executive, 5*, 48–60. doi:10.2307/4165007

Klapper, L. F., Lusardi, A., & Panos, G. A. (2012). Financial literacy and the financial crisis (No. w17930). National Bureau of Economic Research.

Klein, D. A. (1998). *A gestão estratégica do capital intelectual: recursos para a economia baseada em conhecimento.* Qualitymark Editora Ltda.

Klimas, P. (2016). Organizational culture and coopetition: An exploratory study of the features, models and role in the Polish Aviation Industry. *Industrial Marketing Management, 53*, 91–102. doi:10.1016/j.indmarman.2015.11.012

Klir, G. J., & Yuan, B. (1995). *Fuzzy Sets and Fuzzy Logic: Theory and Applications.* Prentice Hall.

Knights, D., & O'Leary, M. (2006). Leadership, ethics and responsibility to the other. *Journal of Business Ethics, 67*(2), 125–137. doi:10.100710551 006 9008 6

Koch, C. (2018). Lead like ... Gareth Southgate. *Director, 71*(3), 28. http://search.ebscohost.com/login.aspx?direct=true&db=bth&AN=128242031&site=eds-live

Koch, P. (1994). *Solitude: A philosophical encounter.* Open Court Publishing.

Koenig, M. E. (1996). Intellectual capital and knowledge management. *IFLA Journal, 22*(4), 299–301. doi:10.1177/034003529602200411

Kohli, A. K., & Jaworski, B. J. (1990). Market orientation: The construct, research propositions, and managerial implications. *Journal of Marketing, 54*(2), 1–18. doi:10.1177/002224299005400201

Kok, P., Van der Weile, T., McKenna, R., & Brown, A. (2001). A corporate social responsibility audit within a quality management framework. *Journal of Business Ethics, 31*(4), 285–297. doi:10.1023/A:1010767001610

Koomson, I., Villano, R. A., & Hadley, D. (2020). Intensifying financial inclusion through the provision of financial literacy training: A gendered perspective. *Applied Economics, 52*(4), 375–387. doi:10.1080/00036846.2019.1645943

Korte, B., & Lovasz, L. (1981). Mathematical structures underlying greedy algorithms. In Fundamentals of Computation Theory. Lecture Notes in Computer Science. Springer-Verlag. doi:10.1007/3-540-10854-8_22

Korte, B., & Lovasz, L. (1983). Structure properties of greedoids. *Combinatorica, 3*(3-4), 359–374. doi:10.1007/BF02579192

Korte, B., Lovasz, L., & Schrader, R. (1991). *Greedoids*. Springer-Verlag. doi:10.1007/978-3-642-58191-5

Korte, B., & Vygen, J. (2012). *Combinatorial optimization: theory and algorithms. Algorithms and Combinatorics* (Vol. 21). Springer-Verlag Berlin Heidelberg. doi:10.1007/978-3-642-24488-9

Košičiarová, I., Kádeková, Z., & Štarchoň, P. (2021). Leadership and motivation as important aspects of the international company's corporate culture. *Sustainability, 13*(7), 3916. doi:10.3390u13073916

Krishnan, H. (1996). Characteristics of memory associations: A consumer-based brand equity perspective. *International Journal of Research in Marketing, 13*(4), 389–405. doi:10.1016/S0167-8116(96)00021-3

Kristianto, Y., Ajmal, M. M., & Sandhu, M. (2012). Adopting TQM approach to achieve customer satisfaction: A flour milling company case study. *The TQM Journal, 24*(1), 29–46. doi:10.1108/17542731211191203

Kruskal, J. B. (1956). On the shortest spanning subtree and the traveling salesman problem. *Proceedings of the American Mathematical Society, 7*(1), 48–50. doi:10.1090/S0002-9939-1956-0078686-7

Kuo, C. M., You, S. H., & Lu, C. Y. (2012). Integration of the Kano and QFD model in health food development: Using black beans as examples. *Quality & Quantity, 48*(1), 225–242. doi:10.100711135-012-9762-8

Kuper, S. (1996). *Football Against the Enemy*. Orion Books.

Kyriakopoulos, K., & Moorman, C. (2004). Tradeoffs in marketing exploitation and explorations Strategies: The overlooked role of marketing orientation. *International Journal of Research in Marketing, 21*(3), 219–240. doi:10.1016/j.ijresmar.2004.01.001

Lança, R. (2015). *Coach to Coach*. Prime Books.

Laporte, G. (2010). A concise guide to the Traveling Salesman Problem. *The Journal of the Operational Research Society, 61*(1), 35–40. doi:10.1057/jors.2009.76

Laporte, G., & Palekar, U. (2002). Some applications of the clustered travelling salesman problem. *The Journal of the Operational Research Society, 53*(9), 972–976. doi:10.1057/palgrave.jors.2601420

Lashitew, A. A., Bals, L., & Tulder, R. V. (2020). Inclusive business at the base of the pyramid: The role of embeddedness for enabling social innovations. *Journal of Business Ethics, 162*(2), 421–448. doi:10.100710551-018-3995-y

Lavidge, R. J., & Steiner, G. A. (1961). A model for predictive measurements of advertising effectiveness. *Journal of Marketing, 25*(6), 59–62. doi:10.1177/002224296102500611

Lawler, E. L. (1985). *The Traveling Salesman Problem: A Guided Tour of Combinatorial Optimization*. John Wiley & Sons.

Lawrence, P. (2009). Equator Principles: Or how I learned to stop worrying and love sustainability. *Impact Assessment and Project Appraisal, 27*(1), 1, 3–6. doi:10.3152/146155109X428996

Le Bas, C., & Lauzikas, M. (2010). Determinants of innovation culture and major impacts on the innovation strategy: The case of the information technology sector in Lithuania. *Social Sciences Studies, 4*(8), 125–139.

Lee, Y., & Kramer, A. (2016). The role of purposeful diversity and inclusion strategy (PDIS) and cultural tightness/looseness in the relationship between national culture and organizational culture. *Human Resource Management Review, 26*(3), 198–208. doi:10.1016/j.hrmr.2016.01.001

Legua J. (2015). *El mercado del Hábitat y el Diseño en México*. Oficina Económica y Comercial de la Embajada de España en México, ICEX España importaciones e inversiones.

Leithy, W. (2017). Organizational culture and organizational performance. *International Journal of Economics & Management Sciences*, *6*(4), 1–66. doi:10.4172/2162-6359.1000442

Levitt, T. (1958). The dangers of social responsibility. *Harvard Business Review*, *36*(5), 41–50.

Liden, R. C., & Antonakis, J. (2009). Considering context in psychological leadership research. *Human Relations*, *62*(11), 1587–1605. doi:10.1177/0018726709346374

Lin, H., & McDonough, E. F. III. (2011). Investigating the role of leadership and organizational culture in fostering innovation ambidexterity. *IEEE Transactions on Engineering Management*, *58*(3), 497–509. doi:10.1109/TEM.2010.2092781

Li, P., Sun, J.-M., Taris, T. W., Xing, L., & Peeters, M. C. W. (2021). Country differences in the relationship between leadership and employee engagement: A meta-analysis. *The Leadership Quarterly*, *32*(1), 101458. doi:10.1016/j.leaqua.2020.101458

Litchfield, R. C., Gilson, L. L., & Gilson, P. W. (2015). Defining creative ideas: Toward a more nuanced approach. *Group & Organization Management*, *40*(2), 238–265. doi:10.1177/1059601115574945

Liu, A., Xiao Y., Ji, X., Wang, K., Tsai, S. B., Lu, H., Cheng, J., Lai, X., & Wang, J. (2018). A novel two-stage integrated model for supplier selection of green fresh product. *Sustainability, 10*(7), 1–23.

Liu, J., & Reeves, C. R. (2001). Constructive and Composite Heuristic Solutions to the P//åCi Scheduling Problem. *European Journal of Operational Research*, *132*(2), 439–452. doi:10.1016/S0377-2217(00)00137-5

Locke, E. A., & Kirkpatrick, S. A. (1995). Promoting creativity in organizations. In C. M. Ford & D. A. Gioia (Eds.), *Creative action in organizations*. Sage Publications. doi:10.4135/9781452243535.n13

Lolowang, N., Troena, E., Djazuli, A., & Aisjah, S. (2019). The effect of leadership and organizational culture on employee performance that is educated by motivation (study on the implementation empowerment programs in Jayapura city). *Problems and Perspectives in Management*, *17*(1), 268–277. doi:10.21511/ppm.17(1).2019.23

Long, C. R., & Averill, J. R. (2003). Solitude: An exploration of benefits of being alone. *Journal for the Theory of Social Behaviour*, *33*(1), 21–44. doi:10.1111/1468-5914.00204

López-Duartea, C., González-Loureirob, M., Vidal-Suáreza, M., & González-Díaza, B. (2016). International strategic alliances and national culture: Mapping the Field and developing a research agenda. *Journal of World Business*, *51*(4), 511–524. doi:10.1016/j.jwb.2016.05.001

Lord, R. G. (2019). Leadership and information processing: A dynamic system, dual-processing perspective. In Leader Thinking Skills (pp. 46–70). Routledge.

Lord, R. G., de Vader, C. L., & Alliger, G. M. (1986). A meta-analysis of the relation between personality traits and leadership perceptions: An application of validity generalization procedures. *The Journal of Applied Psychology*, *71*(3), 402–410. doi:10.1037/0021-9010.71.3.402

Lozano K. & Trinidad J. (2016). Estrategias locales para recuperar la competitividad en el mercado internacional: el caso del sector mueblero de Jalisco, México a partir del 2008. *Panorama de la investigación en las ciencias económico administrativas*, 194.

Luis, D. J., Ruth, C. J., & Zhuofan, Z. (2020). Does stress lead to creativity? The relationship between occupational stress and individual innovative behavior. *Studies in Business & Economics*, *15*(1), 21–30. doi:10.2478be-2020-0003

Lumpkin, G. T., & Dess, G. G. (1996). Clarifying the entrepreneurial orientation construct and linking it to performance. *Academy of Management Review*, *21*(1), 135–172. doi:10.2307/258632

Lundvall, B. Å., & Nielsen, P. (2007). Knowledge management and innovation performance. *International Journal of Manpower*, *28*(3/4), 207–223. doi:10.1108/01437720710755218

Lusch, R. F., Vargo, S. L., & Gustafsson, A. (2016). Fostering a trans-disciplinary perspectives of service ecosystems. *Journal of Business Research*, *69*(8), 2957–2963. doi:10.1016/j.jbusres.2016.02.028

Lyon, T. P., & Montgomery, A. W. (2015). The means and end of greenwash. *Organization & Environment*, *28*(2), 223–249. doi:10.1177/1086026615575332

Lythreatis, S., Mostafa, A. M. S., Pereira, V., Wang, X., & Giudice, M. D. (2021). Servant leadership, CSR perceptions, moral meaningfulness and organizational identification- evidence from the Middle East. *International Business Review*, *30*(5), 101772. doi:10.1016/j.ibusrev.2020.101772

Macdonald, G. (2000). Cross-cultural Methodological Issues in Ethical Research. *Journal of Business Ethics*, *27*(1/2), 89–104. doi:10.1023/A:1006406505398

Macve, R., & Chen, X. (2010). The "equator principles": A success for voluntary codes? *Accounting, Auditing, &. Accountability*, *23*(7), 890–919. doi:10.1108/09513571011080171

Madjar, N., Oldham, G. R., & Pratt, M. G. (2002). There's no place like home? The contributions of work and nonwork creativity support to employees' creative performance. *Academy of Management Journal*, *45*(4), 757–767.

Magalhães, C. (2011). Responsabilidade social e regulada no sector metalomecânico brasileiro e português: um estudo comparativo e cultural. In Responsabilidade Social: Uma visão ibero-americana (pp. 81-109). Almedina.

Maguire, J. (2008). 'Real politic' or 'ethically based': Sport, globalization, migration and nation-state policies. *Sport in Society*, *11*(4), 443–458. doi:10.1080/17430430802019375

Maignan, I., & Ferrell, O. C. (2000). Measuring corporate citizenship in two countries: The case of the United States and France. *Journal of Business Ethics*, *23*(3), 283–297. doi:10.1023/A:1006262325211

Maignan, I., & Ferrell, O. C. (2001a). Antecedents and benefits of corporate citizenship: An investigation of French businesses. *Journal of Business Research*, *51*(1), 37–51. doi:10.1016/S0148-2963(99)00042-9

Maignan, I., & Ferrell, O. C. (2001b). Corporate citizenship as a marketing instrument: Concepts, evidence and research directions. *European Journal of Marketing*, *35*(3-4), 457–484. doi:10.1108/03090560110382110

Maitland, A., Hills, L. A., & Rhind, D. J. (2015). Organisational culture in sport – A systematic review. *Sport Management Review*, *18*(4), 501–516. doi:10.1016/j.smr.2014.11.004

Mankiw, N. G. (2014). *Principles of economics*. Cengage Learning.

Mann, S., & Cadman, R. (2014). Does being bored make us more creative? *Creativity Research Journal*, *26*(2), 165–173. doi:10.1080/10400419.2014.901073

Manzoor, F., Wei, L., Nurunnabi, M., Subhan, Q. A., Shah, S. I. A., & Fallatah, S. (2019). The impact of transformational leadership on job performance and CSR as mediator in SMEs. *Sustainability*, *11*(2), 436. doi:10.3390u11020436

Mardar, M., Zhygunov, D., & Znachek, R. (2016). QFD methodology to develop a new health-conducive grain product. *Eastern-European Journal of Enterprise Technologies*, *2*(11), 42–47. doi:10.15587/1729-4061.2016.65725

Marín, L., Espinal, M., López, S., Maldonado, R., Mira, L., & Pertusa, M. (2011). La percepción de los consumidores sobre la RSC en la región de Murcia, los efectos de la crisis económica. *Servicios de Estudios colegios de economistas Región de Murcia*, *9*, 5857.

Martello, S., & Ries, B. (2015). Advances in Combinatorial Optimization. *Discrete Applied Mathematics*, *196*(C), 1–3. doi:10.1016/j.dam.2015.08.011

Martinez, F., O'Sullivan, P., Smith, M., & Esposito, M. (2017). Perspectives on the role of business in social innovation. *Journal of Management Development*, *36*(5), 681–695. doi:10.1108/JMD-10-2016-0212

Martin, J. (1992). *Cultures in Organizations - three perspectives*. Oxford University Press.

Martins, E. C., & Terblanche, F. (2003). Building organizational culture that stimulates creativity and innovation. *European Journal of Innovation Management*, *6*(1), 64–74. doi:10.1108/14601060310456337

Martins, E., & Martins, N. (2002). An organizational culture model to promote creativity and innovation. *SA Journal of Industrial Psychology*, *28*(4), 58–65. doi:10.4102ajip.v28i4.71

Martins, J. M. (2010). *Gestão do Conhecimento, Criação e transferência de conhecimento*. Edições Sílabo.

Martins, J., Costa, C., Oliveira, T., Gonnalves, R., & Branco, F. (2019). How Smartphone Advertising Influences Consumers' Purchase Intention. *Journal of Business Research*, *94*, 378–387. doi:10.1016/j.jbusres.2017.12.047

Matei, M. (2013). *Responsabilitatea socială a corporaţiilor şi instituţiilor şi dezvoltarea durabilă a României*. Expert Publishing House.

Matos, M. H., & Machado, C. F. (2014). Institutional leaderships: The case of Crédito Agrícola in Portugal. *International Journal of Applied Management Sciences and Engineering*, *1*(1), 67–90. doi:10.4018/ijamse.2014010105

Matten, D., Crane, A., & Chapple, W. (2003). Behind the Mask: Revealing the True Face of Corporate Citizenship. *Journal of Business Ethics*, *45*(1/2), 109–120. doi:10.1023/A:1024128730308

Matthews, S. H., Kelemen, T. K., & Bolino, M. C. (2021). How follower traits and cultural values influence the effects of leadership. *The Leadership Quarterly*, *32*(1), 101497. doi:10.1016/j.leaqua.2021.101497

Maxham, J. G. III, & Netemeyer, R. G. (2003). Firms reap what they sow: The effects of shared values and perceived organizational justice on customers' evaluations of complaint handling. *Journal of Marketing*, *67*(1), 46–62. doi:10.1509/jmkg.67.1.46.18591

Mayer, D., Kuenzi, M., Greenbaum, R., Bardes, M., & Salvador, R. B. (2009). How low does ethical leadership flow? Test of a trickle-down model. *Organizational Behavior and Human Decision Processes*, *108*(1), 1–13. doi:10.1016/j.obhdp.2008.04.002

McCauley, J. L. (2009). *Dynamics of markets: The new financial economics*. Cambridge University Press. doi:10.1017/CBO9780511805448

Mclean, L. D. (2005). Organizational culture's influence on creativity and innovation: A review of the literature and implications for human resource development. *Advances in Developing Human Resources*, *7*(2), 226–246. doi:10.1177/1523422305274528

McLeary, C. N., & Cruise, P. A. (2015). A context-specific model of organizational trust – an examination of cognitive and socio-affective trust determinants in unique cultural settings. *Cross Cultural Management*, *22*(2), 297–302. doi:10.1108/CCM-11-2013-0180

McWilliams, A., & Siegel, D. (2001). Corporate social responsibility: A theory of the firm perspective. *Academy of Management Review*, *26*(1), 117–127.

Meindl, J. R., & Ehrlich, S. B. (1987). The romance of leadership and the evaluation of organizational performance. *Academy of Management Journal*, *30*(1), 1, 91–109. doi:10.2307/255897

Meindl, J. R., Ehrlich, S. B., & Dukerich, J. M. (1985). The Romance of Leadership. *Administrative Science Quarterly*, *30*(1), 78–102. doi:10.2307/2392813

Mendes, R. M., & Miskulin, R. G. S. (2017). A análise de conteúdo como uma metodologia [Content analysis as a methodology]. *Cadernos de Pesquisa, 47(*165), 1044-1066. https://dialnet.unirioja.es/servlet/oaiart?codigo=6132756

Mercier, M., Vinchon, F., Pichot, N., Bonetto, E., Bonnardel, N., Girandola, F., & Lubart, T. (2021). COVID-19: A Boon or a Bane for Creativity? *Frontiers in Psychology*, *11*, 1–10. doi:10.3389/fpsyg.2020.601150 PMID:33536973

Merz, P., & Freisleben, B. (2001). Memetic algorithms for the traveling salesman problem. *Complex Systems*, *13*, 297–345.

Mexico's Economy Ministry. (2017). Inversión Extranjera Directa en México y el mundo. *Carpeta de Información Estadística.* http://www.economia.gob.mx/files/comunidad_negocios/ied/analisis_publicaciones/Otros por ciento20estudios/carpeta_informacion_estadistica_1115.pdf

Meyer, L., Castillo, R., Buchgeister, J., & Tsatsaronis, G. (2009). Application of exergoeconomicand exergoenvironmental analysis to an SOFC system with an allothermal biomass gasifier. *International Journal of Thermodynamics*, *12*(4), 177–186.

Meyerstein, A. (2015). Are the equator principles greenwash or game changers? effectiveness, transparency and future challenges. In S. O. Idowu, & R. Schmidpeter (Eds.), Responsible investment banking: Risk management frameworks, sustainable financial innovation and softlaw standards (pp. 267-284). Springer International Publishing. doi:10.1007/978-3-319-10311-2_16

Michalewicz, Z. (1996). *Genetic algorithms + data structures = evolution programs* (3rd ed.). Springer-Verlag. doi:10.1007/978-3-662-03315-9

Midura, M. (2013). John Vs. Jennifer: A Battle of the Sexes. *Yale Scientific*. Retrieved from https://www.yalescientific.org/2013/02/john-vs-jennifer-a-battle-of-the-sexes/

Mikusová, M., & Horváthová, P. (2019). Prepared for a crisis? Basic elements of crisis management in an organisation. *Economic Research-Ekonomska Istrazivanja*, *32*(1), 1844–1868. doi:10.1080/1331677X.2019.1640625

Miles, J. A., & Naumann, S. E. (2021). Gender, sexual orientation, academic self-concept and leadership. *Higher Education, Skills and Work-Based Learning, 11*(5), 983-1001. doi:10.1108/HESWBL-05-2020-0101

Miles, M. P., & Arnold, D. R. (1991). The relationship between marketing orientation and entrepreneurial orientation. *Entrepreneurship Theory and Practice*, *15*(4), 49–65. doi:10.1177/104225879101500407

Miles, M., & Darroch, J. (2008). A commentary on current research at the marketing and entrepreneurship interface. *Journal of Small Business Management*, *46*(1), 46–49. doi:10.1111/j.1540-627X.2007.00230.x

Miller, D. (1983). Revisited a reflection on EO research and some suggestions for the future. *Entrepreneurship Theory and Practice*, *35*(5), 873–894. doi:10.1111/j.1540-6520.2011.00457.x

Miller, P. W. (2019). 'Tackling' race inequality in school leadership: Positive actions in BAME teacher progression – evidence from three English schools. *Educational Management Administration & Leadership*, *48*(6), 986–1006. doi:10.1177/1741143219873098

Milliman, J., Czaplewski, A. J., & Ferguson, J. (2003). Workplace spirituality and employee work attitudes: An exploratory empirical assessment. *Journal of Organizational Change Management*, *16*(4), 426–447. doi:10.1108/09534810310484172

Mills, J. P., & Boardley, I. D. (2016). Expert Premier League soccer managers' use of transformational leadership behaviours and attitude towards sport integrity: An intrinsic case study. *International Journal of Sports Science & Coaching, 11*(3), 382–394. doi:10.1177/1747954116645011

Mintzberg, H., Simons, R., & Kunal, B. (2002). Beyond Selfishness. *MIT Sloan Management Review, 44*(1), 67–74.

Miron-Spektor, E., & Erez, M. (2017). *Looking at creativity through a paradox lens.* Oxford University Press.

Mirvis, P., Herrera, M., Googins, B., & Albaredad, L. (2016). Corporate social innovation: How firms learn to innovate for the greater good. *Journal of Business Research, 69*(11), 5014–5021. doi:10.1016/j.jbusres.2016.04.073

Mohanty, S., Patra, P. K., Mohapatra, S., & Ray, M. (2017). MPSO: A Novel Meta-Heuristics for Load Balancing in Cloud Computing. *International Journal of Applied Evolutionary Computation, 8*(1), 1–25. doi:10.4018/ijaec.2017010101

Molinillo, S. & Viano-Pastor, A. (2015). El papel del smartphone en la experiencia de compra offline. *Dos Algarves: A Multidisciplinary e-Journal, 26*(2), 68-86.

Moodley, M., Schorn, P. M., Walthew, D. C., & Masinga, P. (2002). Optimising the carbonation process. *Proc S Afr Sug Technol Ass, 76*, 469–476.

Moon, Y. S., & Franke, G. R. (2000). Cultural influences on agency practitioners' ethical perceptions: A comparison of Korea and the U.S. *Journal of Advertising, 26*(1), 51–65. doi:10.1080/00913367.2000.10673603

Moreno-Arrebola, R., Martinez Martinez, A., Zurita Ortega, F., & San Roman Mata, S. (2017). La influencia del liderazgo del entrenador en los deportistas. *European Journal of Child Development. Education and Psychopathology, 5*(2), 73. doi:10.30552/ejpad.v5i2.49

Morgan, G. (1995). *Imagens da organização.* Atlas.

Morgan, T., Anokhin, S., Kretinin, A., & Frishammar, J. (2015). The dark side of the entrepreneurial orientation and market orientation interplay: A new product development perspective. *International Small Business Journal, 33*(7), 731–751. doi:10.1177/0266242614521054

Morina, F., Hysa, E., Ergün, U., Panait, M., & Voica, M. C. (2020). The Effect of Exchange Rate Volatility on Economic Growth: Case of the CEE Countries. *Journal of Risk and Financial Management, 13*(8), 177.

Morris, M. W., & Leung, K. (2010). Creativity east and west: Perspectives and parallels. *Management and Organization Review, 6*(3), 313–327. doi:10.1111/j.1740-8784.2010.00193.x

Mrzljak, V., Oroviæ, J., Poljak, I., & Lorencin, I. (2019). Exergy analysis of steam turbine governing valve from a super critical thermal power plant. In *XXVII International Scientific Conference Trans & Motauto*'19 (pp. 99-102). Scientific-Technical Union of Mechanical Engineering, Industry-4.0.

Mukherjee, A. (2012). CSR Holiday for Banks in India: Dilemma between CSR and Financial Inclusion. *Interdisciplinary Journal of Management and Behavioural Sciences, 1*(4).

Mulki, J. P., Jaramillo, J. F., & Locander, W. B. (2009). Critical role of leadership on ethical climate and salesperson performance. *Journal of Business Ethics, 86*(2), 125–141. doi:10.100710551-008-9839-4

Mulyania, T., Wahyono, M., & Yanto, H. (2021). The Effect of Leadership on Performance Through Work Involvement Based on Engineering Process Values On Educational Quality Assurance School Team in Central Java Indonesia. *Turkish Journal of Computer and Mathematics Education, 12*(13), 7037–7047.

Mumford, M. D., Todd, E. M., Higgs, C., & McIntosh, T. (2017). Cognitive skills and leadership performance: The nine critical skills. *The Leadership Quarterly, 28*(1), 24–39. doi:10.1016/j.leaqua.2016.10.012

Murata, T., Ishibuchi, H., & Tanaka, H. (1996). Genetic algorithms for flowshop scheduling problems. *Computers & Industrial Engineering, 30*(4), 1061–1071. doi:10.1016/0360-8352(96)00053-8

Mustapha, N. D., Jali, M. N., & Muhamad, N. (2021). An Overview of Corporate Social Innovation (CSI). *International Journal of Academic Research in Business & Social Sciences, 11*(7), 691–698. doi:10.6007/IJARBSS/v11-i7/10527

Naranjo-Valencia, J. C., Jiménez-Jiménez, D., & Sanz-Valle, R. (2016). Studying the links between organizational culture, innovation, and performance in Spanish companies. *Revista Latinoamericana de Psicología, 48*(1), 30–41. doi:10.1016/j.rlp.2015.09.009

Narver, J. C., & Slater, S. F. (1990). The effect of a market orientation on business profitability. *Journal of Marketing, 54*(4), 20–34. doi:10.1177/002224299005400403

Naspetti, S., Alberti, F., & Solfanelli, F. (2012). Quality determinants in the organic cheese supply chain: A Quality Function Deployment approach. *New Medit, 11*(4), 61–64.

Naspetti, S., Alberti, F., & Solfanelli, F. (2015). Quality function deployment in the organic animal food sector: Application to poultry meat. *Italian Journal of Animal Science, 14*(3), 544–550. doi:10.4081/ijas.2015.4050

Nawaz, M., Enscore, E. E. Jr, & Ham, I. (1983). A heuristic algorithm for the m-machine, n-job flow-shop sequencing problem. *Omega International Journal of Management Science, 11*(1), 91–95. doi:10.1016/0305-0483(83)90088-9

Nazariana, A., Atkinsonb, P., & Foroudi, P. (2017). Influence of national culture and balanced organizational culture on the hotel industry's performance. *International Journal of Hospitality Management, 63*, 22–32. doi:10.1016/j.ijhm.2017.01.003

Neapolitan, R. E., & Naimipour, K. (2003). *Foundations of Algorithms Using C++ Pseudo-code*. Jones & Bartlett Publishers.

Nembhard, I. M., Burns, L. R., & Shortell, S. M. (2020). Responding to Covid-19: Lessons from management research. *NEJM Catalyst Innovations in Care Delivery, 1*(2), 1–7.

Newbold, D., & Koppel, K. (2018). Carbonated dairy beverages: Challenges and opportunities. *Beverages, 4*(3), 66. doi:10.3390/beverages4030066

Nguyen, H., & Mohamed, S. (2011). Leadership behaviors, organizational culture and knowledge management practices: An empirical investigation. *Journal of Management Development, 30*(2), 206–221. doi:10.1108/02621711111105786

Nicholls, A., Simon, J., & Gabriel, M. (2015). Dimensions of Social Innovation. *New Frontiers in Social Innovation Research*, 1–26.

Nicolau, J. A., & Paranhos, J. (2006). Notas sobre o conceito de inovação. *Textos de Economia, 9*(1), 23–37.

Nijstad, B. A., De Dreu, C. K., Rietzschel, E. F., & Baas, M. (2010). The dual pathway to creativity model: Creative ideation as a function of flexibility and persistence. *European Review of Social Psychology, 21*(1), 34–77. doi:10.1080/10463281003765323

Nishi, T., Ohno, A., Inuiguchi, M., Takahashi, S., & Ueda, K. (2017). A Combined column generation and heuristics for railway short-term rolling stock planning with regular inspection constraints. *Computers & Operations Research, 81*, 14–25. doi:10.1016/j.cor.2016.11.025

Noja, G. G., Thalassinos, E., Cristea, M., & Grecu, I. M. (2021). The Interplay between Board Characteristics, Financial Performance, and Risk Management Disclosure in the Financial Services Sector: New Empirical Evidence from Europe. *Journal of Risk and Financial Management, 14*(2), 79.

Northouse, P. G. (2015). *Leadership: Theory and Practice*. SAGE Publications.

Nulty, D. (2008). The adequacy of response rates to online and paper surveys: What can be done? *Assessment & Evaluation in Higher Education*, *33*(3), 301–314. doi:10.1080/02602930701293231

O'Reilly, C. A. III, Caldwell, D. F., Chatman, J. A., & Doerr, B. (2014). the promise and problems of organizational culture: CEO personality, culture, and firm performance. *Group & Organization Management*, *39*(6), 595–625. doi:10.1177/1059601114550713

Okereke, C. J., Lasode, O. A., & Ohijeagbon, I. O. (2020a). Exergoeconomic analysis of an industrial beverage mixer system. *Heliyon*, *6*. 10.2020.e04402 doi:.1016/j.heliyon

Okereke, C. J., Lasode, O. A., & Ohijeagbon, I. O. (2020b). Exergoeconomic analysis of an industrial beverage mixer system: Process data. *Data in Brief*, *32*, 106125. doi:10.1016/j.dib.2020.106125 PMID:32904336

Okereke, C. J., Ohijeagbon, I. O., & Lasode, O. A. (2019). Energy and exergy analysis of vapor compression refrigeration system with flooded evaporator. *International Journal of Air-Conditioning and Refrigeration*, *27*(04), 1950041. doi:10.1142/S201013251950041X

Olajire, A. A. (2013). A review of mineral carbonation technology in sequestration of CO_2. *Journal of Petroleum Science Engineering*, *109*, 364–392. doi:10.1016/j.petrol.2013.03.013

Oliver, I. M., Smith, D. J., & Holland, J. R. C. (1987), A study of permutation crossover operators on the traveling salesman problem. *Proceedings the 2nd International Conference on Genetic Algorithms*, 224-230.

Oni, O. (2016). *Banking on financial sector sustainability regulations*. University of Waterloo. https://uwspace.uwaterloo.ca/bitstream/handle/10012/10367/Oni_Olawuwo.pdf?isAllowed=y&sequence=1

Osellame, I. (2013). *Corporate social responsibility in Canadian banking a case study on the equator principles*. Academic Press.

Ospina, S., & Foldy, E. (2009). A critical review of race and ethnicity in the leadership literature: Surfacing context, power and the collective dimensions of leadership. *The Leadership Quarterly*, *20*(6), 876–896. doi:10.1016/j.leaqua.2009.09.005

Ostroff, C., Kinicki, A., & Muhammad, R. (2013). Organizational Culture and Climate. In Handbook of Psychology *(Vol 12*, pp. 643–676). John Wiley & Sons, Inc.

Ostrower, F. (1993). *Criatividade e processos de criação* (9th ed.). Vozes.

Palazzo, M., Vollero, A., & Siano, A. (2020). From strategic corporate social responsibility to value creation: An analysis of corporate website communication in the banking sector. *International Journal of Bank Marketing*.

Panait, M., Radulescu, I., & Brezoi, A. (2020). Financial Markets–Under the Sign of CSR. Some Evidences Regarding Financial Education. *LUMEN Proceedings*, *11*, 96-106.

Panait, M., Radulescu, I., & Voica, C. (2018). Financial Inclusion and Vulnerabilities Generated by the International Crisis. *Romanian Journal of Economics*, *47*(2), 71–81.

Papadimitriou, C. D., & Steiglitz, K. (1982). *Combinatorial Optimization: Algorithms and Complexity*. Prentice-Hall.

Park, H., & Kim, J. D. (2020). Transition towards green banking: Role of financial regulators and financial institutions. *Asian Journal of Sustainability and Social Responsibility*, *5*(1), 1–25.

Park, S. H., Ham, S., & Lee, M. A. (2012). How to improve the promotion of Korean beef barbecue, bulgogi, for international customers. An application of quality function deployment. *Appetite*, *59*(2), 324–332. doi:10.1016/j.appet.2012.05.008 PMID:22634190

Patterson, M. G., West, M. A., Shackleton, V. J., Dawson, J. F., Lawthom, R., Maitlis, S., Robinson, D. L., & Wallace, A. M. (2005). Validating the organizational climate measure: Links to managerial practices, productivity and innovation. *Journal of Organizational Behavior*, *26*(4), 379–408. doi:10.1002/job.312

Pereira, A. T., Santos, M. J., & Estevam, D. O. (2021). *Styles of leadership and perceptions of corporate social responsibility book: Sustainable Management for Managers and engineers*. John Wiley & Sons.

Pérez, A., & del Bosque, I. R. (2009). The social role of financial companies as a determinant of consumer behaviour. *International Journal of Bank Marketing*, *27*(6), 467–485.

Pérez, A., & del Bosque, I. R. (2015). Customer values and CSR image in the banking industry. *Journal of Financial Services Marketing*, *20*(1), 46–61.

Pérez, A., López-Gutiérrez, C., García-De Los Salmones, M. D. M., & San-Martín, P. (2020). Stakeholder salience, positive CSR news and the market value of banks. Spanish Journal of Finance and Accounting. *Revista Española de Financiación y Contabilidad*, *49*(4), 483–502.

Peters, A., McEwen, B. S., & Friston, K. (2017). Uncertainty and stress: Why it causes diseases and how it is mastered by the brain. *Progress in Neurobiology*, *156*, 164–188. doi:10.1016/j.pneurobio.2017.05.004 PMID:28576664

Peterson, D. K. (2004). The relationship between perceptions of corporate citizenship and organizational commitment. *Business & Society*, *43*(3), 296–319. doi:10.1177/0007650304268065

Pham, S. T. (2013). Experimental Investigation and Modelling of Carbonation Process in Cement Materials. *The Open Civil Engineering Journal*, *7*(1), 116–125. doi:10.2174/1874149520130815006

Phillips, R., Freeman, R. E., & Wicks, A. C. (2003). What stakeholder theory is not. *Business Ethics Quarterly*, *13*(4), 479–502. doi:10.5840/beq200313434

Pindyck, R. S., Rubinfeld, D. L., & Rabasco, E. (2013). *Microeconomia*. Pearson Italia.

Pinedo, M., Zacharias, C., & Zhu, N. (2015). Scheduling in the service industries: An overview. *Journal of Systems Science and Systems Engineering*, *24*(1), 1–48. doi:10.100711518-015-5266-0

Pinto, J. (2019). Key to effective organizational performance management lies at the intersection of paradox theory and stakeholder theory. *International Journal of Management Reviews*, *21*(2), 185–208. doi:10.1111/ijmr.12199

Plucker, J. A., Beghetto, R. A., & Dow, G. T. (2004). Why isn't creativity more important to educational psychologists? Potentials, pitfalls, and future directions in creativity research. *Educational Psychologist*, *39*(2), 83–96. doi:10.120715326985ep3902_1

Porter, M. E., & Kramer, M. R. (2006). The link between competitive advantage and corporate social responsibility. *Harvard Business Review*, *84*(12), 78–92. PMID:17183795

Porter, M. E., & Kramer, M. R. (2011). The big idea: Creating shared value. *Harvard Business Review*, *89*, 2–17.

Pothukuchi, V., Damanpour, F., Choi, J., Chen, C. C., & Ho Park, S. (2002). National and Organizational Culture Differences and International Joint Venture Performance. *Journal of International Business Studies*, *33*(2), 243–256. doi:10.1057/palgrave.jibs.8491015

Prim, R. C. (1957). Shortest connection networks and some generalizations. *The Bell System Technical Journal*, *36*(6), 1389–1401. doi:10.1002/j.1538-7305.1957.tb01515.x

Puffer, S. M. (1990). Attributions of charismatic leadership: The impact of decision style, outcome, and observer characteristics. *The Leadership Quarterly*, *1*(3), 177–192. doi:10.1016/1048-9843(90)90019-E

Pulleyblank, W. R. (Ed.). (2014). *Progress in combinatorial optimization*. Academic Press.

Puranam, P. (2018). *The microstructure of organizations*. Oxford University Press. doi:10.1093/oso/9780199672363.001.0001

Quélin, B. V., Cabral, S., Lazzarini, S., & Kivleniece, I. (2019). The private scope in public-private collaborations: An institutional and capability-based perspective. *Organization Science*, *30*(4), 831–846. doi:10.1287/orsc.2018.1251

Querol, E., Gonzalez-Regueral, B., & Perez-Benedito, J. L. (2013). *Exergy concept and determination. In Practical Approach to Exergy and Thermoeconomic Analysis of Industrial Process*. Springer. doi:10.1007/978-1-4471-4622-3_1

Qu, L., & Sun, R. (1999). A synergetic approach to genetic algorithms for solving travelling salesman problem. *Information Sciences*, *117*(3-4), 267–283. doi:10.1016/S0020-0255(99)00026-2

Radomska, J., & Wołczek, P. (2020). Integrative perspective on ambidexterity, creativity and networking: Literature overview. *European Research Studies Journal*, *23*(3), 31–49. doi:10.35808/ersj/1623

Raguž, I., & Zekan, S. (2017). Organizational culture and leadership style: key factors in the organizational adaptation process. *Ekonomski Vjesnik/Econviews - Review of Contemporary Business, Entrepreneurship and Economic Issues*, *30*(1). Retrieved from https://hrcak.srce.hr/ojs/index.php/ekonomski-vjesnik/article/view/4603

Raimi, L., Adelopo, A. O., & Yusuf, H. (2019) Corporate social responsibility and sustainable management of solid wastes and effluents in Lagos megacity Nigeria. *Social Responsibility Journal*. Doi.org/10.1108/SRJ-09-2018-0239

Raimi, L., Panait, M., & Hysa, E. (2021). Financial Inclusion in ASEAN Countries–A Gender Gap Perspective and Policy Prescriptions. *LUMEN Proceedings, 15*, 38-55.

Raimi, L. (2020). What are the Motivations for Undertaking Corporate Social Responsibility (CSR) Practices in the Middle-East? A Critical Literature Review. *Global Journal of Human Social Science: E-Economics, 20*(6), 58–64.

Raimi, L., Akhuemonkhan, I. A., & Ogunjirin, O. D. (2015). Corporate social responsibility and entrepreneurship (CSRE): Antidotes to poverty, insecurity and underdevelopment in Nigeria. *Social Responsibility Journal*, *11*(1), 56–81.

Raimi, L., & Isiaka, A. S. (2021). Regional Review: Africa: CSR in Sub-Saharan Africa: Emerging Trends, Perspectives, and Debates. In D. Crowther & S. Seifi (Eds.), *The Palgrave Handbook of Corporate Social Responsibility*. Palgrave Macmillan.

Rajendran, C., & Ziegler, H. (2004). Ant-colony algorithms for permutation flowshop scheduling to minimize makespan/total flowtime of jobs. *European Journal of Operational Research*, *155*(2), 426–438. doi:10.1016/S0377-2217(02)00908-6

Randolph-Seng, B., Cogliser, C. C., Randolph, A. F., Scandura, T. A., Miller, C. D., & Smith-Genthôs, R. (2016). Diversity in leadership: Race in leader-member exchanges. *Leadership and Organization Development Journal*, *37*(6), 750–773. doi:10.1108/LODJ-10-2014-0201

Rashkovskiy, S. A. (2021). Thermodynamics of markets. *Physica A*, *567*, 125699. doi:10.1016/j.physa.2020.125699

Ratten, V. (2020). Coronavirus (covid-19) and entrepreneurship: Changing life and work landscape. *Journal of Small Business and Entrepreneurship*, *32*(5), 503–516. doi:10.1080/08276331.2020.1790167

Ray, S., & Ray, I. (2012). Understanding Democratic Leadership: Some Key Issues And Perception. With Reference To India's Freedom Movement. *Afro Asian Journal of Social Sciences, 3*(1), 1-26.

Reeves, C. R. (1995). A genetic algorithm for flowshop sequencing. *Computers & Operations Research*, *22*(1), 5–13. doi:10.1016/0305-0548(93)E0014-K

Reeves, C. R. (1997). Genetic algorithms for the operations researcher. *INFORMS Journal on Computing*, *9*(3), 231–250. doi:10.1287/ijoc.9.3.231

Reeves, C. R., & Yamada, T. (1998). Genetic algorithms, path relinking, and the flow-shop sequencing prob-lem. *Evolutionary Computation*, *6*(1), 45–60. doi:10.1162/evco.1998.6.1.45 PMID:10021740

Rego, A., Mota, T., Cunha, M., & Seabra, A. (2020). *O treinador como "dominador e amigo"? Uma abordagem ao amor duro na liderança* [The coach as "dominator and friend"? An approach to tough love in leadership]. Católica Porto Business School. Available at https://www.catolicabs.porto.ucp.pt/Files/PortalReady/v000/Docs/PDF/amor_duro.pdf

Rehman, N. U., Hysa, E., & Mao, X. (2020). Does public R&D complement or crowd-out private R&D in pre and post economic crisis of 2008? *Journal of Applied Econometrics*, *23*(1), 349–371.

Reinelt, G. (1991). TSPLIB - a traveling salesman problem library. *ORSA Journal on Computing*, *3*(4), 376–384. doi:10.1287/ijoc.3.4.376

Režek Jambrak, A., Šimunek, M., Grbeš, F., Mandura, A., & Djekic, I. (2018). Analysis of apple beverages treated with high-power ultrasound: A quality function deployment approach. *Journal of the Science of Food and Agriculture*, *98*(6), 2258–2266. doi:10.1002/jsfa.8714 PMID:28981162

Richards, R. (2007). Everyday creativity and the arts. *World Futures*, *63*(7), 500–525. doi:10.1080/02604020701572707

Richards, R., Kinney, D. K., Lunde, I., Benet, M., & Merzel, A. P. (1988). Creativity in manic-depressives, cyclothymes, their normal relatives, and control subjects. *Journal of Abnormal Psychology*, *97*(3), 281–288. doi:10.1037/0021-843X.97.3.281 PMID:3192819

Richins, M. L. (1983). Negative word-of-mouth by dissatisðed consumers: A pilot study. *Journal of Marketing*, *47*(4), 68–78. doi:10.1177/002224298304700107

Richmond, P., Mimkes, J., & Hutzler, S. (2013). *Econophysics and physical economics*. Oxford University Press. doi:10.1093/acprof:oso/9780199674701.001.0001

Riketta, M. (2002). Attitudinal organizational commitment and job performance: A meta-analyis. *Journal of Organizational Behavior*, *23*(3), 257–266. doi:10.1002/job.141

Riley, K. F., Hobson, M. P., & Bence, S. J. (1999). *Mathematical methods for physics and engineering*. Academic Press.

Ritchey, F. J. (2008). Estadística para las Ciencias Sociales. Editorial McGraw Hill.

Rivera-Santos, M., Rufn, C., & Kolk, A. (2012). Bridging the institutional divide: Partnerships in subsistence markets. *Journal of Business Research*, *65*(12), 1721–1727. doi:10.1016/j.jbusres.2012.02.013

Robertson, B. (2011). *Corporate social responsibility programs for emissions-trading risk management in Canadian banks*. Springer Berlin Heidelberg. doi:10.1007/978-3-642-20592-7

Robertson, C. J., & Crittenden, W. F. (2003). Mapping Moral Philosophies: Strategic Implications for Multinational Firms. *Strategic Management Journal*, *24*(4), 385–392. doi:10.1002mj.300

Rogers, D. M., & Rempe, S. B. (2012). Irreversible thermodynamics. *Journal of Physics: Conference Series*, *402*(1), 012014. doi:10.1088/1742-6596/402/1/012014

Rosado Júnior, G. A., Lobato, J. F. P., Echeveste, M., Gerber, A. S., Yang, S., & Rossini, K. (2011). Identification of demand characteristics in the production of sires using a conceptual model of quality function deployment: A case study. *Revista Brasileira de Zootecnia*, *40*(1), 210–220. doi:10.1590/S1516-35982011000100029

Ruiz, A. P. (2018). Reacciones hacia la publicidad de responsabilidad social corporativa de entidades financieras. *UCJC Business & Society Review*, (60), 18-53.

Ruiz, R., & Maroto, C. (2005). A comprehensive review and evaluation of permutation flowshop heuristics. *European Journal of Operational Research, 165*(2), 479–494. doi:10.1016/j.ejor.2004.04.017

Ruiz, R., Maroto, C., & Alcaraz, J. (2006). Two new robust genetic algorithms for the flowshop scheduling problem. *Omega the International Journal of Management Science, 34*(5), 461–476. doi:10.1016/j.omega.2004.12.006

Ruiz, R., & Stützle, T. (2007). A simple and effective iterated greedy algorithm for the permutation flow shop problem. *European Journal of Operational Research, 177*(3), 2033–2049. doi:10.1016/j.ejor.2005.12.009

Runco, M. A., Illies, J. J., & Eisenman, R. (2005). Creativity, originality, and appropriateness: What do explicit instructions tell us about their relationships? *The Journal of Creative Behavior, 39*(2), 137–148. doi:10.1002/j.2162-6057.2005.tb01255.x

Runco, M. A., & Jaeger, G. J. (2012). The standard definition of creativity. *Creativity Research Journal, 24*(1), 92–96. doi:10.1080/10400419.2012.650092

Runco, M. A., & Okuda, S. M. (1988). Problem discovery, divergent thinking, and the creative process. *Journal of Youth and Adolescence, 17*(3), 211–220. doi:10.1007/BF01538162 PMID:24277636

Rupp, D. E., & Mallory, D. B. (2015). Corporate social responsibility: Psychological, person-centric, and progressing. *Annual Review of Organizational Psychology and Organizational Behavior, 2*(1), 211–236. doi:10.1146/annurev-orgpsych-032414-111505

Russell, S. V., Evans, A., Fielding, K. S., & Hill, C. (2016). Turn it off: An action research study of top management influence on energy conservation in the workplace. *Frontiers in Psychology, 7*, 389. doi:10.3389/fpsyg.2016.00389 PMID:27047417

Sahui, J. A. (2008). *Factores que influyen en la conducta del consumidor. Una aproximación desde las ciencias sociales.* Universidad Modelo.

Sahu, O. (2018). Assessment of sugarcane industry: Suitability for production, consumption, and utilization. *Annals of Agrarian Science, 16*(4), 389–395. doi:10.1016/j.aasci.2018.08.001

Saka-Helmhout, A., Chappin, M., & Rodrigues, S. (2021). Corporate Social Innovation in Developing Countries. *Journal of Business Ethics*. Advance online publication. doi:10.100710551-021-04933-x

Salman, Z. (2011). Leadership Styles and Employee Performance (A case Study of Shakarganj Mills Limited Jhang Pakistan). *Interdisciplinary of Contemporary Research in Business, 3*(6), 257–267.

Sánchez, F., Sandoval, A., Rodríguez-Pomeda, J., & Casani, F. (2020). Professional aspirations as indicators of responsible leadership style and corporate social responsibility. Are we training the responsible managers that business and society need? A cross-national study. *Journal of Work and Organizational Psychology, 36*(1), 49–61. doi:10.5093/jwop2020a5

Sánchez, R., Iniesta, M. A., & Holbrook, M. B. (2009). The Conceptualisation and Measurement of Consumer Value in Services. *International Journal of Market Research, 51*(1), 93–113. doi:10.2501/S1470785308200328

Santos, A. B., Fazion, C. B., & De Meroe, G. P. (2011). Inovação: Um estudo sobre a evolução do conceito de Schumpeter. *Caderno de Administração Revista da Faculdade de Administração da FEA, 5*(1), 1–16.

Sanyal, R. (2005). Determinates of bribery in international business: The cultural and economic factors. *Journal of Business Ethics, 59*(1-2), 139–145. doi:10.100710551-005-3406-z

Sarkar, S. (2007). *Innovation, market archetypes and outcome: An integrated framework*. Springer Science & Business Media.

Saslow, W. M. (1999). An economic analogy to thermodynamics. *American Journal of Physics*, *67*(12), 1239–1247. doi:10.1119/1.19110

Saul, J. (2011). *Social innovation, inc.: 5 strategies for driving business growth through social change*. Wiley.

Šavija, B., & Lukovic, M. (2016). Carbonation of cement paste: Understanding, challenges, and opportunities. *Construction & Building Materials*, *117*, 285–301. doi:10.1016/j.conbuildmat.2016.04.138

Sawyer & Dickson, P. R. (1987). Psychological Perspectives on Consumer Response to Sales Promotion. In Research on Sales Promotion: Collected Papers. Cambridge, MA: Marketing Science Institute.

Sayadi, S., Erraach, Y., & López, C. P. (2017). Translating consumer's olive-oil quality-attribute requirements into optimal olive-growing practices: A quality function deployment (QFD) approach. *British Food Journal*, *119*(1), 190–214. doi:10.1108/BFJ-05-2016-0228

Schechter, L. (1984). A Normative Conception of Value, Progressive Grocer. *Executive Report*, 12-14.

Scheidler, S., Edinger-Schons, L. M., Spanjol, J., & Wieseke, J. (2019). Scrooge posing as MotherTeresa: How hypocritical social responsibility strategies hurt employees and firms. *Journal of Business Ethics*, *157*(2), 339–358.

Schein, E. H. (1985). *Organizational culture and leadership*. Jessey Bess.

Schein, E. H. (1996). Culture: The missing concept in organization studies. *Administrative Science Quarterly*, *2*(41), 229–240. doi:10.2307/2393715

Schein, E. H. (2010). *Organizational Culture and Leadership*. Jossey-Bass.

Schepers, D. H. (2011). The equator principles: A promise in progress? *Corporate Governance (Bradford)*, *11*(1), 90–106. doi:10.1108/14720701111108871

Schiavo, G., Korzenowski, A. L., Batista, E. R. S., de Souza, D. L., & Scavarda, A. (2018). Customers' quality demands as directions to the cold chicken supply chain management. *Business Process Management Journal*, *24*(3), 771–785. doi:10.1108/BPMJ-11-2016-0224

Schillebeeckx, S. J., Lin, Y., & George, G. (2019). When do expert teams fail to create impactful inventions? *Journal of Management Studies*, *56*(6), 1073–1104. doi:10.1111/joms.12447

Schindehutte, M., Morris, M. H., & Kocak, A. (2008). Understanding market-driving behavior: The role of entrepreneurship. *Journal of Small Business Management*, *46*(1), 4–26. doi:10.1111/j.1540-627X.2007.00228.x

Schminke, M., Ambrose, M. L., & Neubaum, D. O. (2005). The effect of leader moral development on ethical climate and employee attitudes. *Organizational Behavior and Human Decision Processes*, *97*(2), 135–151. doi:10.1016/j.obhdp.2005.03.006

Schmitt, L. J., & Amini, M. M. (1998). Performance characteristics of alternative genetic algorithmic approaches to the traveling salesman problem using path representation: An empirical study. *European Journal of Operational Research*, *108*(3), 551–570. doi:10.1016/S0377-2217(97)00206-3

Scholtens, B., & Dam, L. (2007). Banking on the equator. are banks that adopted the equator principles different from non-adopters? *World Development*, *35*(8), 1307–1328. doi:10.1016/j.worlddev.2006.10.013

Scholtens, B., & Dam, L. (2007). Cultural values and international differences in business ethics. *Journal of Business Ethics*, *75*(3), 273–284. doi:10.100710551-006-9252-9

Schumpeter, J. A. (1942). *Capitalism, Socialism and Democracy*. Harper & Row.

Schumpeter, J. A. (1988). *Capitalismo, sociedade e democracia*. Abril Cultural.

Sciubba, E. (2005). Exergo-economics: Thermodynamic foundation for a more rational resource use. *International Journal of Energy Research*, *29*(7), 613–636. doi:10.1002/er.1096

Sergeant, A., & Frenkel, S. (2000). When do customer contact employees satisfy customers? *Journal of Service Research*, *3*(1), 18–34. doi:10.1177/109467050031002

Shahzad, F., Luqman, A., Khan, A., & Shabbir, L. (2012). Impact of organizational culture on organizational performance: An overview. *Interdisciplinary Journal of Contemporary Research in Business*, *3*(9), 975–985.

Shalley, C. E., & Gilson, L. L. (2004). What leaders need to know: A review of social and contextual factors that can foster or hinder creativity. *The Leadership Quarterly*, *15*(1), 33–53. doi:10.1016/j.leaqua.2003.12.004

Shalley, C. E., Gilson, L. L., & Blum, T. C. (2000). Matching creativity requirements and the work environment: Effects on satisfaction and intentions to leave. *Academy of Management Journal*, *43*(2), 215–223.

Shalley, C. E., Zhou, J., & Oldham, G. R. (2004). The effects of personal and contextual characteristics on creativity: Where should we go from here? *Journal of Management*, *30*(6), 933–958. doi:10.1016/j.jm.2004.06.007

Shamir, B., House, R. J., & Arthur, M. B. (2018). The motivational effects of charismatic leadership: A self-concept based theory. In I. Katz, G. Eilam-Shamir, R. Kark, & Y. Berson (Eds.), *Leadership Now: Reflections on the Legacy of Boas Shamir* (pp. 9–29). Emerald Publishing Limited. doi:10.1108/S1479-357120180000009009

Shamsi, S., & Omidkhah, M. R. (2012). Optimization of steam pressure levels in a total site using a thermoeconomic method. *Energies*, *5*(3), 702–717. doi:10.3390/en5030702

Shao, Y., Nijstad, B. A., & Täuber, S. (2019). Creativity under workload pressure and integrative complexity: The double-edged sword of paradoxical leadership. *Organizational Behavior and Human Decision Processes*, *155*, 7–19. doi:10.1016/j.obhdp.2019.01.008

Sharkie, R. (2009). Trust in Leadership is vital for Employee Performance. *Management Research News*, *32*(5), 491–498. doi:10.1108/01409170910952985

Sharma, L., & Singh, S. (2013). A Study on the Democratic Style of Leadership. *International Journal of Management & Information Technology*, *3*(2), 54–57. doi:10.24297/ijmit.v3i2.1367

Sheibani, K. (2004). An adaptive greedy genetic algorithm using fuzzy reasoning for the travelling salesman problem. *Proc. Hybrid Metaheuristics, 16th European Conf. on Artificial Intelligence*, 31-49.

Sheibani, K. (2005). *Fuzzy Greedy Evaluation in Search, Optimisation, and Learning* [PhD thesis]. London Metropolitan University, London, UK.

Sheibani, K. (2008). Fuzzy Greedy Search in Combinatorial Optimisation. Tadbir Institute for Operational Research, Systems Design, and Financial Services.

Sheibani, K. (2013). A constructive heuristic for permutation flow-shop scheduling with the makespan criterion: in a particular case where the sum of the processing times of each job is the same on each machine. *Proc. 5th Intl. Conf. on Applied Operational Research*, 18-24.

Sheibani, K. (2006). *Fuzzy greedy search and job-shop problem.* Proc. the UK Planning & Scheduling Special Interest Group.

Sheibani, K. (2010). A fuzzy greedy heuristic for permutation flow-shop scheduling. *The Journal of the Operational Research Society*, *61*(5), 813–818. doi:10.1057/jors.2008.194

Sheibani, K. (2010). An evolutionary approach using fuzzy greedy initialization to permutation flow-shop scheduling with the makespan criterion. *Proc. IEEE Intl. Conf. on Industrial Engineering and Engineering Management*, 638-641. 10.1109/ICIEEM.2010.5646535

Sheibani, K. (2010). The fuzzy greedy search in combinatorial optimization with specific reference to the travelling salesman problem. *Proc. IEEE Intl. Conf. on Industrial Engineering and Engineering Management*, 1367-1370. 10.1109/IEEM.2010.5674393

Sheibani, K. (2011). A hybrid metaheuristic using fuzzy greedy search operator for combinatorial optimization with specific reference to the travelling salesman problem. *Iranian Journal of Operations Research*, *2*(2), 63–71.

Sheibani, K. (2017). Fuzzy greedy search heuristic for combinatorial optimization with specific application to machine scheduling. *International Journal of Management Science and Engineering Management*, *12*(2), 148–153. doi:10.1080/17509653.2016.1172993

Sheibani, K. (2018). A very fast heuristic for combinatorial optimization with specific application to priority rule sequencing in operations management: Fuzzy greedy search. *International Journal of Operations Research and Information Systems*, *9*(3), 77–89. doi:10.4018/IJORIS.2018070104

Sheibani, K. (2019). An annotated bibliography of fuzzy greedy search. *Operational Research and Management Science Letters*, *2*(1), 2–5.

Sheibani, K. (2019). An integrated heuristic for machine sequencing with specific reference to the permutation flow-shop scheduling problem. *International Journal of Strategic Engineering*, *2*(2), 1–8. doi:10.4018/IJoSE.2019070101

Sheibani, K. (2021). Fuzzy greedy search: an algorithmic approach for combinatorial optimisation. In V. González-Prida & M. C. Carnero (Eds.), *Advanced Models and Tools for Effective Decision Making Under Uncertainty and Risk Contexts* (pp. 254–275). IGI Global. doi:10.4018/978-1-7998-3246-1.ch010

Shinyashiki, G. T., Trevizan, M. A., & Mendes, I. A. (2003). Sobre a criação e a gestão do conhecimento organizacional. *Revista Latino-Americana de Enfermagem*, *11*(4), 499–506. doi:10.1590/S0104-11692003000400013 PMID:14748169

Siano, A., Raimi, L., Palazzo, M., & Panait, M. C. (2020). Mobile Banking: An Innovative Solution for Increasing Financial Inclusion in Sub-Saharan African Countries. Evidence from Nigeria. *Sustainability*, *12*(23), 10130.

Siegel, S., & Castellan, N. J. (2012). *Estadística No Paramétrica: Aplicada a las ciencias de la conducta*. Editorial Trillas.

Silva, A. (2016). What is leadership? *Journal of Business Studies Quarterly*, *8*(1), 1–5. PMID:29355200

Silva, S. L. (2002). Informação e competitividade: A contextualização da gestão do conhecimento nos processos organizacionais. *Ciência da Informação*, *31*(2), 142–151. doi:10.1590/S0100-19652002000200015

Simonet, D., & Tett, R. (2013). Five perspectives on the leadership–management relationship: A competency-based evaluation and integration. *Journal of Leadership & Organizational Studies*, *20*(2), 199–213. doi:10.1177/1548051812467205

Simonton, D. K. (1977). Creative productivity, age, and stress: A biographical time-series analysis of 10 classical composers. *Journal of Personality and Social Psychology*, *35*(11), 791–804. doi:10.1037/0022-3514.35.11.791 PMID:915694

Simonton, D. K. (1991). Emergence and realization of genius: The lives and works of 120 classical composers. *Journal of Personality and Social Psychology*, *61*(5), 829–840. doi:10.1037/0022-3514.61.5.829

Simonton, D. K. (1994). Individual differences, developmental changes, and social context. *Behavioral and Brain Sciences*, *17*(3), 552–553. doi:10.1017/S0140525X00035925 PMID:23218117

Simonton, D. K. (1997). Creative productivity: A predictive and explanatory model of career trajectories and landmarks. *Psychological Review*, *104*(1), 66–89. doi:10.1037/0033-295X.104.1.66

Simonton, D. K. (1998). Fickle fashion versus immortal fame: Transhistorical assessments of creative products in the opera house. *Journal of Personality and Social Psychology*, *75*(1), 198–210. doi:10.1037/0022-3514.75.1.198

Simonton, D. K. (2000). Creativity: Cognitive, personal, developmental, and social aspects. *The American Psychologist*, *55*(1), 151–158. doi:10.1037/0003-066X.55.1.151 PMID:11392859

Simonton, D. K. (2004). Psychology's status as a scientific discipline: Its empirical placement within an implicit hierarchy of the sciences. *Review of General Psychology*, *8*(1), 59–67. doi:10.1037/1089-2680.8.1.59

Singh, G., Singh, P. J., Tyagi, V., & Pandey, A. K. (2019). Thermal and exergoeconomic analysis of a dairy food processing plant. *Journal of Thermal Analysis and Calorimetry*, *136*(3), 1365–1382. doi:10.100710973-018-7781-y

Singh, K., Misra, M., & Yadav, J. (2021). Corporate social responsibility and financial inclusion: Evaluating the moderating effect of income. *Managerial and Decision Economics*.

Sioud, A., & Gagné, C. (2018). Enhanced migrating birds optimization algorithm for the permutation flow shop prob lem with sequence dependent setup times. *European Journal of Operational Research*, *264*(1), 66–73. doi:10.1016/j.ejor.2017.06.027

Smith, A., & Hume, E. C. (2005). Linking culture and ethics: A comparison of accountants' ethical belief systems in the individualism/collectivism and power distance contexts. *Journal of Business Ethics*, *62*(3), 209–220. doi:10.100710551-005-4773-1

Sogut, Z., Ilten, N., & Oktay, Z. (2010). Energetic and exergetic performance evaluation of the quadruple effect evaporator unit in tomatoes paste production. *Energy*, *35*(9), 3821–3826. doi:10.1016/j.energy.2010.05.035

Soschinski, C., & Giordani, M., Klann, R., & Brizolla, M. (2021). Influence Of National Culture On Corporate Social Responsibility. *Cross-cultural Management Journal*, *1*, 71–88.

Sousa, G. (2018). *Decidir como um treinador - A ciência e a prática no futebol e na vida* [Deciding like a coach - The science and practice in football and life]. Federação Portuguesa de Futebol.

Srikanth, K., & Puranam, P. (2014). The firm as a coordination system: Evidence from software services offshoring. *Organization Science*, *25*(4), 1253–1271. doi:10.1287/orsc.2013.0886

Stair, R. M. (1998). *Princípios de sistemas de informação: uma abordagem gerencial*. Rio de Janeiro: LTC Editors.

Stancu, A., Panait, M., Raimi, L., & Palazzo, M. (2020, December). Financial Consumers–Promoters of Sustainable Development? Evidences from Europe and Central Asia. In *Proceedings of the International Conference on Economics and Social Sciences* (pp. 841-850). Sciendo.

Steen, D. P. (2006). Carbon dioxide, carbonation and the principles of filling technology. In D. P. Steen & P. R. Ashurst (Eds.), *Carbonated soft drinks: formulation and manufacture* (pp. 112–116). Blackwell Publishing Ltd. doi:10.1002/9780470996034

Steers, R. M., & Shim, W. S. (2013). Strong leaders, strong cultures: Global management lessons from Toyota and Hyundai. *Organizational Dynamics, 42*(3), 217–227. doi:10.1016/j.orgdyn.2013.06.007

Sternberg, R. J. (1985). Implicit theories of intelligence, creativity, and wisdom. *Journal of Personality and Social Psychology, 49*(3), 607–627. doi:10.1037/0022-3514.49.3.607

Sternberg, R. J., O'Hara, L. A., & Lubart, T. I. (1997). Creativity as investment. *California Management Review, 40*(1), 8–21. doi:10.2307/41165919

Stoney, C., & Winstanley, D. (2001). Stakeholding: Confusion or utopia? Mapping the conceptual terrain. *Journal of Management Studies, 38*(5), 603–626. doi:10.1111/1467-6486.00251

Sulaiman, M. A., Oni, A. O., & Fadare, D. A. (2012). Energy and exergy analysis of a vegetable oil Refinery. *Energy and Power Engineering, 4*(5), 358–364. doi:10.4236/epe.2012.45047

Sule, D. R. (1997). *Industrial Scheduling*. PWS Publishing Company.

Sun, L., Wang, D., & Xie, Y. (2021). Energy, exergy and exergoeconomic analysis of two supercritical CO2 cycles for waste heat recovery of gas turbine. *Applied Thermal Engineering, 196*, 117337. doi:10.1016/j.applthermaleng.2021.117337

Suryaningrat, I. B. (2016). Implementation of QFD in food supply chain management: A case of processed cassava product in Indonesia. *International Journal on Advanced Science, Engineering and Information Technology, 6*(3), 302–305. doi:10.18517/ijaseit.6.3.713

Szutowski, D., & Ratajczak, P. (2016). The Relation between CSR and Innovation. Model Approach. *Journal of Entrepreneurship. Management and Innovation, 12*(2), 77–94.

Tabares, S. (2020). Insights from corporate social innovation: A research agenda. *Social Enterprise Journal, 16*(3), 317–338. doi:10.1108/SEJ-08-2019-0057

Taillard, E. (1993). Benchmarks for basic scheduling problems. *European Journal of Operational Research, 64*(2), 278–285. doi:10.1016/0377-2217(93)90182-M

Takai, S., & Kalapurackal, R. M. (2012). Sensitivity analysis of relative worth in quality function deployment matrices. *Concurrent Engineering, Research and Applications, 20*(3), 195–202. doi:10.1177/1063293X12442411

Tang, L., & Liu, J. (2002). A modified genetic algorithm for the flow shop sequencing problem to minimize mean flow time. *Journal of Intelligent Manufacturing, 13*(1), 61–67. doi:10.1023/A:1013681029869

TerraChoice. (2009) *The six sins of greenwashing™: A study of environmental claims in North American consumer markets*. http://sinsofgreenwashing.com/index6b90.pdf

The Telegraph. (2011, July 22). *Women say working abroad improves their career prospects*. Retrieved from https://www.telegraph.co.uk/finance/personalfinance/expat-money/8651745/Women-say-working-abroad-improves-their-career-prospects.html

The Telegraph. (2019a, March 25). *The five barriers to women in business*. Retrieved from https://www.telegraph.co.uk/business/women-entrepreneurs/five-barriers/

The Telegraph. (2019b, March 25). *Ten myths about women in business*. Retrieved from https://www.telegraph.co.uk/business/women-entrepreneurs/ten-myths/

Thomas, A. S., & Mueller, S. L. (2000). A case for comparative entrepreneurship: Assessing the relevance of culture. *Journal of International Business Studies, 31*(2), 287–301. doi:10.1057/palgrave.jibs.8490906

Thome, L., & Saunders, B. (2002). The Socio-Cultural Embeddedness of Individuals' Ethical Reasoning in Organizations. *Journal of Business Ethics*, *35*(1), 1–14. doi:10.1023/A:1012679026061

Thompson, H. (2019). These times. *New Statesman (London, England)*, *5491*(148), 25. http://search.ebscohost.com/login.aspx?direct=true&db=bth&AN=138948208&site=eds-live

Thongrote, C., Wirjantoro, T. I., & Phianmongkhol, A. (2016). Effect of carbonation sources and its addition levels on carbonated mango juice. *International Food Research Journal*, *23*(5), 2159–2165.

Tierney, P., & Farmer, S. M. (2002). Creative self-efficacy: Its potential antecedents and relationship to creative performance. *Academy of Management Journal*, *45*(6), 1137–1148.

Tisza, L., & Manning, I. (1957). Fluctuations and irreversible thermodynamics. *Physical Review*, *105*(6), 1695–1705. doi:10.1103/PhysRev.105.1695

Toohey, P. (2011). *Boredom: A lively history. Yale*. Yale University Press.

Torelli, C. J., Shavitt, S., Cho, Y. I., Holbrook, A. L., Johnson, T. P., & Weiner, S. (2015). Justice or compassion? Cultural differences in power norms affect consumer satisfaction with power-holders. *International Marketing Review*, *32*(3/4), 279–306. doi:10.1108/IMR-09-2013-0222

Torfx. (2016). *Why Women are choosing to Work Abroad to Advance their Careers*. Retrieved from https://blog.torfx.com/emigrating/why-women-are-choosing-to-work-abroad-to-advance-their-careers/

Torres, C., Perez, E., & Valero, A. (2018). A. Guidelines to develop software for thermoeconomic analysis of energy systems. In *Proceedings of ECOS 2018: the 31st International Conference on Efficiency, Cost, Optimization, Simulation and Environmental Impact of Energy Systems* (pp. 435–453). Academic Press.

Tran, Q. H. N. (2021). Organisational culture, leadership behaviour and job satisfaction in the Vietnam context. *The International Journal of Organizational Analysis*, *29*(1), 136–154. doi:10.1108/IJOA-10-2019-1919

Trompenaars, F., & Hampden-Turner, C. (1998). *Riding the Waves of Culture: Understanding Cultural Diversity in Global Business* (2nd ed.). McGraw-Hill.

Truth in Advertising. (2020a). *Earth Day 2020: Companies accused of Greenwashing*. https://www.truthinadvertising.org/six-companies-accused-greenwashing/

Truth in Advertising. (2020b). *Overdraft fee at Sunmark Credit Union*. https://www.truthinadvertising.org/overdraft-fees-at-sunmark-credit-union/

Truth in Advertising. (2020c). *Patelco Credit Union's Overdraft fees*. https://www.truthinadvertising.org/patelco-credit-unions-overdraft-fees/

Truth in Advertising. (2020d). *Overdraft fees at Pathfinder Bank*. https://www.truthinadvertising.org/patelco-credit-unions-overdraft-fees/

Tsai, Y. (2011). Relationship between organizational culture, leadership behavior and job satisfaction. *BMC Health Services Research*, *11*(1), 98. doi:10.1186/1472-6963-11-98 PMID:21569537

Tseng, M. L., Lim, M. K., Wong, W. P., Chen, Y. C., & Zhan, Y. (2018). A framework for evaluating the performance of sustainable service supply chain management under uncertainty. *International Journal of Production Economics*, *195*, 359–372. doi:10.1016/j.ijpe.2016.09.002

Tsirlin, A. M., Kazakov, V., Alimova, N. A., & Ahremenkov, A. A. (2005). Thermodynamic model of capital extraction in economic systems. *Interdisciplinary Description of Complex Systems: INDECS*, *3*(1), 1–16.

Tsirlin, A., & Gagarina, L. (2020). Finite-time thermodynamics in economics. *Entropy (Basel, Switzerland)*, *22*(8), 891. doi:10.3390/e22080891 PMID:33286661

Tuan, L. T., & Venkatesh, S. (2010). Organizational culture and technological innovation adoption in private hospitals. *International Business Research*, *3*(3), 144–153. doi:10.5539/ibr.v3n3p144

Tucker, A. (1995). *Applied Combinatorics*. John Wiley & Sons.

Twati, J. M., & Gammack, J. G. (2006). The impact of organizational culture innovation on the adoption of IS/IT: The case of Libya. *Journal of Enterprise Information Management*, *19*(1-2), 175–191. doi:10.1108/17410390610645076

Uhl-Bien, M., & Arena, M. (2018). Leadership for organizational adaptability: A theoretical synthesis and integrative framework. *The Leadership Quarterly*, *29*(1), 89–104. doi:10.1016/j.leaqua.2017.12.009

UL. (2021). *Sins of Greenwashing*. https://www.ul.com/insights/sins-greenwashing

Uysal, C., & Keçebas, A. (2021). Advanced exergoeconomic analysis with using modified productive structure analysis: An application for a real gas turbine cycle. *Energy*, *223*, 120085. doi:10.1016/j.energy.2021.120085

Uzkurt, C., Kumar, R., & Ensari, N. (2013). Assessing organizational readiness for innovation: An exploratory study on organizational characteristics of innovativeness. *International Journal of Innovation and Technology Management*, *10*(4), 1350018. doi:10.1142/S0219877013500181

Valdez, F., Moreno, F., & Melin, P. (2020). A Comparison of ACO, GA and SA for Solving the TSP Problem. In *Hybrid Intelligent Systems in Control, Pattern Recognition and Medicine* (pp. 181–189). Springer. doi:10.1007/978-3-030-34135-0_13

Valdimarsson, P. (2011). Basic concepts of thermo-economics. In *Short Course on Geothermal Drilling, Resource Development and Power Plants, organized by UNU-GTP and LaGeo* (pp. 1-7). Retrieved 9 1, 2014, from http://www.os.is/gogn/unu-gtp-sc/UNU-GTP-SC-12-41.pdf

Valencia, J. C. N., Valle, R. S., & Jiménez, D. J. (2010). Organizational culture as determinant of product innovation. *European Journal of Innovation Management*, *13*(4), 466–480. doi:10.1108/14601061011086294

Valenzo, M., & Ortiz, C. (2007, July-December). Competitividad de la industria exportadora de mueble en la región centro Occidente de México. *Inceptum*, *2*(3), 155–186.

Valero, A., & Torres, C. (2009). Thermoeconomic Analysis. In C. A. Frangopoulos (Ed.), *Exergy, Energy System Analysis and Optimization–Volume II: Thermoeconomic Analysis Modeling, Simulation and Optimization in Energy Systems* (pp. 1–35). EOLSS Publications.

Van der Have, R., & Rubalcaba, L. (2016). Emerging Area of Innovation Studies? *Research Policy*, *45*(9), 1923–1935. doi:10.1016/j.respol.2016.06.010

Van Doorn, S., Jansen, J. J., Van den Bosch, F. A., & Volberda, H. W. (2013). Entrepreneurial orientation and firm performance: Drawing attention to the senior team. *Journal of Product Innovation Management*, *30*(5), 821–836. doi:10.1111/jpim.12032

Van Marrewijk, M. (2003). Concepts and definitions of CSR and corporate sustainability: Between agency and communion. *Journal of Business Ethics*, *44*(2–3), 95–105. doi:10.1023/A:1023331212247

Vanany, I., Maarif, G. A., & Soon, J. M. (2019). Application of multi-based Quality Function Deployment (QFD) model to improve halal meat industry. *Journal of Islamic Marketing Journal*, *10*(1), 97–124. doi:10.1108/JIMA-10-2017-0119

Varadharajan, T. K., & Rajendran, C. (2005). A multi-objective simulated-annealing algorithm for scheduling in flowshops to minimize the makespan and total flowtime of jobs. *European Journal of Operational Research*, *167*(3), 772–795. doi:10.1016/j.ejor.2004.07.020

Vargo, S. L., & Lusch, R. F. (2017). Service-dominant logic 2025. *International Journal of Research in Marketing*, *34*(1), 46–67. doi:10.1016/j.ijresmar.2016.11.001

Vasile, V., Panait, M., & Apostu, S. A. (2021). Financial Inclusion Paradigm Shift in the Postpandemic Period. Digital-Divide and Gender Gap. *International Journal of Environmental Research and Public Health*, *18*(20), 10938.

Vecchio, R. P. (2002). Leadership and gender advantage. *The Leadership Quarterly*, *13*(6), 643–671. doi:10.1016/S1048-9843(02)00156-X

Vega-Vázquez, M., Cossío-Silva, F.-J., & Revilla-Camacho, M.-Á. (2016). Entrepreneurial orientation–hotel performance: Has market orientation anything to say? *Journal of Business Research*, *69*(11), 1–6. doi:10.1016/j.jbusres.2016.04.085

Verbekea, A., Kanob, L., & Yuanc, W. (2016). Inside the regional multinationals: A new value chain perspective on subsidiary capabilities. *International Business Review*, *25*(3), 785–793. doi:10.1016/j.ibusrev.2016.01.019

Verkasalo, H., López-Nicolás, C., Molina-Castillo, F. J., & Bouman, H. (2010). Analysis of Users and Non-Users of Smartphone Applications. *Telematics and Informatics*, *27*(3), 242–255. doi:10.1016/j.tele.2009.11.001

Verleye, K., Perks, H., Gruber, T., & Voets, J. (2018). The long and winding road: Building legitimacy for complex social innovation in networks. *Journal of Product Innovation Management*, *36*(6), 695–720. doi:10.1111/jpim.12506

Viaene, J., & Januszewska, R. (1999). Quality function deployment in the chocolate industry. *Food Quality and Preference*, *10*(4–5), 377–385. doi:10.1016/S0950-3293(99)00007-5

Vieira, L., Pizzo, G., Contreira, A., Lazier-Leão, T., Moreira, C., Gaion, P., & Nascimento, J. R. Junior. (2017). Associação entre motivação e coesão de grupo no futebol profissional: O relacionamento treinador-atleta é um fator determinante? [Association between motivation and group cohesion in professional football: the coach-athlete relationship is a determining factor?]. *Revista de Psicología del Deporte*, *27*, 51–57.

Vinayak, K., & Kodali, R. (2013). Benchmarking the quality function deployment models. *Benchmarking*, *20*(6), 825–854. doi:10.1108/BIJ-07-2011-0052

Vince, A. (2002). A framework for the greedy algorithm. *Discrete Applied Mathematics*, *121*(1-3), 247–260. doi:10.1016/S0166-218X(01)00362-6

Vodanovich, S. J. (2003). Psychometric measures of boredom: A review of the literature. *The Journal of Psychology*, *137*(6), 569–595. doi:10.1080/00223980309600636 PMID:14992349

Voica, M. C., & Stancu, A. (2021). Corporate Social Responsibility Reporting: Background, Evolution and Sustainability Promoter. *Sustainable Management for Managers and Engineers*, 109-155.

Voica, M. C. (2017). Financial inclusion as a tool for sustainable development. *Romanian Journal of Economics*, *44*(1), 121–129.

Volgger, M., & Huang, S. S. (2019). Scoping irresponsible behaviour in hospitality and tourism: Widening the perspective of CSR. *International Journal of Contemporary Hospitality Management*.

Vollero, A., Palazzo, M., Siano, A., & Elving, W. J. (2016). Avoiding the greenwashing trap: Between CSR communication and stakeholder engagement. *International Journal of Innovation and Sustainable Development*, *10*(2), 120–140.

Vos, J. (2003). Corporate social responsibility and the identification of stakeholders. *Corporate Social Responsibility and Environmental Management, 10*(3), 141–152. doi:10.1002/csr.39

Vroom, V., & Jago, A. (2007). The role of the situation in leadership. *The American Psychologist, 62*(1), 17–24. doi:10.1037/0003-066X.62.1.17 PMID:17209676

Waddock, S., & Smith, N. (2000). Relationships: The real challenge of corporate global citizenship. *Business and Society Review, 105*(1), 47–62. doi:10.1111/0045-3609.00064

Wagstaff, C., & Burton-Wylie, S. (2018). Organisational culture in sport: A conceptual, definitional and methodological review. *Sport and Exercise Psychology Review, 1*(14), 32–52.

Waheed, M. A., Jekayinfa, S. O., Ojediran, J. O., & Imeokparia, O. E. (2008). Energetic analysis of fruit juice processing operations in Nigeria. *Energy, 33*(1), 35–45. doi:10.1016/j.energy.2007.09.001

Waldman, D. A., & Siegel, D. S. (2008). Theoretical and practitioner letters: Defining the socially responsible leader. *The Leadership Quarterly, 19*(1), 117–131. doi:10.1016/j.leaqua.2007.12.008

Walesska, M., & Alvarado, A. (2009). Imagen y reputación corporativa. Estudio empírico de operadoras de telefonía en España. *Teoría y Praxis*, (6), 9–29.

Walker, K., & Wan, F. (2012). The harm of symbolic actions and green-washing: Corporate actions and communications on environmental performance and their financial implications. *Journal of Business Ethics, 109*, 227–242.

Wall, G. (2009). *Exergetics*. Retrieved October 17, 2013, from http://www.exergy.se /ftp/exergetics.pdf

Walters, R. G., & McKenzie, S. B. (1988). A structural equations analysis of the impact of price promotions on store performance. *JMR, Journal of Marketing Research, 25*(1), 51–63. doi:10.1177/002224378802500105

Walumbwa, F. O., & Lawler, J. J. (2003). Building effective organizations: Transformational leadership, collectivist orientation, work-related attitudes, and withdrawal behaviors in three emerging economies. *International Journal of Human Resource Management, 14*(7), 1083–1101. doi:10.1080/0958519032000114219

Wang, L., & Zheng, D. Z. (2003). An effective hybrid heuristic for flow shop scheduling. *International Journal of Advanced Manufacturing Technology, 21*(1), 38–44. doi:10.1007001700300005

Wang, T. Y., Hsiao, H. I., & Sung, W. C. (2019). Quality function deployment modified for the food industry: An example of a granola bar. *Food Science & Nutrition, 7*(5), 1746–1753. doi:10.1002/fsn3.1014 PMID:31139387

Wang, Z., & Klir, G. (2013). *Fuzzy measure theory*. Springer Science & Business Media.

Waples, E. P., & Friedrich, T. L. (2011). Managing creative performance: Important strategies for leaders of creative efforts. *Advances in Developing Human Resources, 13*(3), 366–385. doi:10.1177/1523422311424713

Warhurst, A. (2001). Corporate citizenship and corporate social investment. *Journal of Corporate Citizenship, 1*(1), 57–73. doi:10.9774/GLEAF.4700.2001.sp.00008

Warrick, D. D. (2017). What leaders need to know about organizational culture. *Business Horizons, 60*(3), 395–404. doi:10.1016/j.bushor.2017.01.011

Warrick, D. D., Milliman, J. F., & Ferguson, J. M. (2016). Building high performance cultures. *Organizational Dynamics, 45*(1), 64–70. doi:10.1016/j.orgdyn.2015.12.008

Watts, L. L., Steele, L. M., & Mumford, M. D. (2019). Making sense of pragmatic and charismatic leadership stories: Effects on vision formation. *The Leadership Quarterly, 30*(2), 243–259. doi:10.1016/j.leaqua.2018.09.003

Wayne, T. B., & Kissock, K. J. (2000). Exergy analysis of industrial air compression. *Proceedings from the Twenty-Second National Industrial Energy Technology Conference*, 89–98.

Whitney, H. (1935). On the abstract properties of linear dependence. *American Journal of Mathematics*, *57*(3), 509–533. doi:10.2307/2371182

Wiebe, R., & Gaddy, V. (1940). The Solubility of carbon dioxide in water at various temperatures from 12 to 40° and at pressures to 500 atmospheres. Critical Phenomena. *Journal of the American Chemical Society*, *64*(4), 815–817. doi:10.1021/ja01861a033

Wigand, R., Picot, A., & Reichwald, R. (1997). Information, organization and management. John Wiley.

Williams, O. (2021). Europe's Largest Bank Suspected of Greenwashing. *Forbes*. https://www.forbes.com/sites/oliverwilliams1/2021/01/10/europes-largest-bank-suspected-of-greenwashing/

Williams, T. A., & Shepherd, D. A. (2017). Mixed method social network analysis: Combining inductive concept development, content analysis, and secondary data for quantitative analysis. *Organizational Research Methods*, *20*(2), 268–298.

Wilson, M. (2003). Corporate sustainability: What is it and where does it come from? *Ivey Business Journal*, *67*(6), 1–5.

Wilson, R. J. (1985). *Introduction to Graph Theory* (3rd ed.). Longman.

Winter, S. (1987). Knowledge and competence as strategic assets. In D. Teece (Ed.), *The competitive challenge: strategy for industrial innovation and renewal*. Ballinger Publishing Company.

Wolfinbarger, M., & Gilly, M. (2001). Shopping Online for Freedom, Control, and Fun. *California Management Review*, *43*(2), 34–55. doi:10.2307/41166074

Wong, S. I., & Giessner, S. R. (2016). The thin line between empowering and laissez-faire leadership: An expectancy-match perspective. *Journal of Management*, *44*(2), 757–783. doi:10.1177/0149206315574597

Wood, J. D. (1991). Corporate Social Performance Revised. *Academy of Management Review*, *16*(4), 691–718. doi:10.2307/258977

Wood, J. D., & Logsdon, J. M. (2002). Business citizenship: From domestic to global level of analysis. *Business Ethics Quarterly*, *12*(2), 155–187. doi:10.2307/3857809

Woudstra, N. (2012). *Sustainable energy systems- limitations and challenges based on exergy analysis*. Retrieved September 19, 2014, from http://repository.tudelft.nl/ view/ir/uuid%3A112039e0-89ee-44fa-8d52-17eca3ec41f0/

Wright, C., & Rwabizambuga, A. (2006). Institutional pressures, corporate reputation, and voluntary codes of conduct: An examination of the equator principles. *Business and Society Review*, *111*(1), 89–117.

Wright, T. A., & Cropanzano, R. (2004). The role of psychological well-being in job performance: A fresh look at an age-old quest. *Organizational Dynamics*, *33*(4), 338–351. doi:10.1016/j.orgdyn.2004.09.002

Wu, P. C. S., Yeh, G. Y., & Hsiao, C. (2011). The Effect of Store Image and Service Quality on Brand Image and Purchase Intention for Private Label Brands. *Australasian Marketing Journal*, *19*(1), 30–39. doi:10.1016/j.ausmj.2010.11.001

Xenikou, A., & Simosi, M. (2006). Organizational culture and transformational leadership as predictors of business unit performance. *Journal of Managerial Psychology, 21*(6), 566-579. doi:10.1108/02683940610684409

Xie, G., & Paik, Y. (2019). Cultural differences in creativity and innovation: Are Asian employees truly less creative than western employees? *Asia Pacific Business Review*, *25*(1), 123–147. doi:10.1080/13602381.2018.1535380

Younsi, A., Turcry, P., Rozière, E., Aît-Mokhtar, A., & Loukili, A. (2011). Performance-based design and carbonation of concrete with high fly ash content. *Cement and Concrete Composites*, *33*(10), 993–1000. doi:10.1016/j.cemconcomp.2011.07.005

Yun, Y., & Gen, M. (2003). Performance analysis of adaptive genetic algorithms with fuzzy logic and heuristics. *Fuzzy Optimization and Decision Making*, *2*(2), 161–175. doi:10.1023/A:1023499201829

Zahra, S. A. (1993). Environment, corporate entrepreneurship, and financial performance: A taxonomic approach. *Journal of Business Venturing*, *8*(4), 319–340. doi:10.1016/0883-9026(93)90003-N

Zairi, M., & Al-Mashari, M. (2005). Developing a sustainable culture of innovation management: A perspective approach. *Knowledge and Process Management*, *12*(3), 190–202. doi:10.1002/kpm.229

Zander, L., & Romani, L. (2004). When nationality matters: A study of departmental, hierarchical, professional, gender and age-based employee groupings' leadership preferences across 15 countries. *International Journal of Cross Cultural Management*, *4*(3), 291–315. doi:10.1177/1470595804047811

Zarandi, M. H. F., Asl, A. A. S., Sotudian, S., & Castillo, O. (2020). A state of the art review of intelligent scheduling. *Artificial Intelligence Review*, *53*(1), 501–593. doi:10.100710462-018-9667-6

Zeithaml, V. A. (1988). Consumer perceptions of price, quality, and value: A means-end model and synthesis of evidence. *Journal of Marketing*, *52*(3), 2–22. doi:10.1177/002224298805200302

Zhou, J., & Shalley, C. E. (2003). Research on employee creativity: A critical review and directions for future research. *Research in Personnel and Human Resources Management*, *22*, 165–217. doi:10.1016/S0742-7301(03)22004-1

Zielke, S. (2008). Exploring asymmetric effects in the formation of retail price satisfaction. *Journal of Retailing and Consumer Services*, *15*(5), 335–347. doi:10.1016/j.jretconser.2007.08.005

Zimmermann, H. J. (2001). *Fuzzy Set Theory and its Applications* (4th ed.). Springer-Verlag. doi:10.1007/978-94-010-0646-0

Zwetsloot, G. (2003). From management systems to corporate social responsibility. *Journal of Business Ethics*, *44*(2/3), 201–207. doi:10.1023/A:1023303917699

Zyphur, M. J., Narayanan, J., Koh, G., & Koh, D. (2009). Testosterone–status mismatch lowers collective efficacy in groups: Evidence from a slope-as-predictor multilevel structural equation model. *Organizational Behavior and Human Decision Processes*, *110*(2), 70–79. doi:10.1016/j.obhdp.2009.05.004

Related References

To continue our tradition of advancing information science and technology research, we have compiled a list of recommended IGI Global readings. These references will provide additional information and guidance to further enrich your knowledge and assist you with your own research and future publications.

Abdul Razak, R., & Mansor, N. A. (2021). Instagram Influencers in Social Media-Induced Tourism: Rethinking Tourist Trust Towards Tourism Destination. In M. Dinis, L. Bonixe, S. Lamy, & Z. Breda (Eds.), *Impact of New Media in Tourism* (pp. 135-144). IGI Global. https://doi.org/10.4018/978-1-7998-7095-1.ch009

Abir, T., & Khan, M. Y. (2022). Importance of ICT Advancement and Culture of Adaptation in the Tourism and Hospitality Industry for Developing Countries. In C. Ramos, S. Quinteiro, & A. Gonçalves (Eds.), *ICT as Innovator Between Tourism and Culture* (pp. 30–41). IGI Global. https://doi.org/10.4018/978-1-7998-8165-0.ch003

Abtahi, M. S., Behboudi, L., & Hasanabad, H. M. (2017). Factors Affecting Internet Advertising Adoption in Ad Agencies. *International Journal of Innovation in the Digital Economy*, *8*(4), 18–29. doi:10.4018/IJIDE.2017100102

Afenyo-Agbe, E., & Mensah, I. (2022). Principles, Benefits, and Barriers to Community-Based Tourism: Implications for Management. In I. Mensah & E. Afenyo-Agbe (Eds.), *Prospects and Challenges of Community-Based Tourism and Changing Demographics* (pp. 1–29). IGI Global. doi:10.4018/978-1-7998-7335-8.ch001

Agbo, V. M. (2022). Distributive Justice Issues in Community-Based Tourism. In I. Mensah & E. Afenyo-Agbe (Eds.), *Prospects and Challenges of Community-Based Tourism and Changing Demographics* (pp. 107–129). IGI Global. https://doi.org/10.4018/978-1-7998-7335-8.ch005

Agrawal, S. (2017). The Impact of Emerging Technologies and Social Media on Different Business(es): Marketing and Management. In O. Rishi & A. Sharma (Eds.), *Maximizing Business Performance and Efficiency Through Intelligent Systems* (pp. 37–49). Hershey, PA: IGI Global. doi:10.4018/978-1-5225-2234-8.ch002

Ahmad, A., & Johari, S. (2022). Georgetown as a Gastronomy Tourism Destination: Visitor Awareness Towards Revisit Intention of Nasi Kandar Restaurant. In M. Valeri (Ed.), *New Governance and Management in Touristic Destinations* (pp. 71–83). IGI Global. https://doi.org/10.4018/978-1-6684-3889-3.ch005

Alkhatib, G., & Bayouq, S. T. (2021). A TAM-Based Model of Technological Factors Affecting Use of E-Tourism. *International Journal of Tourism and Hospitality Management in the Digital Age*, *5*(2), 50–67. https://doi.org/10.4018/IJTHMDA.20210701.oa1

Altinay Ozdemir, M. (2021). Virtual Reality (VR) and Augmented Reality (AR) Technologies for Accessibility and Marketing in the Tourism Industry. In C. Eusébio, L. Teixeira, & M. Carneiro (Eds.), *ICT Tools and Applications for Accessible Tourism* (pp. 277-301). IGI Global. https://doi.org/10.4018/978-1-7998-6428-8.ch013

Anantharaman, R. N., Rajeswari, K. S., Angusamy, A., & Kuppusamy, J. (2017). Role of Self-Efficacy and Collective Efficacy as Moderators of Occupational Stress Among Software Development Professionals. *International Journal of Human Capital and Information Technology Professionals*, *8*(2), 45–58. doi:10.4018/IJHCITP.2017040103

Aninze, F., El-Gohary, H., & Hussain, J. (2018). The Role of Microfinance to Empower Women: The Case of Developing Countries. *International Journal of Customer Relationship Marketing and Management*, *9*(1), 54–78. doi:10.4018/IJCRMM.2018010104

Antosova, G., Sabogal-Salamanca, M., & Krizova, E. (2021). Human Capital in Tourism: A Practical Model of Endogenous and Exogenous Territorial Tourism Planning in Bahía Solano, Colombia. In V. Costa, A. Moura, & M. Mira (Eds.), *Handbook of Research on Human Capital and People Management in the Tourism Industry* (pp. 282–302). IGI Global. https://doi.org/10.4018/978-1-7998-4318-4.ch014

Arsenijević, O. M., Orčić, D., & Kastratović, E. (2017). Development of an Optimization Tool for Intangibles in SMEs: A Case Study from Serbia with a Pilot Research in the Prestige by Milka Company. In M. Vemić (Ed.), *Optimal Management Strategies in Small and Medium Enterprises* (pp. 320–347). Hershey, PA: IGI Global. doi:10.4018/978-1-5225-1949-2.ch015

Aryanto, V. D., Wismantoro, Y., & Widyatmoko, K. (2018). Implementing Eco-Innovation by Utilizing the Internet to Enhance Firm's Marketing Performance: Study of Green Batik Small and Medium Enterprises in Indonesia. *International Journal of E-Business Research*, *14*(1), 21–36. doi:10.4018/IJEBR.2018010102

Asero, V., & Billi, S. (2022). New Perspective of Networking in the DMO Model. In M. Valeri (Ed.), *New Governance and Management in Touristic Destinations* (pp. 105–118). IGI Global. https://doi.org/10.4018/978-1-6684-3889-3.ch007

Atiku, S. O., & Fields, Z. (2017). Multicultural Orientations for 21st Century Global Leadership. In N. Baporikar (Ed.), *Management Education for Global Leadership* (pp. 28–51). Hershey, PA: IGI Global. doi:10.4018/978-1-5225-1013-0.ch002

Atiku, S. O., & Fields, Z. (2018). Organisational Learning Dimensions and Talent Retention Strategies for the Service Industries. In N. Baporikar (Ed.), *Global Practices in Knowledge Management for Societal and Organizational Development* (pp. 358–381). Hershey, PA: IGI Global. doi:10.4018/978-1-5225-3009-1.ch017

Atsa'am, D. D., & Kuset Bodur, E. (2021). Pattern Mining on How Organizational Tenure Affects the Psychological Capital of Employees Within the Hospitality and Tourism Industry: Linking Employees' Organizational Tenure With PsyCap. *International Journal of Tourism and Hospitality Management in the Digital Age*, *5*(2), 17–28. https://doi.org/10.4018/IJTHMDA.2021070102

Ávila, L., & Teixeira, L. (2018). The Main Concepts Behind the Dematerialization of Business Processes. In M. Khosrow-Pour, D.B.A. (Ed.), Encyclopedia of Information Science and Technology, Fourth Edition (pp. 888-898). Hershey, PA: IGI Global. https://doi.org/ doi:10.4018/978-1-5225-2255-3.ch076

Ayorekire, J., Mugizi, F., Obua, J., & Ampaire, G. (2022). Community-Based Tourism and Local People's Perceptions Towards Conservation: The Case of Queen Elizabeth Conservation Area, Uganda. In I. Mensah & E. Afenyo-Agbe (Eds.), *Prospects and Challenges of Community-Based Tourism and Changing Demographics* (pp. 56–82). IGI Global. https://doi.org/10.4018/978-1-7998-7335-8.ch003

Baleiro, R. (2022). Tourist Literature and the Architecture of Travel in Olga Tokarczuk and Patti Smith. In R. Baleiro & R. Pereira (Eds.), *Global Perspectives on Literary Tourism and Film-Induced Tourism* (pp. 202-216). IGI Global. https://doi.org/10.4018/978-1-7998-8262-6.ch011

Barat, S. (2021). Looking at the Future of Medical Tourism in Asia. *International Journal of Tourism and Hospitality Management in the Digital Age*, *5*(1), 19–33. https://doi.org/10.4018/IJTHMDA.2021010102

Barbosa, C. A., Magalhães, M., & Nunes, M. R. (2021). Travel Instagramability: A Way of Choosing a Destination? In M. Dinis, L. Bonixe, S. Lamy, & Z. Breda (Eds.), *Impact of New Media in Tourism* (pp. 173-190). IGI Global. https://doi.org/10.4018/978-1-7998-7095-1.ch011

Bari, M. W., & Khan, Q. (2021). Pakistan as a Destination of Religious Tourism. In E. Alaverdov & M. Bari (Eds.), *Global Development of Religious Tourism* (pp. 1-10). IGI Global. https://doi.org/10.4018/978-1-7998-5792-1.ch001

Bartens, Y., Chunpir, H. I., Schulte, F., & Voß, S. (2017). Business/IT Alignment in Two-Sided Markets: A COBIT 5 Analysis for Media Streaming Business Models. In S. De Haes & W. Van Grembergen (Eds.), *Strategic IT Governance and Alignment in Business Settings* (pp. 82–111). Hershey, PA: IGI Global. doi:10.4018/978-1-5225-0861-8.ch004

Bashayreh, A. M. (2018). Organizational Culture and Organizational Performance. In W. Lee & F. Sabetzadeh (Eds.), *Contemporary Knowledge and Systems Science* (pp. 50–69). Hershey, PA: IGI Global. doi:10.4018/978-1-5225-5655-8.ch003

Bechthold, L., Lude, M., & Prügl, R. (2021). Crisis Favors the Prepared Firm: How Organizational Ambidexterity Relates to Perceptions of Organizational Resilience. In A. Zehrer, G. Glowka, K. Schwaiger, & V. Ranacher-Lackner (Eds.), *Resiliency Models and Addressing Future Risks for Family Firms in the Tourism Industry* (pp. 178–205). IGI Global. https://doi.org/10.4018/978-1-7998-7352-5.ch008

Bedford, D. A. (2018). Sustainable Knowledge Management Strategies: Aligning Business Capabilities and Knowledge Management Goals. In N. Baporikar (Ed.), *Global Practices in Knowledge Management for Societal and Organizational Development* (pp. 46–73). Hershey, PA: IGI Global. doi:10.4018/978-1-5225-3009-1.ch003

Bekjanov, D., & Matyusupov, B. (2021). Influence of Innovative Processes in the Competitiveness of Tourist Destination. In J. Soares (Ed.), *Innovation and Entrepreneurial Opportunities in Community Tourism* (pp. 243–263). IGI Global. https://doi.org/10.4018/978-1-7998-4855-4.ch014

Bharwani, S., & Musunuri, D. (2018). Reflection as a Process From Theory to Practice. In M. Khosrow-Pour, D.B.A. (Ed.), Encyclopedia of Information Science and Technology, Fourth Edition (pp. 1529-1539). Hershey, PA: IGI Global. doi:10.4018/978-1-5225-2255-3.ch132

Bhatt, G. D., Wang, Z., & Rodger, J. A. (2017). Information Systems Capabilities and Their Effects on Competitive Advantages: A Study of Chinese Companies. *Information Resources Management Journal, 30*(3), 41–57. doi:10.4018/IRMJ.2017070103

Bhushan, M., & Yadav, A. (2017). Concept of Cloud Computing in ESB. In R. Bhadoria, N. Chaudhari, G. Tomar, & S. Singh (Eds.), *Exploring Enterprise Service Bus in the Service-Oriented Architecture Paradigm* (pp. 116–127). Hershey, PA: IGI Global. doi:10.4018/978-1-5225-2157-0.ch008

Bhushan, S. (2017). System Dynamics Base-Model of Humanitarian Supply Chain (HSCM) in Disaster Prone Eco-Communities of India: A Discussion on Simulation and Scenario Results. *International Journal of System Dynamics Applications*, *6*(3), 20–37. doi:10.4018/IJSDA.2017070102

Binder, D., & Miller, J. W. (2021). A Generations' Perspective on Employer Branding in Tourism. In V. Costa, A. Moura, & M. Mira (Eds.), *Handbook of Research on Human Capital and People Management in the Tourism Industry* (pp. 152–174). IGI Global. https://doi.org/10.4018/978-1-7998-4318-4.ch008

Birch Freeman, A. A., Mensah, I., & Antwi, K. B. (2022). Smiling vs. Frowning Faces: Community Participation for Sustainable Tourism in Ghanaian Communities. In I. Mensah & E. Afenyo-Agbe (Eds.), *Prospects and Challenges of Community-Based Tourism and Changing Demographics* (pp. 83–106). IGI Global. https://doi.org/10.4018/978-1-7998-7335-8.ch004

Biswas, A., & De, A. K. (2017). On Development of a Fuzzy Stochastic Programming Model with Its Application to Business Management. In S. Trivedi, S. Dey, A. Kumar, & T. Panda (Eds.), *Handbook of Research on Advanced Data Mining Techniques and Applications for Business Intelligence* (pp. 353–378). Hershey, PA: IGI Global. doi:10.4018/978-1-5225-2031-3.ch021

Boragnio, A., & Faracce Macia, C. (2021). "Taking Care of Yourself at Home": Use of E-Commerce About Food and Care During the COVID-19 Pandemic in the City of Buenos Aires. In M. Korstanje (Ed.), *Socio-Economic Effects and Recovery Efforts for the Rental Industry: Post-COVID-19 Strategies* (pp. 45–71). IGI Global. https://doi.org/10.4018/978-1-7998-7287-0.ch003

Borges, V. D. (2021). Happiness: The Basis for Public Policy in Tourism. In A. Perinotto, V. Mayer, & J. Soares (Eds.), *Rebuilding and Restructuring the Tourism Industry: Infusion of Happiness and Quality of Life* (pp. 1–25). IGI Global. https://doi.org/10.4018/978-1-7998-7239-9.ch001

Bücker, J., & Ernste, K. (2018). Use of Brand Heroes in Strategic Reputation Management: The Case of Bacardi, Adidas, and Daimler. In A. Erdemir (Ed.), *Reputation Management Techniques in Public Relations* (pp. 126–150). Hershey, PA: IGI Global. doi:10.4018/978-1-5225-3619-2.ch007

Buluk Eşitti, B. (2021). COVID-19 and Alternative Tourism: New Destinations and New Tourism Products. In M. Demir, A. Dalgıç, & F. Ergen (Eds.), *Handbook of Research on the Impacts and Implications of COVID-19 on the Tourism Industry* (pp. 786–805). IGI Global. https://doi.org/10.4018/978-1-7998-8231-2.ch038

Bureš, V. (2018). Industry 4.0 From the Systems Engineering Perspective: Alternative Holistic Framework Development. In R. Brunet-Thornton & F. Martinez (Eds.), *Analyzing the Impacts of Industry 4.0 in Modern Business Environments* (pp. 199–223). Hershey, PA: IGI Global. doi:10.4018/978-1-5225-3468-6.ch011

Buzady, Z. (2017). Resolving the Magic Cube of Effective Case Teaching: Benchmarking Case Teaching Practices in Emerging Markets – Insights from the Central European University Business School, Hungary. In D. Latusek (Ed.), *Case Studies as a Teaching Tool in Management Education* (pp. 79–103). Hershey, PA: IGI Global. doi:10.4018/978-1-5225-0770-3.ch005

Camillo, A. (2021). *Legal Matters, Risk Management, and Risk Prevention: From Forming a Business to Legal Representation.* IGI Global. doi:10.4018/978-1-7998-4342-9.ch004

Căpusneanu, S., & Topor, D. I. (2018). Business Ethics and Cost Management in SMEs: Theories of Business Ethics and Cost Management Ethos. In I. Oncioiu (Ed.), *Ethics and Decision-Making for Sustainable Business Practices* (pp. 109–127). Hershey, PA: IGI Global. doi:10.4018/978-1-5225-3773-1.ch007

Chan, R. L., Mo, P. L., & Moon, K. K. (2018). Strategic and Tactical Measures in Managing Enterprise Risks: A Study of the Textile and Apparel Industry. In K. Strang, M. Korstanje, & N. Vajjhala (Eds.), *Research, Practices, and Innovations in Global Risk and Contingency Management* (pp. 1–19). Hershey, PA: IGI Global. doi:10.4018/978-1-5225-4754-9.ch001

Charlier, S. D., Burke-Smalley, L. A., & Fisher, S. L. (2018). Undergraduate Programs in the U.S: A Contextual and Content-Based Analysis. In J. Mendy (Ed.), *Teaching Human Resources and Organizational Behavior at the College Level* (pp. 26–57). Hershey, PA: IGI Global. doi:10.4018/978-1-5225-2820-3.ch002

Chumillas, J., Güell, M., & Quer, P. (2022). The Use of ICT in Tourist and Educational Literary Routes: The Role of the Guide. In C. Ramos, S. Quinteiro, & A. Gonçalves (Eds.), *ICT as Innovator Between Tourism and Culture* (pp. 15–29). IGI Global. https://doi.org/10.4018/978-1-7998-8165-0.ch002

Dahlberg, T., Kivijärvi, H., & Saarinen, T. (2017). IT Investment Consistency and Other Factors Influencing the Success of IT Performance. In S. De Haes & W. Van Grembergen (Eds.), *Strategic IT Governance and Alignment in Business Settings* (pp. 176–208). Hershey, PA: IGI Global. doi:10.4018/978-1-5225-0861-8.ch007

Damnjanović, A. M. (2017). Knowledge Management Optimization through IT and E-Business Utilization: A Qualitative Study on Serbian SMEs. In M. Vemić (Ed.), *Optimal Management Strategies in Small and Medium Enterprises* (pp. 249–267). Hershey, PA: IGI Global. doi:10.4018/978-1-5225-1949-2.ch012

Daneshpour, H. (2017). Integrating Sustainable Development into Project Portfolio Management through Application of Open Innovation. In M. Vemić (Ed.), *Optimal Management Strategies in Small and Medium Enterprises* (pp. 370–387). Hershey, PA: IGI Global. doi:10.4018/978-1-5225-1949-2.ch017

Daniel, A. D., & Reis de Castro, V. (2018). Entrepreneurship Education: How to Measure the Impact on Nascent Entrepreneurs. In A. Carrizo Moreira, J. Guilherme Leitão Dantas, & F. Manuel Valente (Eds.), *Nascent Entrepreneurship and Successful New Venture Creation* (pp. 85–110). Hershey, PA: IGI Global. doi:10.4018/978-1-5225-2936-1.ch004

David, R., Swami, B. N., & Tangirala, S. (2018). Ethics Impact on Knowledge Management in Organizational Development: A Case Study. In N. Baporikar (Ed.), *Global Practices in Knowledge Management for Societal and Organizational Development* (pp. 19–45). Hershey, PA: IGI Global. doi:10.4018/978-1-5225-3009-1.ch002

De Uña-Álvarez, E., & Villarino-Pérez, M. (2022). Fostering Ecocultural Resources, Identity, and Tourism in Inland Territories (Galicia, NW Spain). In G. Fernandes (Ed.), *Challenges and New Opportunities for Tourism in Inland Territories: Ecocultural Resources and Sustainable Initiatives* (pp. 1-16). IGI Global. https://doi.org/10.4018/978-1-7998-7339-6.ch001

Delias, P., & Lakiotaki, K. (2018). Discovering Process Horizontal Boundaries to Facilitate Process Comprehension. *International Journal of Operations Research and Information Systems*, *9*(2), 1–31. doi:10.4018/IJORIS.2018040101

Denholm, J., & Lee-Davies, L. (2018). Success Factors for Games in Business and Project Management. In *Enhancing Education and Training Initiatives Through Serious Games* (pp. 34–68). Hershey, PA: IGI Global. doi:10.4018/978-1-5225-3689-5.ch002

Deshpande, M. (2017). Best Practices in Management Institutions for Global Leadership: Policy Aspects. In N. Baporikar (Ed.), *Management Education for Global Leadership* (pp. 1–27). Hershey, PA: IGI Global. doi:10.4018/978-1-5225-1013-0.ch001

Deshpande, M. (2018). Policy Perspectives for SMEs Knowledge Management. In N. Baporikar (Ed.), *Knowledge Integration Strategies for Entrepreneurship and Sustainability* (pp. 23–46). Hershey, PA: IGI Global. doi:10.4018/978-1-5225-5115-7.ch002

Dezdar, S. (2017). ERP Implementation Projects in Asian Countries: A Comparative Study on Iran and China. *International Journal of Information Technology Project Management*, *8*(3), 52–68. doi:10.4018/IJITPM.2017070104

Domingos, D., Respício, A., & Martinho, R. (2017). Reliability of IoT-Aware BPMN Healthcare Processes. In C. Reis & M. Maximiano (Eds.), *Internet of Things and Advanced Application in Healthcare* (pp. 214–248). Hershey, PA: IGI Global. doi:10.4018/978-1-5225-1820-4.ch008

Dosumu, O., Hussain, J., & El-Gohary, H. (2017). An Exploratory Study of the Impact of Government Policies on the Development of Small and Medium Enterprises in Developing Countries: The Case of Nigeria. *International Journal of Customer Relationship Marketing and Management*, *8*(4), 51–62. doi:10.4018/IJCRMM.2017100104

Durst, S., Bruns, G., & Edvardsson, I. R. (2017). Retaining Knowledge in Smaller Building and Construction Firms. *International Journal of Knowledge and Systems Science*, *8*(3), 1–12. doi:10.4018/IJKSS.2017070101

Edvardsson, I. R., & Durst, S. (2017). Outsourcing, Knowledge, and Learning: A Critical Review. *International Journal of Knowledge-Based Organizations*, *7*(2), 13–26. doi:10.4018/IJKBO.2017040102

Edwards, J. S. (2018). Integrating Knowledge Management and Business Processes. In M. Khosrow-Pour, D.B.A. (Ed.), Encyclopedia of Information Science and Technology, Fourth Edition (pp. 5046-5055). Hershey, PA: IGI Global. doi:10.4018/978-1-5225-2255-3.ch437

Eichelberger, S., & Peters, M. (2021). Family Firm Management in Turbulent Times: Opportunities for Responsible Tourism. In A. Zehrer, G. Glowka, K. Schwaiger, & V. Ranacher-Lackner (Eds.), *Resiliency Models and Addressing Future Risks for Family Firms in the Tourism Industry* (pp. 103–124). IGI Global. https://doi.org/10.4018/978-1-7998-7352-5.ch005

Eide, D., Hjalager, A., & Hansen, M. (2022). Innovative Certifications in Adventure Tourism: Attributes and Diffusion. In R. Augusto Costa, F. Brandão, Z. Breda, & C. Costa (Eds.), *Planning and Managing the Experience Economy in Tourism* (pp. 161-175). IGI Global. https://doi.org/10.4018/978-1-7998-8775-1.ch009

Ejiogu, A. O. (2018). Economics of Farm Management. In *Agricultural Finance and Opportunities for Investment and Expansion* (pp. 56–72). Hershey, PA: IGI Global. doi:10.4018/978-1-5225-3059-6.ch003

Ekanem, I., & Abiade, G. E. (2018). Factors Influencing the Use of E-Commerce by Small Enterprises in Nigeria. *International Journal of ICT Research in Africa and the Middle East*, *7*(1), 37–53. doi:10.4018/IJICTRAME.2018010103

Ekanem, I., & Alrossais, L. A. (2017). Succession Challenges Facing Family Businesses in Saudi Arabia. In P. Zgheib (Ed.), *Entrepreneurship and Business Innovation in the Middle East* (pp. 122–146). Hershey, PA: IGI Global. doi:10.4018/978-1-5225-2066-5.ch007

El Faquih, L., & Fredj, M. (2017). Ontology-Based Framework for Quality in Configurable Process Models. *Journal of Electronic Commerce in Organizations*, *15*(2), 48–60. doi:10.4018/JECO.2017040104

Faisal, M. N., & Talib, F. (2017). Building Ambidextrous Supply Chains in SMEs: How to Tackle the Barriers? *International Journal of Information Systems and Supply Chain Management*, *10*(4), 80–100. doi:10.4018/IJISSCM.2017100105

Fernandes, T. M., Gomes, J., & Romão, M. (2017). Investments in E-Government: A Benefit Management Case Study. *International Journal of Electronic Government Research*, *13*(3), 1–17. doi:10.4018/IJEGR.2017070101

Figueira, L. M., Honrado, G. R., & Dionísio, M. S. (2021). Human Capital Management in the Tourism Industry in Portugal. In V. Costa, A. Moura, & M. Mira (Eds.), *Handbook of Research on Human Capital and People Management in the Tourism Industry* (pp. 1–19). IGI Global. doi:10.4018/978-1-7998-4318-4.ch001

Gao, S. S., Oreal, S., & Zhang, J. (2018). Contemporary Financial Risk Management Perceptions and Practices of Small-Sized Chinese Businesses. In I. Management Association (Ed.), Global Business Expansion: Concepts, Methodologies, Tools, and Applications (pp. 917-931). Hershey, PA: IGI Global. doi:10.4018/978-1-5225-5481-3.ch041

Garg, R., & Berning, S. C. (2017). Indigenous Chinese Management Philosophies: Key Concepts and Relevance for Modern Chinese Firms. In B. Christiansen & G. Koc (Eds.), *Transcontinental Strategies for Industrial Development and Economic Growth* (pp. 43–57). Hershey, PA: IGI Global. doi:10.4018/978-1-5225-2160-0.ch003

Gencer, Y. G. (2017). Supply Chain Management in Retailing Business. In U. Akkucuk (Ed.), *Ethics and Sustainability in Global Supply Chain Management* (pp. 197–210). Hershey, PA: IGI Global. doi:10.4018/978-1-5225-2036-8.ch011

Gera, R., Arora, S., & Malik, S. (2021). Emotional Labor in the Tourism Industry: Strategies, Antecedents, and Outcomes. In V. Costa, A. Moura, & M. Mira (Eds.), *Handbook of Research on Human Capital and People Management in the Tourism Industry* (pp. 73–91). IGI Global. https://doi.org/10.4018/978-1-7998-4318-4.ch004

Giacosa, E. (2018). The Increasing of the Regional Development Thanks to the Luxury Business Innovation. In L. Carvalho (Ed.), *Handbook of Research on Entrepreneurial Ecosystems and Social Dynamics in a Globalized World* (pp. 260–273). Hershey, PA: IGI Global. doi:10.4018/978-1-5225-3525-6.ch011

Glowka, G., Tusch, M., & Zehrer, A. (2021). The Risk Perception of Family Business Owner-Manager in the Tourism Industry: A Qualitative Comparison of the Intra-Firm Senior and Junior Generation. In A. Zehrer, G. Glowka, K. Schwaiger, & V. Ranacher-Lackner (Eds.), *Resiliency Models and Addressing Future Risks for Family Firms in the Tourism Industry* (pp. 126–153). IGI Global. https://doi.org/10.4018/978-1-7998-7352-5.ch006

Glykas, M., & George, J. (2017). Quality and Process Management Systems in the UAE Maritime Industry. *International Journal of Productivity Management and Assessment Technologies*, *5*(1), 20–39. doi:10.4018/IJPMAT.2017010102

Glykas, M., Valiris, G., Kokkinaki, A., & Koutsoukou, Z. (2018). Banking Business Process Management Implementation. *International Journal of Productivity Management and Assessment Technologies*, *6*(1), 50–69. doi:10.4018/IJPMAT.2018010104

Gomes, J., & Romão, M. (2017). The Balanced Scorecard: Keeping Updated and Aligned with Today's Business Trends. *International Journal of Productivity Management and Assessment Technologies*, *5*(2), 1–15. doi:10.4018/IJPMAT.2017070101

Gomes, J., & Romão, M. (2017). Aligning Information Systems and Technology with Benefit Management and Balanced Scorecard. In S. De Haes & W. Van Grembergen (Eds.), *Strategic IT Governance and Alignment in Business Settings* (pp. 112–131). Hershey, PA: IGI Global. doi:10.4018/978-1-5225-0861-8.ch005

Goyal, A. (2021). Communicating and Building Destination Brands With New Media. In M. Dinis, L. Bonixe, S. Lamy, & Z. Breda (Eds.), *Impact of New Media in Tourism* (pp. 1-20). IGI Global. https://doi.org/10.4018/978-1-7998-7095-1.ch001

Grefen, P., & Turetken, O. (2017). Advanced Business Process Management in Networked E-Business Scenarios. *International Journal of E-Business Research, 13*(4), 70–104. doi:10.4018/IJEBR.2017100105

Guasca, M., Van Broeck, A. M., & Vanneste, D. (2021). Tourism and the Social Reintegration of Colombian Ex-Combatants. In J. da Silva, Z. Breda, & F. Carbone (Eds.), *Role and Impact of Tourism in Peacebuilding and Conflict Transformation* (pp. 66-86). IGI Global. https://doi.org/10.4018/978-1-7998-5053-3.ch005

Haider, A., & Saetang, S. (2017). Strategic IT Alignment in Service Sector. In S. Rozenes & Y. Cohen (Eds.), *Handbook of Research on Strategic Alliances and Value Co-Creation in the Service Industry* (pp. 231–258). Hershey, PA: IGI Global. doi:10.4018/978-1-5225-2084-9.ch012

Hajilari, A. B., Ghadaksaz, M., & Fasghandis, G. S. (2017). Assessing Organizational Readiness for Implementing ERP System Using Fuzzy Expert System Approach. *International Journal of Enterprise Information Systems, 13*(1), 67–85. doi:10.4018/IJEIS.2017010105

Haldorai, A., Ramu, A., & Murugan, S. (2018). Social Aware Cognitive Radio Networks: Effectiveness of Social Networks as a Strategic Tool for Organizational Business Management. In H. Bansal, G. Shrivastava, G. Nguyen, & L. Stanciu (Eds.), *Social Network Analytics for Contemporary Business Organizations* (pp. 188–202). Hershey, PA: IGI Global. doi:10.4018/978-1-5225-5097-6.ch010

Hall, O. P. Jr. (2017). Social Media Driven Management Education. *International Journal of Knowledge-Based Organizations, 7*(2), 43–59. doi:10.4018/IJKBO.2017040104

Hanifah, H., Halim, H. A., Ahmad, N. H., & Vafaei-Zadeh, A. (2017). Innovation Culture as a Mediator Between Specific Human Capital and Innovation Performance Among Bumiputera SMEs in Malaysia. In N. Ahmad, T. Ramayah, H. Halim, & S. Rahman (Eds.), *Handbook of Research on Small and Medium Enterprises in Developing Countries* (pp. 261–279). Hershey, PA: IGI Global. doi:10.4018/978-1-5225-2165-5.ch012

Hartlieb, S., & Silvius, G. (2017). Handling Uncertainty in Project Management and Business Development: Similarities and Differences. In Y. Raydugin (Ed.), *Handbook of Research on Leveraging Risk and Uncertainties for Effective Project Management* (pp. 337–362). Hershey, PA: IGI Global. doi:10.4018/978-1-5225-1790-0.ch016

Hass, K. B. (2017). Living on the Edge: Managing Project Complexity. In Y. Raydugin (Ed.), *Handbook of Research on Leveraging Risk and Uncertainties for Effective Project Management* (pp. 177–201). Hershey, PA: IGI Global. doi:10.4018/978-1-5225-1790-0.ch009

Hawking, P., & Carmine Sellitto, C. (2017). Developing an Effective Strategy for Organizational Business Intelligence. In M. Tavana (Ed.), *Enterprise Information Systems and the Digitalization of Business Functions* (pp. 222–237). Hershey, PA: IGI Global. doi:10.4018/978-1-5225-2382-6.ch010

Hawking, P., & Sellitto, C. (2017). A Fast-Moving Consumer Goods Company and Business Intelligence Strategy Development. *International Journal of Enterprise Information Systems, 13*(2), 22–33. doi:10.4018/IJEIS.2017040102

Hawking, P., & Sellitto, C. (2017). Business Intelligence Strategy: Two Case Studies. *International Journal of Business Intelligence Research, 8*(2), 17–30. doi:10.4018/IJBIR.2017070102

Hee, W. J., Jalleh, G., Lai, H., & Lin, C. (2017). E-Commerce and IT Projects: Evaluation and Management Issues in Australian and Taiwanese Hospitals. *International Journal of Public Health Management and Ethics, 2*(1), 69–90. doi:10.4018/IJPHME.2017010104

Hernandez, A. A. (2018). Exploring the Factors to Green IT Adoption of SMEs in the Philippines. *Journal of Cases on Information Technology, 20*(2), 49–66. doi:10.4018/JCIT.2018040104

Hollman, A., Bickford, S., & Hollman, T. (2017). Cyber InSecurity: A Post-Mortem Attempt to Assess Cyber Problems from IT and Business Management Perspectives. *Journal of Cases on Information Technology, 19*(3), 42–70. doi:10.4018/JCIT.2017070104

Ibrahim, F., & Zainin, N. M. (2021). Exploring the Technological Impacts: The Case of Museums in Brunei Darussalam. *International Journal of Tourism and Hospitality Management in the Digital Age, 5*(1), 1–18. https://doi.org/10.4018/IJTHMDA.2021010101

Igbinakhase, I. (2017). Responsible and Sustainable Management Practices in Developing and Developed Business Environments. In Z. Fields (Ed.), *Collective Creativity for Responsible and Sustainable Business Practice* (pp. 180–207). Hershey, PA: IGI Global. doi:10.4018/978-1-5225-1823-5.ch010

Iwata, J. J., & Hoskins, R. G. (2017). Managing Indigenous Knowledge in Tanzania: A Business Perspective. In P. Jain & N. Mnjama (Eds.), *Managing Knowledge Resources and Records in Modern Organizations* (pp. 198–214). Hershey, PA: IGI Global. doi:10.4018/978-1-5225-1965-2.ch012

Jain, P. (2017). Ethical and Legal Issues in Knowledge Management Life-Cycle in Business. In P. Jain & N. Mnjama (Eds.), *Managing Knowledge Resources and Records in Modern Organizations* (pp. 82–101). Hershey, PA: IGI Global. doi:10.4018/978-1-5225-1965-2.ch006

James, S., & Hauli, E. (2017). Holistic Management Education at Tanzanian Rural Development Planning Institute. In N. Baporikar (Ed.), *Management Education for Global Leadership* (pp. 112–136). Hershey, PA: IGI Global. doi:10.4018/978-1-5225-1013-0.ch006

Janošková, M., Csikósová, A., & Čulková, K. (2018). Measurement of Company Performance as Part of Its Strategic Management. In R. Leon (Ed.), *Managerial Strategies for Business Sustainability During Turbulent Times* (pp. 309–335). Hershey, PA: IGI Global. doi:10.4018/978-1-5225-2716-9.ch017

Jean-Vasile, A., & Alecu, A. (2017). Theoretical and Practical Approaches in Understanding the Influences of Cost-Productivity-Profit Trinomial in Contemporary Enterprises. In A. Jean Vasile & D. Nicolò (Eds.), *Sustainable Entrepreneurship and Investments in the Green Economy* (pp. 28–62). Hershey, PA: IGI Global. doi:10.4018/978-1-5225-2075-7.ch002

Joia, L. A., & Correia, J. C. (2018). CIO Competencies From the IT Professional Perspective: Insights From Brazil. *Journal of Global Information Management, 26*(2), 74–103. doi:10.4018/JGIM.2018040104

Juma, A., & Mzera, N. (2017). Knowledge Management and Records Management and Competitive Advantage in Business. In P. Jain & N. Mnjama (Eds.), *Managing Knowledge Resources and Records in Modern Organizations* (pp. 15–28). Hershey, PA: IGI Global. doi:10.4018/978-1-5225-1965-2.ch002

K., I., & A, V. (2018). Monitoring and Auditing in the Cloud. In K. Munir (Ed.), *Cloud Computing Technologies for Green Enterprises* (pp. 318-350). Hershey, PA: IGI Global. https://doi.org/ doi:10.4018/978-1-5225-3038-1.ch013

Kabra, G., Ghosh, V., & Ramesh, A. (2018). Enterprise Integrated Business Process Management and Business Intelligence Framework for Business Process Sustainability. In A. Paul, D. Bhattacharyya, & S. Anand (Eds.), *Green Initiatives for Business Sustainability and Value Creation* (pp. 228–238). Hershey, PA: IGI Global. doi:10.4018/978-1-5225-2662-9.ch010

Kaoud, M. (2017). Investigation of Customer Knowledge Management: A Case Study Research. *International Journal of Service Science, Management, Engineering, and Technology*, *8*(2), 12–22. doi:10.4018/IJSSMET.2017040102

Katuu, S. (2018). A Comparative Assessment of Enterprise Content Management Maturity Models. In N. Gwangwava & M. Mutingi (Eds.), *E-Manufacturing and E-Service Strategies in Contemporary Organizations* (pp. 93–118). Hershey, PA: IGI Global. doi:10.4018/978-1-5225-3628-4.ch005

Khan, M. Y., & Abir, T. (2022). The Role of Social Media Marketing in the Tourism and Hospitality Industry: A Conceptual Study on Bangladesh. In C. Ramos, S. Quinteiro, & A. Gonçalves (Eds.), *ICT as Innovator Between Tourism and Culture* (pp. 213–229). IGI Global. https://doi.org/10.4018/978-1-7998-8165-0.ch013

Kinnunen, S., Ylä-Kujala, A., Marttonen-Arola, S., Kärri, T., & Baglee, D. (2018). Internet of Things in Asset Management: Insights from Industrial Professionals and Academia. *International Journal of Service Science, Management, Engineering, and Technology*, *9*(2), 104–119. doi:10.4018/IJSSMET.2018040105

Klein, A. Z., Sabino de Freitas, A., Machado, L., Freitas, J. C. Jr, Graziola, P. G. Jr, & Schlemmer, E. (2017). Virtual Worlds Applications for Management Education. In L. Tomei (Ed.), *Exploring the New Era of Technology-Infused Education* (pp. 279–299). Hershey, PA: IGI Global. doi:10.4018/978-1-5225-1709-2.ch017

Kővári, E., Saleh, M., & Steinbachné Hajmásy, G. (2022). The Impact of Corporate Digital Responsibility (CDR) on Internal Stakeholders' Satisfaction in Hungarian Upscale Hotels. In M. Valeri (Ed.), *New Governance and Management in Touristic Destinations* (pp. 35–51). IGI Global. https://doi.org/10.4018/978-1-6684-3889-3.ch003

Kożuch, B., & Jabłoński, A. (2017). Adopting the Concept of Business Models in Public Management. In M. Lewandowski & B. Kożuch (Eds.), *Public Sector Entrepreneurship and the Integration of Innovative Business Models* (pp. 10–46). Hershey, PA: IGI Global. doi:10.4018/978-1-5225-2215-7.ch002

Kumar, J., Adhikary, A., & Jha, A. (2017). Small Active Investors' Perceptions and Preferences Towards Tax Saving Mutual Fund Schemes in Eastern India: An Empirical Note. *International Journal of Asian Business and Information Management*, *8*(2), 35–45. doi:10.4018/IJABIM.2017040103

Latusi, S., & Fissore, M. (2021). Pilgrimage Routes to Happiness: Comparing the Camino de Santiago and Via Francigena. In A. Perinotto, V. Mayer, & J. Soares (Eds.), *Rebuilding and Restructuring the Tourism Industry: Infusion of Happiness and Quality of Life* (pp. 157–182). IGI Global. https://doi.org/10.4018/978-1-7998-7239-9.ch008

Lavassani, K. M., & Movahedi, B. (2017). Applications Driven Information Systems: Beyond Networks toward Business Ecosystems. *International Journal of Innovation in the Digital Economy*, *8*(1), 61–75. doi:10.4018/IJIDE.2017010104

Lazzareschi, V. H., & Brito, M. S. (2017). Strategic Information Management: Proposal of Business Project Model. In G. Jamil, A. Soares, & C. Pessoa (Eds.), *Handbook of Research on Information Management for Effective Logistics and Supply Chains* (pp. 59–88). Hershey, PA: IGI Global. doi:10.4018/978-1-5225-0973-8.ch004

Lechuga Sancho, M. P., & Martín Navarro, A. (2022). Evolution of the Literature on Social Responsibility in the Tourism Sector: A Systematic Literature Review. In G. Fernandes (Ed.), *Challenges and New Opportunities for Tourism in Inland Territories: Ecocultural Resources and Sustainable Initiatives* (pp. 169–186). IGI Global. https://doi.org/10.4018/978-1-7998-7339-6.ch010

Lederer, M., Kurz, M., & Lazarov, P. (2017). Usage and Suitability of Methods for Strategic Business Process Initiatives: A Multi Case Study Research. *International Journal of Productivity Management and Assessment Technologies*, *5*(1), 40–51. doi:10.4018/IJPMAT.2017010103

Lee, I. (2017). A Social Enterprise Business Model and a Case Study of Pacific Community Ventures (PCV). In V. Potocan, M. Üngan, & Z. Nedelko (Eds.), *Handbook of Research on Managerial Solutions in Non-Profit Organizations* (pp. 182–204). Hershey, PA: IGI Global. doi:10.4018/978-1-5225-0731-4.ch009

Leon, L. A., Seal, K. C., Przasnyski, Z. H., & Wiedenman, I. (2017). Skills and Competencies Required for Jobs in Business Analytics: A Content Analysis of Job Advertisements Using Text Mining. *International Journal of Business Intelligence Research*, *8*(1), 1–25. doi:10.4018/IJBIR.2017010101

Levy, C. L., & Elias, N. I. (2017). SOHO Users' Perceptions of Reliability and Continuity of Cloud-Based Services. In M. Moore (Ed.), *Cybersecurity Breaches and Issues Surrounding Online Threat Protection* (pp. 248–287). Hershey, PA: IGI Global. doi:10.4018/978-1-5225-1941-6.ch011

Levy, M. (2018). Change Management Serving Knowledge Management and Organizational Development: Reflections and Review. In N. Baporikar (Ed.), *Global Practices in Knowledge Management for Societal and Organizational Development* (pp. 256–270). Hershey, PA: IGI Global. doi:10.4018/978-1-5225-3009-1.ch012

Lewandowski, M. (2017). Public Organizations and Business Model Innovation: The Role of Public Service Design. In M. Lewandowski & B. Kożuch (Eds.), *Public Sector Entrepreneurship and the Integration of Innovative Business Models* (pp. 47–72). Hershey, PA: IGI Global. doi:10.4018/978-1-5225-2215-7.ch003

Lhannaoui, H., Kabbaj, M. I., & Bakkoury, Z. (2017). A Survey of Risk-Aware Business Process Modelling. *International Journal of Risk and Contingency Management, 6*(3), 14–26. doi:10.4018/IJRCM.2017070102

Li, J., Sun, W., Jiang, W., Yang, H., & Zhang, L. (2017). How the Nature of Exogenous Shocks and Crises Impact Company Performance?: The Effects of Industry Characteristics. *International Journal of Risk and Contingency Management, 6*(4), 40–55. doi:10.4018/IJRCM.2017100103

Lopez-Fernandez, M., Perez-Perez, M., Serrano-Bedia, A., & Cobo-Gonzalez, A. (2021). Small and Medium Tourism Enterprise Survival in Times of Crisis: "El Capricho de Gaudí. In D. Toubes & N. Araújo-Vila (Eds.), *Risk, Crisis, and Disaster Management in Small and Medium-Sized Tourism Enterprises* (pp. 103–129). IGI Global. doi:10.4018/978-1-7998-6996-2.ch005

Mahajan, A., Maidullah, S., & Hossain, M. R. (2022). Experience Toward Smart Tour Guide Apps in Travelling: An Analysis of Users' Reviews on Audio Odigos and Trip My Way. In R. Augusto Costa, F. Brandão, Z. Breda, & C. Costa (Eds.), *Planning and Managing the Experience Economy in Tourism* (pp. 255-273). IGI Global. https://doi.org/10.4018/978-1-7998-8775-1.ch014

Malega, P. (2017). Small and Medium Enterprises in the Slovak Republic: Status and Competitiveness of SMEs in the Global Markets and Possibilities of Optimization. In M. Vemić (Ed.), *Optimal Management Strategies in Small and Medium Enterprises* (pp. 102–124). Hershey, PA: IGI Global. doi:10.4018/978-1-5225-1949-2.ch006

Malewska, K. M. (2017). Intuition in Decision-Making on the Example of a Non-Profit Organization. In V. Potocan, M. Ünğan, & Z. Nedelko (Eds.), *Handbook of Research on Managerial Solutions in Non-Profit Organizations* (pp. 378–399). Hershey, PA: IGI Global. doi:10.4018/978-1-5225-0731-4.ch018

Maroofi, F. (2017). Entrepreneurial Orientation and Organizational Learning Ability Analysis for Innovation and Firm Performance. In N. Baporikar (Ed.), *Innovation and Shifting Perspectives in Management Education* (pp. 144–165). Hershey, PA: IGI Global. doi:10.4018/978-1-5225-1019-2.ch007

Marques, M., Moleiro, D., Brito, T. M., & Marques, T. (2021). Customer Relationship Management as an Important Relationship Marketing Tool: The Case of the Hospitality Industry in Estoril Coast. In M. Dinis, L. Bonixe, S. Lamy, & Z. Breda (Eds.), Impact of New Media in Tourism (pp. 39-56). IGI Global. https://doi.org/ doi:10.4018/978-1-7998-7095-1.ch003

Martins, P. V., & Zacarias, M. (2017). A Web-based Tool for Business Process Improvement. *International Journal of Web Portals, 9*(2), 68–84. doi:10.4018/IJWP.2017070104

Matthies, B., & Coners, A. (2017). Exploring the Conceptual Nature of e-Business Projects. *Journal of Electronic Commerce in Organizations, 15*(3), 33–63. doi:10.4018/JECO.2017070103

Mayer, V. F., Fraga, C. C., & Silva, L. C. (2021). Contributions of Neurosciences to Studies of Well-Being in Tourism. In A. Perinotto, V. Mayer, & J. Soares (Eds.), *Rebuilding and Restructuring the Tourism Industry: Infusion of Happiness and Quality of Life* (pp. 108–128). IGI Global. https://doi.org/10.4018/978-1-7998-7239-9.ch006

McKee, J. (2018). Architecture as a Tool to Solve Business Planning Problems. In M. Khosrow-Pour, D.B.A. (Ed.), Encyclopedia of Information Science and Technology, Fourth Edition (pp. 573-586). Hershey, PA: IGI Global. doi:10.4018/978-1-5225-2255-3.ch050

McMurray, A. J., Cross, J., & Caponecchia, C. (2018). The Risk Management Profession in Australia: Business Continuity Plan Practices. In N. Bajgoric (Ed.), *Always-On Enterprise Information Systems for Modern Organizations* (pp. 112–129). Hershey, PA: IGI Global. doi:10.4018/978-1-5225-3704-5.ch006

Meddah, I. H., & Belkadi, K. (2018). Mining Patterns Using Business Process Management. In R. Hamou (Ed.), *Handbook of Research on Biomimicry in Information Retrieval and Knowledge Management* (pp. 78–89). Hershey, PA: IGI Global. doi:10.4018/978-1-5225-3004-6.ch005

Melian, A. G., & Camprubí, R. (2021). The Accessibility of Museum Websites: The Case of Barcelona. In C. Eusébio, L. Teixeira, & M. Carneiro (Eds.), *ICT Tools and Applications for Accessible Tourism* (pp. 234–255). IGI Global. https://doi.org/10.4018/978-1-7998-6428-8.ch011

Mendes, L. (2017). TQM and Knowledge Management: An Integrated Approach Towards Tacit Knowledge Management. In D. Jaziri-Bouagina & G. Jamil (Eds.), *Handbook of Research on Tacit Knowledge Management for Organizational Success* (pp. 236–263). Hershey, PA: IGI Global. doi:10.4018/978-1-5225-2394-9.ch009

Menezes, V. D., & Cavagnaro, E. (2021). Communicating Sustainable Initiatives in the Hotel Industry: The Case of the Hotel Jakarta Amsterdam. In F. Brandão, Z. Breda, R. Costa, & C. Costa (Eds.), *Handbook of Research on the Role of Tourism in Achieving Sustainable Development Goals* (pp. 224-234). IGI Global. https://doi.org/10.4018/978-1-7998-5691-7.ch013

Menezes, V. D., & Cavagnaro, E. (2021). Communicating Sustainable Initiatives in the Hotel Industry: The Case of the Hotel Jakarta Amsterdam. In F. Brandão, Z. Breda, R. Costa, & C. Costa (Eds.), *Handbook of Research on the Role of Tourism in Achieving Sustainable Development Goals* (pp. 224-234). IGI Global. https://doi.org/10.4018/978-1-7998-5691-7.ch013

Mitas, O., Bastiaansen, M., & Boode, W. (2022). If You're Happy, I'm Happy: Emotion Contagion at a Tourist Information Center. In R. Augusto Costa, F. Brandão, Z. Breda, & C. Costa (Eds.), *Planning and Managing the Experience Economy in Tourism* (pp. 122-140). IGI Global. https://doi.org/10.4018/978-1-7998-8775-1.ch007

Mnjama, N. M. (2017). Preservation of Recorded Information in Public and Private Sector Organizations. In P. Jain & N. Mnjama (Eds.), *Managing Knowledge Resources and Records in Modern Organizations* (pp. 149–167). Hershey, PA: IGI Global. doi:10.4018/978-1-5225-1965-2.ch009

Mokoqama, M., & Fields, Z. (2017). Principles of Responsible Management Education (PRME): Call for Responsible Management Education. In Z. Fields (Ed.), *Collective Creativity for Responsible and Sustainable Business Practice* (pp. 229–241). Hershey, PA: IGI Global. doi:10.4018/978-1-5225-1823-5.ch012

Monteiro, A., Lopes, S., & Carbone, F. (2021). Academic Mobility: Bridging Tourism and Peace Education. In J. da Silva, Z. Breda, & F. Carbone (Eds.), *Role and Impact of Tourism in Peacebuilding and Conflict Transformation* (pp. 275-301). IGI Global. https://doi.org/10.4018/978-1-7998-5053-3.ch016

Muniapan, B. (2017). Philosophy and Management: The Relevance of Vedanta in Management. In P. Ordóñez de Pablos (Ed.), *Managerial Strategies and Solutions for Business Success in Asia* (pp. 124–139). Hershey, PA: IGI Global. doi:10.4018/978-1-5225-1886-0.ch007

Murad, S. E., & Dowaji, S. (2017). Using Value-Based Approach for Managing Cloud-Based Services. In A. Turuk, B. Sahoo, & S. Addya (Eds.), *Resource Management and Efficiency in Cloud Computing Environments* (pp. 33–60). Hershey, PA: IGI Global. doi:10.4018/978-1-5225-1721-4.ch002

Mutahar, A. M., Daud, N. M., Thurasamy, R., Isaac, O., & Abdulsalam, R. (2018). The Mediating of Perceived Usefulness and Perceived Ease of Use: The Case of Mobile Banking in Yemen. *International Journal of Technology Diffusion, 9*(2), 21–40. doi:10.4018/IJTD.2018040102

Naidoo, V. (2017). E-Learning and Management Education at African Universities. In N. Baporikar (Ed.), *Management Education for Global Leadership* (pp. 181–201). Hershey, PA: IGI Global. doi:10.4018/978-1-5225-1013-0.ch009

Naidoo, V., & Igbinakhase, I. (2018). Opportunities and Challenges of Knowledge Retention in SMEs. In N. Baporikar (Ed.), *Knowledge Integration Strategies for Entrepreneurship and Sustainability* (pp. 70–94). Hershey, PA: IGI Global. doi:10.4018/978-1-5225-5115-7.ch004

Naumov, N., & Costandachi, G. (2021). Creativity and Entrepreneurship: Gastronomic Tourism in Mexico. In J. Soares (Ed.), *Innovation and Entrepreneurial Opportunities in Community Tourism* (pp. 90–108). IGI Global. https://doi.org/10.4018/978-1-7998-4855-4.ch006

Nayak, S., & Prabhu, N. (2017). Paradigm Shift in Management Education: Need for a Cross Functional Perspective. In N. Baporikar (Ed.), *Management Education for Global Leadership* (pp. 241–255). Hershey, PA: IGI Global. doi:10.4018/978-1-5225-1013-0.ch012

Nedelko, Z., & Potocan, V. (2017). Management Solutions in Non-Profit Organizations: Case of Slovenia. In V. Potocan, M. Üngan, & Z. Nedelko (Eds.), *Handbook of Research on Managerial Solutions in Non-Profit Organizations* (pp. 1–22). Hershey, PA: IGI Global. doi:10.4018/978-1-5225-0731-4.ch001

Nedelko, Z., & Potocan, V. (2017). Priority of Management Tools Utilization among Managers: International Comparison. In V. Wang (Ed.), *Encyclopedia of Strategic Leadership and Management* (pp. 1083–1094). Hershey, PA: IGI Global. doi:10.4018/978-1-5225-1049-9.ch075

Nedelko, Z., Raudeliūnienė, J., & Črešnar, R. (2018). Knowledge Dynamics in Supply Chain Management. In N. Baporikar (Ed.), *Knowledge Integration Strategies for Entrepreneurship and Sustainability* (pp. 150–166). Hershey, PA: IGI Global. doi:10.4018/978-1-5225-5115-7.ch008

Nguyen, H. T., & Hipsher, S. A. (2018). Innovation and Creativity Used by Private Sector Firms in a Resources-Constrained Environment. In S. Hipsher (Ed.), *Examining the Private Sector's Role in Wealth Creation and Poverty Reduction* (pp. 219–238). Hershey, PA: IGI Global. doi:10.4018/978-1-5225-3117-3.ch010

Obicci, P. A. (2017). Risk Sharing in a Partnership. In *Risk Management Strategies in Public-Private Partnerships* (pp. 115–152). Hershey, PA: IGI Global. doi:10.4018/978-1-5225-2503-5.ch004

Obidallah, W. J., & Raahemi, B. (2017). Managing Changes in Service Oriented Virtual Organizations: A Structural and Procedural Framework to Facilitate the Process of Change. *Journal of Electronic Commerce in Organizations*, *15*(1), 59–83. doi:10.4018/JECO.2017010104

Ojo, O. (2017). Impact of Innovation on the Entrepreneurial Success in Selected Business Enterprises in South-West Nigeria. *International Journal of Innovation in the Digital Economy*, *8*(2), 29–38. doi:10.4018/IJIDE.2017040103

Okdinawati, L., Simatupang, T. M., & Sunitiyoso, Y. (2017). Multi-Agent Reinforcement Learning for Value Co-Creation of Collaborative Transportation Management (CTM). *International Journal of Information Systems and Supply Chain Management*, *10*(3), 84–95. doi:10.4018/IJISSCM.2017070105

Olivera, V. A., & Carrillo, I. M. (2021). Organizational Culture: A Key Element for the Development of Mexican Micro and Small Tourist Companies. In J. Soares (Ed.), *Innovation and Entrepreneurial Opportunities in Community Tourism* (pp. 227–242). IGI Global. doi:10.4018/978-1-7998-4855-4.ch013

Ossorio, M. (2022). Corporate Museum Experiences in Enogastronomic Tourism. In R. Augusto Costa, F. Brandão, Z. Breda, & C. Costa (Eds.), Planning and Managing the Experience Economy in Tourism (pp. 107-121). IGI Global. https://doi.org/ doi:10.4018/978-1-7998-8775-1.ch006

Ossorio, M. (2022). Enogastronomic Tourism in Times of Pandemic. In G. Fernandes (Ed.), *Challenges and New Opportunities for Tourism in Inland Territories: Ecocultural Resources and Sustainable Initiatives* (pp. 241–255). IGI Global. https://doi.org/10.4018/978-1-7998-7339-6.ch014

Özekici, Y. K. (2022). ICT as an Acculturative Agent and Its Role in the Tourism Context: Introduction, Acculturation Theory, Progress of the Acculturation Theory in Extant Literature. In C. Ramos, S. Quinteiro, & A. Gonçalves (Eds.), *ICT as Innovator Between Tourism and Culture* (pp. 42–66). IGI Global. https://doi.org/10.4018/978-1-7998-8165-0.ch004

Pal, K. (2018). Building High Quality Big Data-Based Applications in Supply Chains. In A. Kumar & S. Saurav (Eds.), *Supply Chain Management Strategies and Risk Assessment in Retail Environments* (pp. 1–24). Hershey, PA: IGI Global. doi:10.4018/978-1-5225-3056-5.ch001

Palos-Sanchez, P. R., & Correia, M. B. (2018). Perspectives of the Adoption of Cloud Computing in the Tourism Sector. In J. Rodrigues, C. Ramos, P. Cardoso, & C. Henriques (Eds.), *Handbook of Research on Technological Developments for Cultural Heritage and eTourism Applications* (pp. 377–400). Hershey, PA: IGI Global. doi:10.4018/978-1-5225-2927-9.ch018

Papadopoulou, G. (2021). Promoting Gender Equality and Women Empowerment in the Tourism Sector. In F. Brandão, Z. Breda, R. Costa, & C. Costa (Eds.), Handbook of Research on the Role of Tourism in Achieving Sustainable Development Goals (pp. 152-174). IGI Global. https://doi.org/ doi:10.4018/978-1-7998-5691-7.ch009

Papp-Váry, Á. F., & Tóth, T. Z. (2022). Analysis of Budapest as a Film Tourism Destination. In R. Baleiro & R. Pereira (Eds.), *Global Perspectives on Literary Tourism and Film-Induced Tourism* (pp. 257-279). IGI Global. https://doi.org/10.4018/978-1-7998-8262-6.ch014

Patiño, B. E. (2017). New Generation Management by Convergence and Individual Identity: A Systemic and Human-Oriented Approach. In N. Baporikar (Ed.), *Innovation and Shifting Perspectives in Management Education* (pp. 119–143). Hershey, PA: IGI Global. doi:10.4018/978-1-5225-1019-2.ch006

Patro, C. S. (2021). Digital Tourism: Influence of E-Marketing Technology. In M. Dinis, L. Bonixe, S. Lamy, & Z. Breda (Eds.), *Impact of New Media in Tourism* (pp. 234-254). IGI Global. https://doi.org/10.4018/978-1-7998-7095-1.ch014

Pawliczek, A., & Rössler, M. (2017). Knowledge of Management Tools and Systems in SMEs: Knowledge Transfer in Management. In A. Bencsik (Ed.), *Knowledge Management Initiatives and Strategies in Small and Medium Enterprises* (pp. 180–203). Hershey, PA: IGI Global. doi:10.4018/978-1-5225-1642-2.ch009

Pejic-Bach, M., Omazic, M. A., Aleksic, A., & Zoroja, J. (2018). Knowledge-Based Decision Making: A Multi-Case Analysis. In R. Leon (Ed.), *Managerial Strategies for Business Sustainability During Turbulent Times* (pp. 160–184). Hershey, PA: IGI Global. doi:10.4018/978-1-5225-2716-9.ch009

Perano, M., Hysa, X., & Calabrese, M. (2018). Strategic Planning, Cultural Context, and Business Continuity Management: Business Cases in the City of Shkoder. In A. Presenza & L. Sheehan (Eds.), *Geopolitics and Strategic Management in the Global Economy* (pp. 57–77). Hershey, PA: IGI Global. doi:10.4018/978-1-5225-2673-5.ch004

Pereira, R., Mira da Silva, M., & Lapão, L. V. (2017). IT Governance Maturity Patterns in Portuguese Healthcare. In S. De Haes & W. Van Grembergen (Eds.), *Strategic IT Governance and Alignment in Business Settings* (pp. 24–52). Hershey, PA: IGI Global. doi:10.4018/978-1-5225-0861-8.ch002

Pérez-Uribe, R. I., Torres, D. A., Jurado, S. P., & Prada, D. M. (2018). Cloud Tools for the Development of Project Management in SMEs. In R. Perez-Uribe, C. Salcedo-Perez, & D. Ocampo-Guzman (Eds.), *Handbook of Research on Intrapreneurship and Organizational Sustainability in SMEs* (pp. 95–120). Hershey, PA: IGI Global. doi:10.4018/978-1-5225-3543-0.ch005

Petrisor, I., & Cozmiuc, D. (2017). Global Supply Chain Management Organization at Siemens in the Advent of Industry 4.0. In L. Saglietto & C. Cezanne (Eds.), *Global Intermediation and Logistics Service Providers* (pp. 123–142). Hershey, PA: IGI Global. doi:10.4018/978-1-5225-2133-4.ch007

Pierce, J. M., Velliaris, D. M., & Edwards, J. (2017). A Living Case Study: A Journey Not a Destination. In N. Silton (Ed.), *Exploring the Benefits of Creativity in Education, Media, and the Arts* (pp. 158–178). Hershey, PA: IGI Global. doi:10.4018/978-1-5225-0504-4.ch008

Pipia, S., & Pipia, S. (2021). Challenges of Religious Tourism in the Conflict Region: An Example of Jerusalem. In E. Alaverdov & M. Bari (Eds.), *Global Development of Religious Tourism* (pp. 135-148). IGI Global. https://doi.org/10.4018/978-1-7998-5792-1.ch009

Poulaki, P., Kritikos, A., Vasilakis, N., & Valeri, M. (2022). The Contribution of Female Creativity to the Development of Gastronomic Tourism in Greece: The Case of the Island of Naxos in the South Aegean Region. In M. Valeri (Ed.), *New Governance and Management in Touristic Destinations* (pp. 246–258). IGI Global. https://doi.org/10.4018/978-1-6684-3889-3.ch015

Radosavljevic, M., & Andjelkovic, A. (2017). Multi-Criteria Decision Making Approach for Choosing Business Process for the Improvement: Upgrading of the Six Sigma Methodology. In J. Stanković, P. Delias, S. Marinković, & S. Rochhia (Eds.), *Tools and Techniques for Economic Decision Analysis* (pp. 225–247). Hershey, PA: IGI Global. doi:10.4018/978-1-5225-0959-2.ch011

Radovic, V. M. (2017). Corporate Sustainability and Responsibility and Disaster Risk Reduction: A Serbian Overview. In M. Camilleri (Ed.), *CSR 2.0 and the New Era of Corporate Citizenship* (pp. 147–164). Hershey, PA: IGI Global. doi:10.4018/978-1-5225-1842-6.ch008

Raghunath, K. M., Devi, S. L., & Patro, C. S. (2018). Impact of Risk Assessment Models on Risk Factors: A Holistic Outlook. In K. Strang, M. Korstanje, & N. Vajjhala (Eds.), *Research, Practices, and Innovations in Global Risk and Contingency Management* (pp. 134–153). Hershey, PA: IGI Global. doi:10.4018/978-1-5225-4754-9.ch008

Raman, A., & Goyal, D. P. (2017). Extending IMPLEMENT Framework for Enterprise Information Systems Implementation to Information System Innovation. In M. Tavana (Ed.), *Enterprise Information Systems and the Digitalization of Business Functions* (pp. 137–177). Hershey, PA: IGI Global. doi:10.4018/978-1-5225-2382-6.ch007

Rao, Y., & Zhang, Y. (2017). The Construction and Development of Academic Library Digital Special Subject Databases. In L. Ruan, Q. Zhu, & Y. Ye (Eds.), *Academic Library Development and Administration in China* (pp. 163–183). Hershey, PA: IGI Global. doi:10.4018/978-1-5225-0550-1.ch010

Ravasan, A. Z., Mohammadi, M. M., & Hamidi, H. (2018). An Investigation Into the Critical Success Factors of Implementing Information Technology Service Management Frameworks. In K. Jakobs (Ed.), *Corporate and Global Standardization Initiatives in Contemporary Society* (pp. 200–218). Hershey, PA: IGI Global. doi:10.4018/978-1-5225-5320-5.ch009

Rezaie, S., Mirabedini, S. J., & Abtahi, A. (2018). Designing a Model for Implementation of Business Intelligence in the Banking Industry. *International Journal of Enterprise Information Systems, 14*(1), 77–103. doi:10.4018/IJEIS.2018010105

Richards, V., Matthews, N., Williams, O. J., & Khan, Z. (2021). The Challenges of Accessible Tourism Information Systems for Tourists With Vision Impairment: Sensory Communications Beyond the Screen. In C. Eusébio, L. Teixeira, & M. Carneiro (Eds.), *ICT Tools and Applications for Accessible Tourism* (pp. 26–54). IGI Global. https://doi.org/10.4018/978-1-7998-6428-8.ch002

Rodrigues de Souza Neto, V., & Marques, O. (2021). Rural Tourism Fostering Welfare Through Sustainable Development: A Conceptual Approach. In A. Perinotto, V. Mayer, & J. Soares (Eds.), *Rebuilding and Restructuring the Tourism Industry: Infusion of Happiness and Quality of Life* (pp. 38–57). IGI Global. https://doi.org/10.4018/978-1-7998-7239-9.ch003

Romano, L., Grimaldi, R., & Colasuonno, F. S. (2017). Demand Management as a Success Factor in Project Portfolio Management. In L. Romano (Ed.), *Project Portfolio Management Strategies for Effective Organizational Operations* (pp. 202–219). Hershey, PA: IGI Global. doi:10.4018/978-1-5225-2151-8.ch008

Rubio-Escuderos, L., & García-Andreu, H. (2021). Competitiveness Factors of Accessible Tourism E-Travel Agencies. In C. Eusébio, L. Teixeira, & M. Carneiro (Eds.), *ICT Tools and Applications for Accessible Tourism* (pp. 196–217). IGI Global. https://doi.org/10.4018/978-1-7998-6428-8.ch009

Rucci, A. C., Porto, N., Darcy, S., & Becka, L. (2021). Smart and Accessible Cities?: Not Always – The Case for Accessible Tourism Initiatives in Buenos Aries and Sydney. In C. Eusébio, L. Teixeira, & M. Carneiro (Eds.), *ICT Tools and Applications for Accessible Tourism* (pp. 115–145). IGI Global. https://doi.org/10.4018/978-1-7998-6428-8.ch006

Ruhi, U. (2018). Towards an Interdisciplinary Socio-Technical Definition of Virtual Communities. In M. Khosrow-Pour, D.B.A. (Ed.), Encyclopedia of Information Science and Technology, Fourth Edition (pp. 4278-4295). Hershey, PA: IGI Global. doi:10.4018/978-1-5225-2255-3.ch371

Ryan, L., Catena, M., Ros, P., & Stephens, S. (2021). Designing Entrepreneurial Ecosystems to Support Resource Management in the Tourism Industry. In V. Costa, A. Moura, & M. Mira (Eds.), *Handbook of Research on Human Capital and People Management in the Tourism Industry* (pp. 265–281). IGI Global. https://doi.org/10.4018/978-1-7998-4318-4.ch013

Sabuncu, I. (2021). Understanding Tourist Perceptions and Expectations During Pandemic Through Social Media Big Data. In M. Demir, A. Dalgıç, & F. Ergen (Eds.), *Handbook of Research on the Impacts and Implications of COVID-19 on the Tourism Industry* (pp. 330–350). IGI Global. https://doi.org/10.4018/978-1-7998-8231-2.ch016

Safari, M. R., & Jiang, Q. (2018). The Theory and Practice of IT Governance Maturity and Strategies Alignment: Evidence From Banking Industry. *Journal of Global Information Management, 26*(2), 127–146. doi:10.4018/JGIM.2018040106

Sahoo, J., Pati, B., & Mohanty, B. (2017). Knowledge Management as an Academic Discipline: An Assessment. In B. Gunjal (Ed.), *Managing Knowledge and Scholarly Assets in Academic Libraries* (pp. 99–126). Hershey, PA: IGI Global. doi:10.4018/978-1-5225-1741-2.ch005

Saini, D. (2017). Relevance of Teaching Values and Ethics in Management Education. In N. Baporikar (Ed.), *Management Education for Global Leadership* (pp. 90–111). Hershey, PA: IGI Global. doi:10.4018/978-1-5225-1013-0.ch005

Sambhanthan, A. (2017). Assessing and Benchmarking Sustainability in Organisations: An Integrated Conceptual Model. *International Journal of Systems and Service-Oriented Engineering, 7*(4), 22–43. doi:10.4018/IJSSOE.2017100102

Sambhanthan, A., & Potdar, V. (2017). A Study of the Parameters Impacting Sustainability in Information Technology Organizations. *International Journal of Knowledge-Based Organizations, 7*(3), 27–39. doi:10.4018/IJKBO.2017070103

Sánchez-Fernández, M. D., & Manríquez, M. R. (2018). The Entrepreneurial Spirit Based on Social Values: The Digital Generation. In P. Isaias & L. Carvalho (Eds.), *User Innovation and the Entrepreneurship Phenomenon in the Digital Economy* (pp. 173–193). Hershey, PA: IGI Global. doi:10.4018/978-1-5225-2826-5.ch009

Sanchez-Ruiz, L., & Blanco, B. (2017). Process Management for SMEs: Barriers, Enablers, and Benefits. In M. Vemić (Ed.), *Optimal Management Strategies in Small and Medium Enterprises* (pp. 293–319). Hershey, PA: IGI Global. doi:10.4018/978-1-5225-1949-2.ch014

Sanz, L. F., Gómez-Pérez, J., & Castillo-Martinez, A. (2018). Analysis of the European ICT Competence Frameworks. In V. Ahuja & S. Rathore (Eds.), *Multidisciplinary Perspectives on Human Capital and Information Technology Professionals* (pp. 225–245). Hershey, PA: IGI Global. doi:10.4018/978-1-5225-5297-0.ch012

Sarvepalli, A., & Godin, J. (2017). Business Process Management in the Classroom. *Journal of Cases on Information Technology*, *19*(2), 17–28. doi:10.4018/JCIT.2017040102

Saxena, G. G., & Saxena, A. (2021). Host Community Role in Medical Tourism Development. In M. Singh & S. Kumaran (Eds.), *Growth of the Medical Tourism Industry and Its Impact on Society: Emerging Research and Opportunities* (pp. 105–127). IGI Global. https://doi.org/10.4018/978-1-7998-3427-4.ch006

Saygili, E. E., Ozturkoglu, Y., & Kocakulah, M. C. (2017). End Users' Perceptions of Critical Success Factors in ERP Applications. *International Journal of Enterprise Information Systems*, *13*(4), 58–75. doi:10.4018/IJEIS.2017100104

Saygili, E. E., & Saygili, A. T. (2017). Contemporary Issues in Enterprise Information Systems: A Critical Review of CSFs in ERP Implementations. In M. Tavana (Ed.), *Enterprise Information Systems and the Digitalization of Business Functions* (pp. 120–136). Hershey, PA: IGI Global. doi:10.4018/978-1-5225-2382-6.ch006

Schwaiger, K. M., & Zehrer, A. (2021). The COVID-19 Pandemic and Organizational Resilience in Hospitality Family Firms: A Qualitative Approach. In A. Zehrer, G. Glowka, K. Schwaiger, & V. Ranacher-Lackner (Eds.), *Resiliency Models and Addressing Future Risks for Family Firms in the Tourism Industry* (pp. 32–49). IGI Global. https://doi.org/10.4018/978-1-7998-7352-5.ch002

Scott, N., & Campos, A. C. (2022). Cognitive Science of Tourism Experiences. In R. Augusto Costa, F. Brandão, Z. Breda, & C. Costa (Eds.), Planning and Managing the Experience Economy in Tourism (pp. 1-21). IGI Global. https://doi.org/ doi:10.4018/978-1-7998-8775-1.ch001

Seidenstricker, S., & Antonino, A. (2018). Business Model Innovation-Oriented Technology Management for Emergent Technologies. In M. Khosrow-Pour, D.B.A. (Ed.), Encyclopedia of Information Science and Technology, Fourth Edition (pp. 4560-4569). Hershey, PA: IGI Global. doi:10.4018/978-1-5225-2255-3.ch396

Selvi, M. S. (2021). Changes in Tourism Sales and Marketing Post COVID-19. In M. Demir, A. Dalgıç, & F. Ergen (Eds.), *Handbook of Research on the Impacts and Implications of COVID-19 on the Tourism Industry* (pp. 437–460). IGI Global. doi:10.4018/978-1-7998-8231-2.ch021

Senaratne, S., & Gunarathne, A. D. (2017). Excellence Perspective for Management Education from a Global Accountants' Hub in Asia. In N. Baporikar (Ed.), *Management Education for Global Leadership* (pp. 158–180). Hershey, PA: IGI Global. doi:10.4018/978-1-5225-1013-0.ch008

Sensuse, D. I., & Cahyaningsih, E. (2018). Knowledge Management Models: A Summative Review. *International Journal of Information Systems in the Service Sector, 10*(1), 71–100. doi:10.4018/IJISSS.2018010105

Seth, M., Goyal, D., & Kiran, R. (2017). Diminution of Impediments in Implementation of Supply Chain Management Information System for Enhancing its Effectiveness in Indian Automobile Industry. *Journal of Global Information Management, 25*(3), 1–20. doi:10.4018/JGIM.2017070101

Seyal, A. H., & Rahman, M. N. (2017). Investigating Impact of Inter-Organizational Factors in Measuring ERP Systems Success: Bruneian Perspectives. In M. Tavana (Ed.), *Enterprise Information Systems and the Digitalization of Business Functions* (pp. 178–204). Hershey, PA: IGI Global. doi:10.4018/978-1-5225-2382-6.ch008

Shaqrah, A. A. (2018). Analyzing Business Intelligence Systems Based on 7s Model of McKinsey. *International Journal of Business Intelligence Research, 9*(1), 53–63. doi:10.4018/IJBIR.2018010104

Sharma, A. J. (2017). Enhancing Sustainability through Experiential Learning in Management Education. In N. Baporikar (Ed.), *Management Education for Global Leadership* (pp. 256–274). Hershey, PA: IGI Global. doi:10.4018/978-1-5225-1013-0.ch013

Shetty, K. P. (2017). Responsible Global Leadership: Ethical Challenges in Management Education. In N. Baporikar (Ed.), *Innovation and Shifting Perspectives in Management Education* (pp. 194–223). Hershey, PA: IGI Global. doi:10.4018/978-1-5225-1019-2.ch009

Sinthupundaja, J., & Kohda, Y. (2017). Effects of Corporate Social Responsibility and Creating Shared Value on Sustainability. *International Journal of Sustainable Entrepreneurship and Corporate Social Responsibility, 2*(1), 27–38. doi:10.4018/IJSECSR.2017010103

Škarica, I., & Hrgović, A. V. (2018). Implementation of Total Quality Management Principles in Public Health Institutes in the Republic of Croatia. *International Journal of Productivity Management and Assessment Technologies, 6*(1), 1–16. doi:10.4018/IJPMAT.2018010101

Skokic, V. (2021). How Small Hotel Owners Practice Resilience: Longitudinal Study Among Small Family Hotels in Croatia. In A. Zehrer, G. Glowka, K. Schwaiger, & V. Ranacher-Lackner (Eds.), *Resiliency Models and Addressing Future Risks for Family Firms in the Tourism Industry* (pp. 50–73). IGI Global. doi:10.4018/978-1-7998-7352-5.ch003

Smuts, H., Kotzé, P., Van der Merwe, A., & Loock, M. (2017). Framework for Managing Shared Knowledge in an Information Systems Outsourcing Context. *International Journal of Knowledge Management, 13*(4), 1–30. doi:10.4018/IJKM.2017100101

Sousa, M. J., Cruz, R., Dias, I., & Caracol, C. (2017). Information Management Systems in the Supply Chain. In G. Jamil, A. Soares, & C. Pessoa (Eds.), *Handbook of Research on Information Management for Effective Logistics and Supply Chains* (pp. 469–485). Hershey, PA: IGI Global. doi:10.4018/978-1-5225-0973-8.ch025

Spremic, M., Turulja, L., & Bajgoric, N. (2018). Two Approaches in Assessing Business Continuity Management Attitudes in the Organizational Context. In N. Bajgoric (Ed.), *Always-On Enterprise Information Systems for Modern Organizations* (pp. 159–183). Hershey, PA: IGI Global. doi:10.4018/978-1-5225-3704-5.ch008

Steenkamp, A. L. (2018). Some Insights in Computer Science and Information Technology. In *Examining the Changing Role of Supervision in Doctoral Research Projects: Emerging Research and Opportunities* (pp. 113–133). Hershey, PA: IGI Global. doi:10.4018/978-1-5225-2610-0.ch005

Stipanović, C., Rudan, E., & Zubović, V. (2022). Reaching the New Tourist Through Creativity: Sustainable Development Challenges in Croatian Coastal Towns. In M. Valeri (Ed.), *New Governance and Management in Touristic Destinations* (pp. 231–245). IGI Global. https://doi.org/10.4018/978-1-6684-3889-3.ch014

Tabach, A., & Croteau, A. (2017). Configurations of Information Technology Governance Practices and Business Unit Performance. *International Journal of IT/Business Alignment and Governance*, *8*(2), 1–27. doi:10.4018/IJITBAG.2017070101

Talaue, G. M., & Iqbal, T. (2017). Assessment of e-Business Mode of Selected Private Universities in the Philippines and Pakistan. *International Journal of Online Marketing*, *7*(4), 63–77. doi:10.4018/IJOM.2017100105

Tam, G. C. (2017). Project Manager Sustainability Competence. In *Managerial Strategies and Green Solutions for Project Sustainability* (pp. 178–207). Hershey, PA: IGI Global. doi:10.4018/978-1-5225-2371-0.ch008

Tambo, T. (2018). Fashion Retail Innovation: About Context, Antecedents, and Outcome in Technological Change Projects. In I. Management Association (Ed.), Fashion and Textiles: Breakthroughs in Research and Practice (pp. 233-260). Hershey, PA: IGI Global. https://doi.org/ doi:10.4018/978-1-5225-3432-7.ch010

Tantau, A. D., & Frăţilă, L. C. (2018). Information and Management System for Renewable Energy Business. In *Entrepreneurship and Business Development in the Renewable Energy Sector* (pp. 200–244). Hershey, PA: IGI Global. doi:10.4018/978-1-5225-3625-3.ch006

Teixeira, N., Pardal, P. N., & Rafael, B. G. (2018). Internationalization, Financial Performance, and Organizational Challenges: A Success Case in Portugal. In L. Carvalho (Ed.), *Handbook of Research on Entrepreneurial Ecosystems and Social Dynamics in a Globalized World* (pp. 379–423). Hershey, PA: IGI Global. doi:10.4018/978-1-5225-3525-6.ch017

Teixeira, P., Teixeira, L., Eusébio, C., Silva, S., & Teixeira, A. (2021). The Impact of ICTs on Accessible Tourism: Evidence Based on a Systematic Literature Review. In C. Eusébio, L. Teixeira, & M. Carneiro (Eds.), *ICT Tools and Applications for Accessible Tourism* (pp. 1–25). IGI Global. doi:10.4018/978-1-7998-6428-8.ch001

Trad, A., & Kalpić, D. (2018). The Business Transformation Framework, Agile Project and Change Management. In M. Khosrow-Pour, D.B.A. (Ed.), Encyclopedia of Information Science and Technology, Fourth Edition (pp. 620-635). Hershey, PA: IGI Global. https://doi.org/ doi:10.4018/978-1-5225-2255-3.ch054

Trad, A., & Kalpić, D. (2018). The Business Transformation and Enterprise Architecture Framework: The Financial Engineering E-Risk Management and E-Law Integration. In B. Sergi, F. Fidanoski, M. Ziolo, & V. Naumovski (Eds.), *Regaining Global Stability After the Financial Crisis* (pp. 46–65). Hershey, PA: IGI Global. doi:10.4018/978-1-5225-4026-7.ch003

Trengereid, V. (2022). Conditions of Network Engagement: The Quest for a Common Good. In R. Augusto Costa, F. Brandão, Z. Breda, & C. Costa (Eds.), *Planning and Managing the Experience Economy in Tourism* (pp. 69-84). IGI Global. https://doi.org/10.4018/978-1-7998-8775-1.ch004

Turulja, L., & Bajgoric, N. (2018). Business Continuity and Information Systems: A Systematic Literature Review. In N. Bajgoric (Ed.), *Always-On Enterprise Information Systems for Modern Organizations* (pp. 60–87). Hershey, PA: IGI Global. doi:10.4018/978-1-5225-3704-5.ch004

Vargas-Hernández, J. G. (2017). Professional Integrity in Business Management Education. In N. Baporikar (Ed.), *Management Education for Global Leadership* (pp. 70–89). Hershey, PA: IGI Global. doi:10.4018/978-1-5225-1013-0.ch004

Varnacı Uzun, F. (2021). The Destination Preferences of Foreign Tourists During the COVID-19 Pandemic and Attitudes Towards: Marmaris, Turkey. In M. Demir, A. Dalgıç, & F. Ergen (Eds.), *Handbook of Research on the Impacts and Implications of COVID-19 on the Tourism Industry* (pp. 285–306). IGI Global. https://doi.org/10.4018/978-1-7998-8231-2.ch014

Vasista, T. G., & AlAbdullatif, A. M. (2017). Role of Electronic Customer Relationship Management in Demand Chain Management: A Predictive Analytic Approach. *International Journal of Information Systems and Supply Chain Management, 10*(1), 53–67. doi:10.4018/IJISSCM.2017010104

Vieru, D., & Bourdeau, S. (2017). Survival in the Digital Era: A Digital Competence-Based Multi-Case Study in the Canadian SME Clothing Industry. *International Journal of Social and Organizational Dynamics in IT, 6*(1), 17–34. doi:10.4018/IJSODIT.2017010102

Vijayan, G., & Kamarulzaman, N. H. (2017). An Introduction to Sustainable Supply Chain Management and Business Implications. In M. Khan, M. Hussain, & M. Ajmal (Eds.), *Green Supply Chain Management for Sustainable Business Practice* (pp. 27–50). Hershey, PA: IGI Global. doi:10.4018/978-1-5225-0635-5.ch002

Vlachvei, A., & Notta, O. (2017). Firm Competitiveness: Theories, Evidence, and Measurement. In A. Vlachvei, O. Notta, K. Karantininis, & N. Tsounis (Eds.), *Factors Affecting Firm Competitiveness and Performance in the Modern Business World* (pp. 1–42). Hershey, PA: IGI Global. doi:10.4018/978-1-5225-0843-4.ch001

Wang, C., Schofield, M., Li, X., & Ou, X. (2017). Do Chinese Students in Public and Private Higher Education Institutes Perform at Different Level in One of the Leadership Skills: Critical Thinking?: An Exploratory Comparison. In V. Wang (Ed.), *Encyclopedia of Strategic Leadership and Management* (pp. 160–181). Hershey, PA: IGI Global. doi:10.4018/978-1-5225-1049-9.ch013

Wang, J. (2017). Multi-Agent based Production Management Decision System Modelling for the Textile Enterprise. *Journal of Global Information Management, 25*(4), 1–15. doi:10.4018/JGIM.2017100101

Wiedemann, A., & Gewald, H. (2017). Examining Cross-Domain Alignment: The Correlation of Business Strategy, IT Management, and IT Business Value. *International Journal of IT/Business Alignment and Governance, 8*(1), 17–31. doi:10.4018/IJITBAG.2017010102

Wolf, R., & Thiel, M. (2018). Advancing Global Business Ethics in China: Reducing Poverty Through Human and Social Welfare. In S. Hipsher (Ed.), *Examining the Private Sector's Role in Wealth Creation and Poverty Reduction* (pp. 67–84). Hershey, PA: IGI Global. doi:10.4018/978-1-5225-3117-3.ch004

Yablonsky, S. (2018). Innovation Platforms: Data and Analytics Platforms. In *Multi-Sided Platforms (MSPs) and Sharing Strategies in the Digital Economy: Emerging Research and Opportunities* (pp. 72–95). Hershey, PA: IGI Global. doi:10.4018/978-1-5225-5457-8.ch003

Yaşar, B. (2021). The Impact of COVID-19 on Volatility of Tourism Stocks: Evidence From BIST Tourism Index. In M. Demir, A. Dalgıç, & F. Ergen (Eds.), *Handbook of Research on the Impacts and Implications of COVID-19 on the Tourism Industry* (pp. 23–44). IGI Global. https://doi.org/10.4018/978-1-7998-8231-2.ch002

Yusoff, A., Ahmad, N. H., & Halim, H. A. (2017). Agropreneurship among Gen Y in Malaysia: The Role of Academic Institutions. In N. Ahmad, T. Ramayah, H. Halim, & S. Rahman (Eds.), *Handbook of Research on Small and Medium Enterprises in Developing Countries* (pp. 23–47). Hershey, PA: IGI Global. doi:10.4018/978-1-5225-2165-5.ch002

Zacher, D., & Pechlaner, H. (2021). Resilience as an Opportunity Approach: Challenges and Perspectives for Private Sector Participation on a Community Level. In A. Zehrer, G. Glowka, K. Schwaiger, & V. Ranacher-Lackner (Eds.), *Resiliency Models and Addressing Future Risks for Family Firms in the Tourism Industry* (pp. 75–102). IGI Global. https://doi.org/10.4018/978-1-7998-7352-5.ch004

Zanin, F., Comuzzi, E., & Costantini, A. (2018). The Effect of Business Strategy and Stock Market Listing on the Use of Risk Assessment Tools. In *Management Control Systems in Complex Settings: Emerging Research and Opportunities* (pp. 145–168). Hershey, PA: IGI Global. doi:10.4018/978-1-5225-3987-2.ch007

Zgheib, P. W. (2017). Corporate Innovation and Intrapreneurship in the Middle East. In P. Zgheib (Ed.), *Entrepreneurship and Business Innovation in the Middle East* (pp. 37–56). Hershey, PA: IGI Global. doi:10.4018/978-1-5225-2066-5.ch003

About the Contributors

Carolina Machado received her PhD degree in Management Sciences (Organizational and Policies Management area / Human Resources Management) from the University of Minho in 1999, Master degree in Management (Strategic Human Resource Management) from Technical University of Lisbon in 1994, and Degree in Business Administration from University of Minho in 1989. Teaching in the Human Resources Management subjects since 1989 at University of Minho, she is an Associate Professor with Habilitation, with experience and research interest areas in the field of Human Resource Management, International Human Resource Management, Human Resource Management in SMEs, Training and Development, Emotional Intelligence, Management Change, Knowledge Management and Management/HRM in the Digital Age/Business Analytics. She is Head of the Human Resources Management Work Group at the School of Economics and Management at University of Minho, Coordinator of Advanced Training Courses at the Interdisciplinary Centre of Social Sciences, Member of the Interdisciplinary Centre of Social Sciences (CICS.NOVA.UMinho), University of Minho, as well as Chief Editor of the International Journal of Applied Management Sciences and Engineering (IJAMSE), Guest Editor of journals, books Editor and book Series Editor, as well as reviewer in different international prestigious journals. In addition, she has also published both as editor/co-editor and as author/co-author several books, book chapters and articles in journals and conferences.

J. Paulo Davim is a Full Professor at the University of Aveiro, Portugal. He is also distinguished as honorary professor in several universities/colleges/institutes in China, India and Spain. He received his Ph.D. degree in Mechanical Engineering in 1997, M.Sc. degree in Mechanical Engineering (materials and manufacturing processes) in 1991, Mechanical Engineering degree (5 years) in 1986, from the University of Porto (FEUP), the Aggregate title (Full Habilitation) from the University of Coimbra in 2005 and the D.Sc. (Higher Doctorate) from London Metropolitan University in 2013. He is Senior Chartered Engineer by the Portuguese Institution of Engineers with an MBA and Specialist titles in Engineering and Industrial Management as well as in Metrology. He is also Eur Ing by FEANI-Brussels and Fellow (FIET) of IET-London. He has more than 30 years of teaching and research experience in Manufacturing, Materials, Mechanical and Industrial Engineering, with special emphasis in Machining & Tribology. He has also interest in Management, Engineering Education and Higher Education for Sustainability. He has guided large numbers of postdoc, Ph.D. and master's students as well as has coordinated and participated in several financed research projects. He has received several scientific awards and honors. He has worked as evaluator of projects for ERC-European Research Council and other international research agencies as well as examiner of Ph.D. thesis for many universities in different countries. He is the Editor in Chief of several international journals, Guest Editor of journals, books Editor, book Series Editor and Scientific

Advisory for many international journals and conferences. Presently, he is an Editorial Board member of 30 international journals and acts as reviewer for more than 100 prestigious Web of Science journals. In addition, he has also published as editor (and co-editor) more than 200 books and as author (and co-author) more than 15 books, 100 book chapters and 500 articles in journals and conferences (more than 280 articles in journals indexed in Web of Science core collection/h-index 62+/13500+ citations, SCOPUS/h-index 68+/16500+ citations, Google Scholar/h-index 87+/27000+ citations). He has listed in World´s Top 2% Scientists by Stanford University study.

* * *

Paulina Aburto de la Llave has a bachelor's degree in International Business and Administration and a master's degree in International Business and Markets, both from the Universidad Panamericana. She is currently head of the Foreign Trade Department of Syngenta in Mexico City, a leading transnational in agricultural science and technology, in particular seeds and crop protection products, based in Basel, Switzerland.

Connie Atristain-Suárez holds undergraduate and graduate degrees in Administration from the Instituto Tecnológico Autónomo de México (ITAM) and a Doctoral Degree in Administrative Sciences from the EGADE Business School. She is a full-time Professor and Researcher at ECEE at the Universidad Panamericana and teaches in undergraduate and graduate programs. She has specialized in competitiveness, strategic planning and managerial efficiency of SMEs in Mexico, as well as the Family Business. She has been conferred the National Researcher Level-I of the Mexican National System of Researchers by CONACYT.

Diana Fernandes received her Master degree in International Business in 2017, and her Degree in International Relations in 2015, both from University of Minho. She currently develops professional activity in HR Services. In the meanwhile she dedicates to academic research, being her main research interest areas in the field of International Business, Human Resources Management, Green Economy and Management Change.

Ana Paula Ferreira, PhD in Business Sciences from the University of Minho in 2008, she joined the School of Economics and Management (SEM) of this University in 1997. Since then, she teaches and researches in the areas of Organizational Behavior and Human Resources Management. She took over the course direction of the Management bachelor degree and, currently, in the Human Resource Management MSc. She also belongs to the direction of the Management and Business MSc. She belongs to the SEM School Board and Pedagogical Board. Between 2015 and 2017 she was an executive member of UMinhoExec – Executive Business Education. Between 1991 and 1997, she worked as an HRM and Training consultant, as well as a HR technician at the Banking sector. She has published in international journals, in the areas of careers, training and psychological contract, as well as book chapters in international (Palgrave, Springer, De Gruyter) and national (RH Editora) publishers.

Juliani Elis Duarte Gotardi is an undergraduate in Biological Sciences at Faculdades Adamantinenses Integradas, MBA in Environmental and Business Management and master's student of the Graduate Program in Agribusiness and Development at São Paulo State University (UNESP), Tupã Campus, Brazil.

Eglantina Hysa is a Lecturer of Economics at the Epoka University where she teaches International Economy, Development and Growth, and Microeconomics courses. Her research is concentrated in the fields of development economics, social aspects, and quality in higher education institutions.

Abiodun Isiaka (PhD, Wilfrid Laurier University) is an Assistant Professor at University of Regina, Saskatchewan, Canada. Her research interests are in financial reporting and disclosure, corporate governance, and corporate social responsibility.

Priscilla Ayleen Bustos Mac-Lean is an Assistant Professor in School of Science and Engineering at Sao Paulo State University (Unesp), Tupã Campus, Brazil, lecturing on Biosystems Engineering course. Doctor in Animal Quality and Productivity by the Postgraduate Program in Animal Science from the University of São Paulo (2012). Master in Animal Production by the Postgraduate Program in Animal Science (2008) and undergraduated in animal science (2005) from Universidade Estadual de Maringá (UEM). Her research interests include the areas of improvement of the animal production chain evaluating good practices related to product quality and quantity, ambience and animal biometeorology and welfare, working mainly in the productive chains of dairy cattle and sericulture.

Carla Magalhães is a PhD in Business Sciences (Human Resources), by University of Minho, in partnership with Getulio Vargas Foundation (Rio de Janeiro). Master in Public Administration (Human Resources and Marketing) by Getulio Vargas Foundation. Associate Professor at Lusófona University of Porto, with the following positions: Director of the 1st study cycle on Human Resources Management and Development and Director of the Advanced Lifelong Training Unit - LUFAV.ULP. Invited Assistant Professor at School of Economics and Management of University of Minho. Integrated researcher at the Center for Population Studies, Economy and Society (CEPESE) and associate member at the Center for Transdisciplinary Research for Entrepreneurship & Ecosystem Innovation (TRIE).

Idehai Olayemi Ohijeagbon obtained B.Eng, M.Eng and P.hD degrees in Mechanical Engineering. His research interests include Energy conversion systems and efficiency, Sustainable energy utilization of production systems, Product development, and Manufacturing processes. He is a Professor of Mechanical Engineering and engage in teaching and research activities at the University of Ilorin. He had served as Head, Department of Mechanical Engineering, University of Ilorin, and he is currently the Director of Laboratory to Products (LABTOP) at the University of Ilorin.

Chukwemeka Joshua Okereke completed his PhD degree in Mechanical Engineering from the University of Ilorin. Has more than one decade of experience in manufacturing operations, maintenance, and project management. Research interest include industrial process improvement and optimisation, computational and experimental fluid dynamics; heat exchange and heat networking; multi-phase flow analysis; refrigeration and ventilation system analysis.

Tiago Vilaça Oliveira received his Master's degree in Human Resources Management in 2021, and his Degree in Management in 2019, both from the School of Economics and Management of the University of Minho. He develops professional activity in HR Management and is involved in the management of a family owned SME. Simultaneously, he is a Football Coach and holds the UEFA B license. He is currently Assistant Coach in the senior football team of Forjães Sport Club. He focuses on topics related to management in football, high performance teams, leadership and organizational culture.

Mirela Panait currently holds the position of PhD Professor at Faculty of Economic Sciences, Petroleum-Gas University of Ploiesti (Romania). She is also scientific researcher at Institute of National Economy from Romanian Academy. Her work experience derives from and includes: teaching courses and seminaries; research activities; participating in scientific conferences. Regarding research activities, these are represented by publishing seven books and over 100 scientific papers in the field of interest: corporate social responsibility, sustainable development, foreign direct investment and financial market. She has published articles in leading international journals, including Land Use Policy, Energies, Sustainability and PlosOne. She is reviewer for journals and conferences like Sustainability, Energies, Journal of Risk and Financial Management. She is also guest editor or associate editor for prestigious journals from MDPI group or IGI Global publishing house.

Alessandra Panzella is a master student in Human Resource Management at the University of Minho, Braga, Portugal, with research interests in HRM, IHRM, Emotional Intelligence, Strategic HRM, HR Analytics.

Eduardo Satolo is an Associate Professor in Faculty of Science and Engineering at São Paulo State University (UNESP), Tupã Campus, lecturing on Business Administration course and Agribussiness and Development Post Graduation Program. Post-Doctorate in Production Engineering from UFF (2018), doctor in Industrial Engineering from the Universidade Metodista de Piracicaba (UNIMEP). Has a master's in Production Engineering from UNIMEP is a graduate in Production Engineering from UNIMEP.

Kaveh Sheibani holds a PhD in Operational Research from London Metropolitan University, UK. His main research interests lie in exploring search methodologies for hard combinatorial optimisation problems and their efficiency across a wide variety of applications, such as scheduling, operations management, timetabling, manufacturing, healthcare, telecommunications, and logistics.

Andrea Uribe-Palau is a Marketing Bachelor student at the Universidad Panamericana.

Index

I

K

L

M

N

O

P

Q

S

T

V

W

Premier Reference Source
Stabilizing Currency and Preserving Economic Sovereignty Using the Grondona System

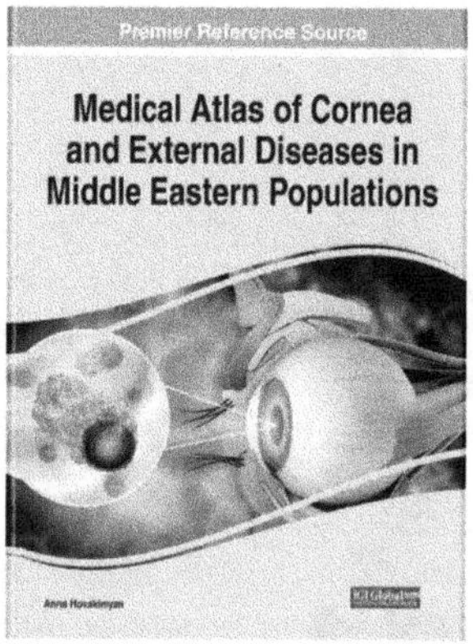
Premier Reference Source
Medical Atlas of Cornea and External Diseases in Middle Eastern Populations

Premier Reference Source
Examining Biophilia and Societal Indifference to Environmental Protection

Premier Reference Source
Using Narratives and Storytelling to Promote Cultural Diversity on College Campuses

www.ingramcontent.com/pod-product-compliance
Lightning Source LLC
LaVergne TN
LVHW062139300926
849161LV00010B/2575